S0-BQI-367

# Cellular Receptors

# Cellular Receptors

## *for Hormones and Neurotransmitters*

*Edited by*
Dennis Schulster
*Hormones Division,*
*National Institute for Biological Standards and Control,*
*Holly Hill, Hampstead,*
*London NW3 6RB, England*

and

Alexander Levitzki
*Department of Biological Chemistry,*
*The Institute of Life Sciences,*
*Hebrew University,*
*Jerusalem, Israel*



JOHN WILEY & SONS

Chichester · New York · Brisbane · Toronto

***British Library Cataloguing in Publication Data:***

Cellular receptors for hormones and neurotransmitters.
1. Hormone receptors 2. Neurotransmitter receptors
I. Schulster, Dennis II. Levitzki, Alexander
591.1′9′27 QP188.5.H67 79-41216

ISBN 0 471 27682 0

Typeset by Preface Ltd., Salisbury, Wilts. and printed by Page Bros. (Norwich) Ltd., Norwich.

# List of Contributors

DR GRETA AGUILERA, Endocrinology and Reproduction Research Branch, National Institute of Child Health and Human Development, National Institute of Health, Bethesda, Maryland 20014, USA.

PROF. LUTZ BIRNBAUMER, Department of Cell Biology, Baylor College Med., Houston, Texas 77030, USA.

DR SHMARYAHU BLUMBERG, Department of Biophysics, The Weizmann Institute of Science, Rehovot, Israel.

PROF. TOM BLUNDELL, Laboratory of Molecular Biology, Department of Crystallography, Birkbeck College, University of London, Malet Street, London WC1E 7HX, UK.

DR KEVIN J. CATT, Endocrinology and Reproduction Research Branch, National Institute of Child Health and Human Development, National Institutes of Health, Bethesda, Maryland 20014, USA.

DR ALEX N. EBERLE, Institute of Molecular Biology and Biophysics, Swiss Federal Institute of Technology (ETHZ), CH-8093 Zürich, Switzerland.

PROF. JACK PETER GREEN, Department of Pharmacology, Mount Sinai School of Medicine of the City University of New York, New York NY 10029, USA.

DR LINDSAY B. HOUGH, Department of Pharmacology, Mount Sinai School of Medicine of the City University of New York, New York NY 10029, USA.

DR MILES HOUSLAY, Department of Biochemistry, University of Manchester Institute of Science and Technology, PO Box 88, Manchester M60 1QD, UK.

DR RAVI IYENGAR, Department of Cell Biology, Baylor College Med., Houston, Texas 77030, USA.

PROF. SERGE JARD, Collège de France, Laboratoire de Physiologie Cellulaire, 11 Place Marcellin, Berthelot 75231, Paris Cedex 05, France.

DR LOTHAR KUEHN, Biochemische Abteilung, Diabetes-Forschungsinstitut an der Universität Düsseldorf, Hennekamp 65, Düsseldorf, W. Germany.

PROF. ALEXANDER LEVITZKI, Department of Biological Chemistry, The Institute of Life Sciences, The Hebrew University of Jerusalem, Jerusalem, Israel.

PROF. W. IAN P. MAINWARING, Department of Biochemistry, University of Leeds, 9 Hyde Terrace, Leeds LS2 9LS, UK.

DR ROBERT H. MICHELL, Department of Biochemistry, University of Birmingham, P.O. Box 363, Birmingham B15 2TT, U.K.

DR RICHARD J. MILLER, The University of Chicago, Department of Pharmacological and Physiological Sciences, 947 East 58th Street, Chicago, Illinois 60637, USA.

DR JOAV M. PRIVES, Department of Anatomical Sciences, Health Sciences Center, State University of New York at Stony Brook, Stony Brook, Long Island, NY 11794, USA.

DR YORAM SALOMON, Department of Hormone Research, The Weizmann Institute of Science, Rehovot, Israel.

DR DENNIS SCHULSTER, Hormones Division, National Institute for Biological Standards and Control, Holly Hill, Hampstead, London NW3 6RB, UK.

PROF. ROBERT SCHWYZER, Institute of Molecular Biology and Biophysics, Swiss Federal Institute of Technology (ETHZ), CH-8093 Zürich, Switzerland.

DR JAMSHED R. TATA, F.R.S., National Institute for Medical Research, Mill Hill, London NW7 1AA, UK.

DR VIVIAN I. TEICHBERG, Department of Neurobiology, The Weizmann Institute of Science, Rehovot, Israel.

DR MICHAEL WALLIS, School of Biological Sciences, University of Sussex, Falmer, Brighton BN1 9QG, UK.

DR GRAHAM WARREN, European Molecular Biology Laboratory, Postfach 10.2209, 6900 Heidelberg, W. Germany.

# Contents

# Preface

The field of 'cellular receptors' embraces a very broad spectrum of biomolecular interactions. Not all aspects can be covered in depth in a relatively short book of this nature. The objective has been to provide an adequate, up-to-date background to the subject as a whole, with sufficient key references to enable the reader to pursue details in those areas of specific interest. The book is aimed primarily at the advanced student who wishes to gain some basic knowledge in what is known as 'receptorology', rather than the research investigator working on particular receptors. Nevertheless, it is hoped that investigators working in one area of receptor research may find it valuable to have a compilation of information from work on other receptors and other aspects of the field. It is clear that studies on different receptors have progressed at vastly different rates and along different lines. The reasons are often related to the particular systems, ligands and assay methodologies involved, usually at an early stage in the work. Now, may well be an appropriate time for some 'cross-fertilization' of idea and approaches, and it is now the objective of this book to seed such an interaction.

This text represents a departure from the traditional style of presenting such information. The first part of the book is devoted to the general properties of receptors and the methodologies currently available to study them. Following this general section individual receptors, or families of receptors are discussed in detail. All of the specialist chapters have been written by selected authors, who are recognized authorities in their own field. The range of countries from which the contributors emanate, attests to this international expertise. Such multi-authored texts are usually reserved for conference proceedings in which a collection of presented papers are bound together in the form of a book. These are of undoubted value to the well-established research worker, but are often bewildering and much less useful to the student, and research worker new to the field. The attempt here has been to provide a unified account of the field (in which the fundamental and methodological aspects are covered in Section I), but in the later sections to retain the obvious advantages of presentation by experts actively engaged in their specific topics.

The excellent cooperation of these authors in the preparation and editing of their manuscripts is gratefully acknowledged, as is the helpful collaboration of the publishers. Our thanks are also due to Dr. E. Giorgi and Mr D. Montague for their help in the preparation of this text.

DENNIS SCHULSTER,
*London, England*

ALEXANDER LEVITZKI,
*Jerusalem, Israel*

Feb. 1979

# Abbreviations

| | |
|---|---|
| A | Amperes |
| A.II | Angiotensin II |
| Å | Ångström units ($10^{-8}$ cm) |
| 9-AAP | 9-aminoacridino-propranolol |
| ACh | Acetylcholine |
| ACTH | Adrenocorticotropic hormone, corticotropin |
| ADH | Antidiuretic hormone |
| ADP | Adenosine diphosphate |
| ATP | Adenosine 5′-triphosphate |
| ATPase | Adenosine triphosphatase |
| Ala | Alanine |
| Amino acids | For one-letter symbols for amino acid abbreviations, see Appendix to this list |
| Arg | Arginine |
| Asp | Aspartic acid |
| $\alpha$-BTX | $\alpha$-bungarotoxin |
| $c_a$ | Antagonist concentration |
| °C | Degree Celsius |
| CBG | Glucocorticoid-binding $\alpha_2$-globulin |
| CLIP | Corticotropin-like intermediate lobe peptide |
| CM-cellulose | *O*-(carboxymethyl) cellulose |
| CNS | Central nervous system |
| CTP | Cytosine 5′-triphosphate |
| $Ca^{2+}$-ATPase | Calcium activated adenosine triphosphatase |
| C-component or C | Catalytic unit of the adenylate cyclase system |
| C′ | Activated catalytic unit of the adenylate cyclase system |
| C-fragment | $\beta$-lipotropin ($C_{61}$–$C_{91}$) fragment |
| $[Ca^{2+}]_{cyt}$ | Calcium pool in the cytoplasm |
| Calmodulin | $Ca^{2+}$-dependent modulator protein, or receptor protein |
| cat$^-$ | Cells containing no measurable adenylate cyclase activity |
| Ci | Curie ($2.22 \times 10^6$ dpm) |

| | |
|---|---|
| cpm | Counts per minute |
| C-terminal | Free $\alpha$-carboxyl group of a protein |
| cyclic AMP; cAMP | Cyclic adenosine 3′:5′ monophosphate |
| cyclic GMP:cGMP | Cyclic guanosine 3′:5′ monophosphate |
| *d* | Dextrorotatory |
| D | Dextro |
| DABA | 2,4-diamino-butyric acid |
| DAPN | Dansyl analogue of propranolol |
| DEAE-Sephadex | *O*-(diaminodiethylaminoethyl)-Sephadex |
| *d*-LSD | Lysergic acid diethylamide |
| DNA | Deoxyribonucleic acid |
| DPOC | Dioleylphosphatidylcholine |
| dpm | Disintegrations per minute |
| DR | Dose ratio; a ratio of the $ED_{50}$ values of agonist in the presence and absence of antagonist |
| DTT | Dithiothreitol |
| e | Intrinsic activity |
| E | Enzyme |
| $ED_{50}$ | Median effective dose (half maximal response) |
| EDTA | Ethylenediaminotetraacetic acid |
| EGTA | Ethyleneglycol-bis(-aminoethylether-*N′*,*N′*)tetraacetic acid |
| END | Endorphin(s) |
| ENK | Enkephalin(s) |
| Epc | End plate current |
| Epp | End plate potential |
| f | Femto; $10^{-15}$ |
| $f/f_0$ | Frictional ratio |
| $f_1$ | Fraction of occupied receptor |
| $f_2$ | Fraction of occupied receptor in the presence of H, the low affinity ligand |
| F | $f_2/f_1$ see page 18 |
| FITC | Fluorescein isothiocyanate |
| FSH | Follicle stimulating hormone, follitropin |
| g | Gram |
| g | Unit of gravitational field (9.81 m.s$^{-2}$) |
| GABA | $\gamma$-amino-butyric acid |
| GABA-T | $\gamma$-amino-butyric acid transaminase |
| GAD | Glutamic acid decarboxylase |
| GDP | Guanosine diphosphate |
| Gpp(NH)p | see p(NH)ppG |
| GTP | Guanosine triphosphate |
| GTPase | Guanosine triphosphatase |

| | |
|---|---|
| GTP$\gamma$S | Guanosine-5′-*O*-(3-thiotriphosphate) |
| G-component | GTP binding component of the adenylate cyclase system |
| Gln | Glycine |
| Glu | Glutamine |
| h | Hour |
| [$^3$H] | Tritium ($\beta$ radiation) |
| H | Low affinity ligand *or* hormone (Chapter 2) |
| $[H]_{0.5}$ | Ligand concentration required to displace 50% of the bound radioactive ligand *or* for 50% saturation of the response activity measured |
| H (subscript) | Refers to low affinity ligand (Chapter 2) |
| H | Heavy form of acetylcholine receptor in *Torpedo* (Chapter 18) |
| $H_1$, $H_2$ | Histamine receptors with different relative activities to a series of agonists |
| $HATP^{3-}$ | Adenosine 5′-triphosphate (ionized form) |
| hCG | Human chorionic gonadotropin |
| hFSH | Human follicle stimulating hormone |
| HRC | Hormone-receptor–catalytic unit complex |
| HRC′ | Hormone-receptor–activated catalytic unit complex |
| HSA | Human serum albumin |
| HTC | Hepatoma cells |
| HYP | Hydroxybenzylpindolol |
| His | Histidine |
| *i* (subscript) | Refers to the *i*th component (Chapter 2) |
| $^{125}I$, $^{131}I$ | Radioactive iodine ($\gamma$ and $\beta$ radiations) |
| I | Labelled antagonist (Chapter 2) |
| $[I_1]$ | Concentration of free labelled antagonist |
| $[I_2]$ | Concentration of free labelled antagonist in the presence of H, the low affinity ligand |
| I (subscript) | Refers to labelled antagonist (Chapter 2) |
| $IC_{50}$ | Concentration which causes 50% inhibition of binding of [$^3$H] naloxone |
| ITP | Inosine-5′-triphosphate |
| IgE | Antibody |
| Ile | Isoleucine |
| i.u. | International units |
| *k* (with subscript) | Rate constant |
| $k_a$ | Rate constant of association; $M^{-1}$ $sec^{-1}$ |
| $k_{-1}$ | Rate constant of dissociation; $sec^{-1}$ |
| $k_1$ | Rate constant of association; $M^{-1}$ $sec^{-1}$ |
| K | Association constant; litres/mole; $M^{-1}$ (Chapter 10) |

| | |
|---|---|
| $K = [X_{0.5}]^n$ | In the general case of multiple ligands, K is related to the ligand which yields 50% saturation by exp($n$), where $n$ is the Hill coefficient (Chapter 2) |
| K (with subscripts $1, i \ldots N$) | Thermodynamic dissociation constant; moles/litre; M (Chapter 2) |
| $K_A$; $K_a$ | Association constant; litres/mole; $M^{-1}$ |
| $K_B$ | Apparent dissociation constant of the antagonist |
| $K_D$: $K_D'$; $K_{D_{app}}$; $K_d$ | Apparent dissociation constant; moles/litre; M |
| $K_{act}$ | Apparent activation constant, i.e. concentration of agonist that causes half-maximal increase in the rate of adenylate cyclase activity |
| $K_i$ | Dissociation constant for binding-inhibitor complex |
| $K_m$ | Michaelis constant, substrate concentration for half-maximal velocity |
| $k_{off}$ | Inverse of half life |
| $K_t$ | Concentration of solute which causes half saturation of uptake |
| $K^+$-ATPase | Potassium-activated adenosine triphosphatase |
| l | Litre |
| *l* | Laevorotatory |
| L | Laevo |
| L | Light form of acetylcholine receptor in *Torpedo* (Chapter 18) |
| L | HRC′/HRC the ratio of the concentration of the hormone-receptor complex with the activated and the non-activated catalytic unit (Chapter 2) |
| L* | Free radioligand |
| L*R | Bound radioligand |
| LH | Luteinizing hormone, lutropin |
| LHRH | Luteinizing hormone releasing hormone |
| LPH | Lipotropic hormone, lipotropin |
| Leu | Leucine |
| Li | Authentic light chain of the immunoglobulin |
| Lys | Lysine |
| m | Metre |
| m | Milli; $10^{-3}$ |
| min | Minutes |
| M | Moles/litre, molar |
| MBTA | 4-(*N*-maleimido) benzyltrimethylammonium |
| mho | Reciprocal ohm = amp. volt$^{-1}$ |
| mol | Moles |
| mRNA | Messenger ribonucleic acid |

| | |
|---|---|
| MSA | Multiplication stimulating activity |
| MSH | Melanocyte stimulating hormone; melanotropin; intermedin |
| M.W. | Molecular weight |
| Mepps | Miniature end plate potentials |
| Met | Methionine |
| MeV | Megavolts |
| Mg ATP | Magnesium-adenosine triphosphate complex |
| Mn ATP | Manganese-adenosine triphosphate complex |
| n | Nano; $10^{-9}$ |
| $n$ | Hill coefficient |
| $n_H$ | Hill coefficient at 50% ligand saturation |
| $N$ | Number of binding sites, maximal binding (Chapter 2) |
| $N_X$ | Average number of ligand molecules bound to the receptor |
| $N$ (subscript) | Refers to the $N$th component (Chapter 2) |
| N-terminal | Free $\alpha$-amino group of a protein |
| $NAD^+$ | Nicotinamine adenine dinucleotide |
| NADPH | Nicotinamine adenine dinucleotide phosphate (reduced) |
| $Na^+$-ATPase | Sodium activated adenosine triphosphatase |
| NEM | *N*-ethylmaleimide |
| NPS | *O*-nitrophenylsulphenyl |
| p | Pico; $10^{-12}$ |
| $^{32}P$ | Radioactive phosphorus ($\beta$ radiation) |
| $pA_2$ value | The intercept with the abscissa in the Schild plot of log(DR-1) versus log antagonist concentration |
| PBG | Progesterone-binding $\alpha_1$-globulin |
| $PGE_1$ | Prostaglandin $E_1$ |
| pH | $-\log[H^+]$ |
| Phe | Phenylalanine |
| $P_i$ | Inorganic orthophosphate —$H_2PO_3$ |
| pI | Isoelectric point |
| p(NH)ppG | Guanosine 5′-($\beta$,$\gamma$-imido)-triphosphate |
| PPi | Inorganic pyrophosphate —$HPO_3.H_2PO_3$ |
| Pro | Proline |
| Ps | Pseudohermaphrodite (rat) |
| PTH | Parathyroid hormone |
| R | Receptor |
| $[R]$ | Free receptor concentration |
| $[R_T]$ | Total receptor concentration |
| RC | Receptor coupled to catalytic unit C |
| RH | Receptor–low affinity ligand complex |

| | |
|---|---|
| $RI_1$ | Receptor–labelled antagonist complex |
| $RI_2$ | Receptor–labelled antagonist complex in the presence of H, the low affinity ligand |
| RX | Receptor–ligand complex |
| R | Regulatory unit bearing specific binding sites for cAMP (Chapter 5) |
| $R_2cAMP_2$ | Occupied binding sites of R above |
| RC | Unoccupied binding sites of R above |
| $R_2C_2$ | Holoenzyme (inactive) of adenylate cyclase |
| res. | Residue(s) of amino acids |
| $RM_R$ | Microsomal vesicles |
| $RM_{RK}$ | Microsomal vesicles washed with 0.5M KCl |
| RNA | Ribonucleic acid |
| rRNA | Ribosomal ribonucleic acid |
| s | Second(s) |
| $[^{35}S]$ | Radioactive sulphur ($\beta$ radiation) |
| $S_{20,w}$, S | Sedimentation constant ($10^{-13}$ $cm^2$ $s^{-1}$ $dyn^{-1}$) |
| SBG | Sex steroid-binding $\beta$-globulin |
| SDS | Sodium dodecyl sulphate |
| SDS-PAGE | Sodium dodecyl sulphate-polyacrylamide gel electrophoresis |
| SP | Substance P (P for preparation) |
| Sar | Sarcosine |
| Ser | Serine |
| Steroids | For abbreviations of common steroids, see Appendix, at the end of this list |
| $t_{1/2}$ | Half life (turnover time) |
| $T_3$ | Triiodothyronine |
| $T_4$ | Thyroxine |
| TBG | Thyroxine-binding $\beta$-globulin |
| TBPA | Thyroxine-binding pre-albumin |
| TMV | Tobacco mosaic virus |
| TSH | Thyroid stimulating hormone, thyrotropin |
| Tfm | Testicular feminization (mouse) |
| Tris | Tris(hydroxymethyl)aminomethane |
| Trp | Tryptophan |
| Tyr | Tyrosine |
| UTP | Uridine triphosphate |
| V | Volts |
| $V_{max}$ | Maximal rate of reaction |
| Val | Valine |
| X | Ligand |
| [X] | Concentration of free ligand |

| | |
|---|---|
| $\bar{Y}_x$ | The fraction of binding sites occupied by receptor |
| $\mu$ | Micro; $10^{-6}$ |
| $\Psi$ (with subscript) | The inverse of the thermodynamic dissociation constant K (with subscript) |

## AMINO ACIDS: ONE-LETTER SYMBOLS

| | |
|---|---|
| A | alanine |
| C | cysteine |
| D | aspartic acid |
| E | glutamic acid |
| F | phenylalanine |
| G | glycine |
| H | histidine |
| I | isoleucine |
| K | lysine |
| L | leucine |
| M | methionine |
| N | asparagine |
| P | proline |
| Q | glutamine |
| R | arginine |
| S | serine |
| T | threonine |
| V | valine |
| W | tryptophan |
| Y | tyrosine |

## STEROIDS

| | |
|---|---|
| Aldosterone | 18,11-Hemiacetal of 11$\beta$,21-dihydroxy-3,20-dioxo-4-pregnen-18-al |
| 5$\alpha$-Androstanediol | 5$\alpha$-Androstane-3$\alpha$($\beta$), 17$\beta$-diol |
| 5$\alpha$-Androstanedione | 5$\alpha$-Androstane-3,17-dione |
| Androstenedione | 4-Androstene-3,17-dione |
| Cholecalciferol | 9,10-Seco-5,7,10(19)-cholestatrien-3$\beta$-ol |
| Cholesterol | 5-Cholesten-3$\beta$-ol |
| Cortexolone (11-Deoxy-17$\alpha$-corticosterone) | 21,17$\alpha$-Dihydroxy-4-pregnene-3,20 dione |
| Corticosterone | 11$\beta$, 21-Dihydroxy-4-pregnene-3;20-dione |
| Cortisol | 11$\beta$,17$\alpha$,21-Trihydroxy-4-pregnene-3,20-dione |

| | |
|---|---|
| 7-Dehydrocholesterol | 5,7-Cholestadien-3β-ol |
| 11-Deoxycorticosterone | 21-Hydroxy-4-pregnene-3,20-dione |
| Dexamethasone | 16α-Methyl-9α-fluoro-11β,17α,21-trihydroxy-1,4-pregnadiene-3,20-dione |
| 5α-Dihydrocortisol | 11β,17α,21-Trihydroxy-5α-pregnane-3,20-dione |
| 5α-Dihydrotestosterone | 17β-Hydroxy-5α-androstan-3-one |
| 5β-Dihydrotestosterone | 17β-Hydroxy-5β-androstan-3-one |
| 1α,25-Dihydroxycholecalciferol | 9,10-Seco-5,7,10(19)-cholestatriene-1α,3β,25-triol |
| 7,17-Dimethyl-19-nor-testosterone | 7α,17α-Dimethyl-17β-hydroxy-19-nor-4-androsten-3-one |
| Epicortisol | 11α,17α,21-Trihydroxy-4-pregnene-3,20-dione |
| Epioestriol | 1,3,5(10)-estratriene-3,16β,17β-triol |
| 9α-Fluorocortisol | 9α-Fluoro-11β,17α,21-trihydroxy-4-pregnene-3,20-dione |
| 9α-Fluoroprednisolone | 9α-Fluoro-11β,17α,21-trihydroxy-1,4-pregnadiene-3,20-dione |
| 25-Hydroxy-cholecalciferol | 9,10-Seco-5,7,10(19)-cholestatriene-3β,25-diol |
| 17α-Isoaldosterone | 18,11-Hemiacetal of 11β,21-dihydroxy-3,20-dioxo-17α-4-pregnen-18-al |
| Oestradiol-17α | 1,3,5(10)-Estratriene-3,17α-diol |
| Oestradiol-17β | 1,3,5(10)-Estratriene-3,17β-diol |
| Oestriol | 1,3,5(10)-Estratriene-3,16α,17β-triol |
| Oestrone | 3-Hydroxy-1,3,5(10)-estratrien-17-one |
| Prednisolone | 11β,17α,21-Trihydroxy-1,4-pregnadiene-3,20-dione |
| Progesterone | 4-Pregnene-3,20-dione |
| Testosterone | 17β-Hydroxy-4-androsten-3-one |
| Triamcinolone | 9α-Fluoro-11β,16α,17α,21-tetrahydroxy-1,4-pregnadiene-3,20-dione |
| Triamcinolone acetonide | 16α,17α-Cyclic acetal with acetone of 9α-Fluoro-11β,16α,17α,21-tetrahydroxy-1,4-pregnadiene-3,20-dione |

SECTION I

# Fundamental concepts in receptor research

Cellular Receptors for Hormones and Neurotransmitters
Edited by D. Schulster and A. Levitzki


CHAPTER 1

# Introduction

Dennis Schulster
*Hormones Division, National Institute for Biological Standards and Control, Holly Hill, Hampstead, London NW3 6RB*
Alexander Levitzki
*Department of Biological Chemistry, Hebrew University, Jerusalem, Israel*

A number of research fields in biology have focused in the recent decade on the concept of receptors. It has been recognized that hormones, neurotransmitters, and drugs must first interact with specific receptors in order to elicit their biochemical, physiological, endocrinological, and pharmacological effects; the basic phenomenology of receptor activity has been the subject of intense investigation in these fields. More recently however, mainly in the last decade, the details of the ligand–receptor interactions have become more amenable to investigation. Thus for example, it has been recognized that a variety of neurotransmitters and hormones induce the activation of adenylate cyclase upon interaction with the receptor, and in many laboratories the mechanistic details of the receptor–cyclase interrelationships have become the main subject of study. Similarly, the mode of coupling of certain receptors to the movement of specific ions, such as $Ca^{2+}$ and $Na^{+}$, across a membrane has become a central issue in a number of related fields such as neurophysiology, biochemistry, and pharmacology. With the well-founded concepts of regulatory biochemistry and the sophisticated physicochemical tools now available to the researcher, it has become possible to explore receptor related phenomenon in much finer detail. This book attempts to summarize the fundamental concepts and present state of knowledge in the field of cellular receptors, and to introduce the non-expert reader to current ideas in receptor research.

Necessarily, in order to confine this book to a reasonable size, the coverage has had to be limited both in depth and scope, and for further details the reader is referred to key references in each of the chapters. Other useful general texts or conference proceedings in the receptor field include References 1–8.

This book has been divided into four sections. The first of these (Section I) covers fundamental, general, and methodological aspects and is intended to provide the basis for subsequent sections.

Cellular receptors for hormones and neurotransmitters may be subdivided broadly into two main types.

1. Those localized inside the cell, both in the cytoplasm and the nucleus: the intracellular receptors (Section II).
2. Those sited within the cell membrane: the cell membrane–surface receptors (Sections III and IV).

Clearly the water soluble intracellular receptors for steroids and thyroid hormones, are of a completely different character from the hydrophobic molecules localized in the cell membrane, that act as receptors for the other hormones and neurotransmitters. Nevertheless, the principles of selective and specific binding, related directly to distinctive cellular responses, operate for both kinds of receptors.

It could be said that the concept of receptors dates back to the immunological studies of Paul Ehrlich on toxin: anti-toxin interactions and the ideas of Sir Henry Dale.[9] Much of our present knowledge of receptors is founded on pharmacological studies into the selective action of drugs, since it was in this area that soluble pure ligands were readily available. Quantitative studies on drug action by Clark, over 50 years ago, laid the foundations for the 'receptor theory' by showing that binding to receptors was reversible (and thus non-covalent), obeyed the law of mass action, and that specificity was observed at very low ligand concentrations ($10^{-9}$–$10^{-10}$ M). Another major development was the availability of structural analogues—opening the way to establish structure–function relationships. Studies on structurally related drug analogues (see Chapter 8 onwards) have established that structural requirements for biological activity are precise and exacting, and that ligands structurally related to a stimulatory molecule (or agonist) could have an inhibitory (or antagonist) effect. From these observations rational drug and hormone design has emerged as a practical possibility, with an enormous impact on the pharmaceutical aspects of the field.

Inherent in the evaluation of the receptor function, is its coupling to a response signal. Only when it is coupled to a response can we be sure that we are dealing with a '*receptor*, rather than with some non-specific 'binding molecule'. No cellular recognition system yet exists in which the transduction mechanism through the membrane is fully understood. Nevertheless, the cyclic nucleotides and $Ca^{2+}$ (and other ions) have emerged as the two main types of secondary messenger molecules for communicating between the cell surface and the intracellular metabolic machinery.

The intracellular events regulated by cyclic adenosine

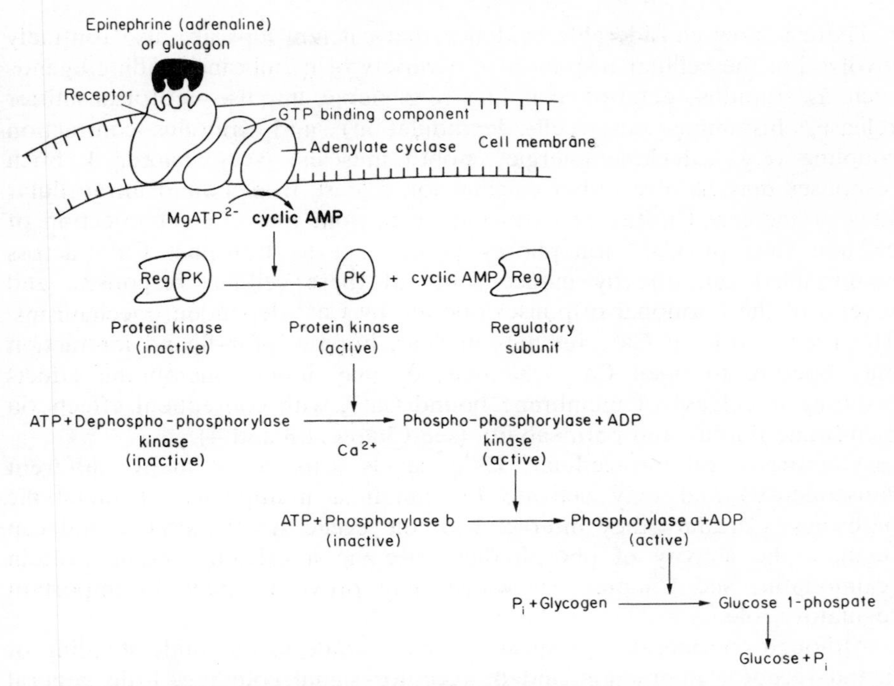


Figure 1.1. Amplification cascade for the breakdown of glycogen in the liver cell by epinephrine (adrenaline) or glucagon. The amplification factor (blood hormone concentration: blood glucose concentration) is over 1 million-fold. Adrenaline acts similarly on the muscle cell. The interaction of receptors with adenylate cyclase has been found to involve a specific GTP binding component (or regulatory protein); other as yet, unidentified, components may be involved

3′:5′-monophosphate (cyclic AMP) are perhaps best understood in the systems in which it was first detected by Earl Sutherland and coworkers[10]—the glucagon and epinephrine (adrenaline) regulated breakdown of glycogen by mammalian liver and muscle cells. This involves the degradative effects of phosphorylase enzymes which are activated in a cascade manner by kinases (Figure 1.1). Cyclic AMP has a regulatory role in an enormous number of different cellular systems and, apart from a few deficient mutant cell types, virtually all cells synthesize it.[10]

Cyclic AMP is formed from $MgATP^{2-}$ via the membrane localized enzyme adenylate cyclase; cyclic AMP breakdown is regulated by membranal and cytoplasmic phosphodiesterases. In keeping with its ubiquitous and important nature, the ligand–receptor–cyclic AMP system may be regulated at a variety of levels including stimulation and inhibition of both adenylate cyclase and phosphodiesterase. These and other aspects of adenylate cyclase regulation are dealt with in further detail in Chapter 4.

There is now considerable evidence that calcium ions are also routinely involved in the cellular responses to a variety of membrane binding ligands such as stimulus–secretion coupling (e.g. nerve impulse–neurotransmitter release, histamine–mast cell degranulation) and stimulus–contraction coupling (e.g. catecholaminergic smooth muscles)—see Chapter 4. Such responses may involve either calcium ion release from bound intracellular sites or increased influx of extracellular calcium ions. Direct injection of calcium ions or $Ca^{2+}$ ionophores (which act to transport $Ca^{2+}$ across membranes) can directly induce some of these cellular responses, and several of the hormonal responses operate by $Ca^{2+}$-dependent mechanisms. The precise role of $Ca^{2+}$ remains unclear, but receptor–ligand interaction may operate to open $Ca^{2+}$ channels, or may induce membrane effects resulting in release of membrane bound $Ca^{2+}$, with consequent effects on membrane fluidity and permeability (see Chapters 3 and 4).

Alterations in intracellular $Ca^{2+}$ levels can have many different consequences and may activate or inhibit a multiplicity of metabolic pathways. Certainly they interact with the cyclic AMP pathway and can regulate the activity of phosphodiesterase via a calcium-binding protein (calmodulin; see Chapter 4), which may prove to play an important regulatory role.

Although considerable progress has been made in our understanding of ligand–receptor interaction and of receptor–signal coupling, little general progress has as yet been made in elucidating the structures of receptors and of the components in the biochemical apparatus coupled to them. Indeed, much current research is devoted to the study of these particular aspects. At present, of all the receptors considered in this book, the nicotinic acetylcholine receptor is probably the best characterized due to a unique combination of favourable circumstances, and its multi-subunit nature has been demonstrated (see Chapter 18). The recent advances gained following structural and reconstitutional studies on the nicotinic acetylcholine receptor, immunoreceptors, and toxin receptors reveal the fruitfulness of using this line of approach for furthering our understanding of the mechanism of action of hormones and neurotransmitters.

## I. REFERENCES


1. Beers, R. F., and Bassett, E. G. *Miles Int. Symposium Series No. 9 Cell membrane receptors for viruses, antigens and antibodies, polypeptide hormones and small molecules*, Raven Press, New York, (1976).
2. Bradshaw, R. A., Frazier, W. A., Merrell, R. C., Gottlieb, D. I., and Hogue-Angeletti, R. A. (eds.) *Surface Membrane Receptors: interface between cells and their environment*, Plenum Press, New York and London, (1976).
3. Cuatrecasas, P., and Greaves, M. F. *Receptors and recognition, Series A.* Chapman and Hall, London, Vol. 1 (1976); Vol. 2 (1976); Vol. 3 (1977); Vol. 4 (1977); Vol. 5 (1978); Vol. 6 (1978).

4. Gualtieri, F., Gianella, M., and Melchiorre, C. (eds.) *Recent Advances in Receptor Chemistry*, Elsevier, North Holland Biomedical Press, Amsterdam, (1979).
5. Nicolson, G. L., Raftery, M. A., Rodbell, M., and Fox, C. F. (eds.) *Progress in Clinical and Biological Research, Vol. 8. Cell surface receptors. Proc. of UCN–UCLA Conference, California 1975*, Alan Liss Inc., New York, (1976).
6. O'Malley, B. W., and Birnbaumer, L. (eds.) *Receptors and Hormone Action*, Academic Press, New York, Vols. 1–3, (1978).
7. Rickenberg, H. V. (ed.) *International Review of Biochemistry, Biochemistry and Mode of action of hormones II*, University Park Press, Baltimore, Vol. 20, (1978).
8. Smythies, J. R., and Bradley, R. J. (eds.) *Modern Pharmacology–Toxicology Series, Receptors in Pharmacology*, Marcel Dekker, Basel, Vol. 11, (1978).
9. Dale, H. (ed.) *The Collected Papers of Paul Ehrlich* Pergamon Press, London, Vol. 2, (1957).
10. Robison, G. A., Butcher, R. W., and Sutherland, E. W. *Cyclic AMP.* Academic Press, New York, (1971).

Cellular Receptors for Hormones and Neurotransmitters
Edited by D. Schulster and A. Levitzki


CHAPTER 2

# Quantitative aspects of ligand binding to receptors

Alexander Levitzki
*Department of Biological Chemistry, Hebrew University, Jerusalem, Israel*

## I. INTRODUCTION

The interaction of a ligand with its receptor is the first step in a cascade of events which leads to the observed biological response of the cell. The binding of a ligand to its receptor usually triggers a number of events such as membrane potential changes, permeability changes, alterations in the intracellular level of certain biochemicals, etc. It is not always easy to determine which of the phenomena induced by the interaction of the ligand with its receptor is directly related to the biological signal. In recent years techniques have been developed to study ligand–receptor interactions. It is now quite well established that receptors to hormones and neurotransmitters are localized on the cell surface, except for intracellular receptors for steroid hormones and tri-iodothyronine ($T_3$)and thyroxine ($T_4$). In this chapter we discuss some general concepts of ligand–receptor interactions. Little is known about the mechanisms which account for the coupling between ligand binding and. the primary biochemical response that elicits the biological signal. We shall, however, discuss some general principles which must be dealt with, when this aspect of receptor activity is investigated.

## II. MODES OF LIGAND BINDING

### (i) The general case

The binding of a ligand to a receptor possessing *N* binding sites is described by the equilibria:

$$\mathrm{R} + \mathrm{X} \overset{\mathrm{K}_1}{\rightleftharpoons} \mathrm{RX} \qquad \mathrm{K}_1 = \frac{[\mathrm{R}][\mathrm{X}]}{[\mathrm{RX}]} \tag{1}$$

$$\mathrm{RX}_{i-1} + \mathrm{X} \overset{\mathrm{K}_i}{\rightleftharpoons} \mathrm{RX}_i \qquad \mathrm{K}_i = \frac{[\mathrm{RX}_{i-1}][\mathrm{X}]}{[\mathrm{RX}_i]} \tag{2}$$

$$\mathrm{RX}_{N-1} + \mathrm{X} \overset{\mathrm{K}_N}{\rightleftharpoons} \mathrm{RX}_N \qquad K_N = \frac{[\mathrm{RX}_{N-1}][\mathrm{X}]}{[\mathrm{RX}_N]} \tag{3}$$

where R is the receptor, X is the ligand, and $K_1$ through $K_N$ are the thermodynamic dissociation constants (usually expressed in units of mol $l^{-1}$, M). The dissociation constant is a useful parameter as it gives a measure of the concentration of reactants necessary to form half maximal amounts of complex. The reciprocal of the dissociation constant is the association constant $K_A$ (expressed as 1 $mol^{-1}$ or $M^{-1}$).

The average number of ligand molecules bound to the receptor molecule is given by:

$$N_x = \frac{[\mathrm{RX}] + 2[\mathrm{RX}_2] + \cdots + i[\mathrm{RX}_i] + \cdots + N[\mathrm{RX}_N]}{[\mathrm{R}] + [\mathrm{RX}] + [\mathrm{RX}_2] + \cdots + [\mathrm{RX}_i] + \cdots + [\mathrm{RX}_N]} \tag{4}$$

Inserting equations (1) through (3) into equation (4), one obtains:

$$N_x = \frac{\dfrac{[X]}{K_1} + \dfrac{2[X]^2}{K_1K_2} + \cdots + \dfrac{i[X]^i}{K_1K_2\cdots K_i} + \cdots + \dfrac{N[X]^N}{K_1K_2\cdots K_i\cdots K_N}}{1 + \dfrac{[X]}{K_1} + \dfrac{[X]^2}{K_1K_2} + \cdots + \dfrac{[X]^i}{K_1K_2\cdots K_i} + \cdots + \dfrac{[X]^N}{K_1K_2\cdots K_i\cdots K_N}} \tag{5}$$

If one defines:

$$\psi_1 = \frac{1}{K_1} \tag{6}$$

$$\psi_2 = \frac{1}{K_1K_2} \tag{7}$$

$$\vdots$$

$$\psi_i = \frac{1}{K_1K_2\cdots K_i} \tag{8}$$

$$\vdots$$

$$\psi_N = \frac{1}{K_1K_2\cdots K_i\cdots K_N} \tag{9}$$

one can rewrite equation (5) in the form:

$$N_x = \frac{\sum_{i=1}^{N} i\psi_i[X]^i}{1 + \sum_{i=1}^{N} \psi_i[X]^i} \tag{10}$$

In many cases the investigator is interested in the fraction of sites occupied $\bar{Y}_x$:

$$Y_x = \frac{N_x}{N} = \frac{1}{N} \times \frac{\sum_{i=1}^{N} i\psi_i[X]^i}{1 + \sum_{i=1}^{N} \psi_i[X]^i} \tag{11}$$

Equations (5) through (7) are valid for the most general case where no assumptions are made on the mode of binding. Equations (5), (10), and (11) can be used to fit any binding data, and one can obtain an estimate of the thermodynamic parameters that characterize the binding.

**(ii) The simple cases**

In the simplest situation the binding of ligand can be described by a single dissociation constant, as no heterogeneity in the receptor population is found and no cooperativity among receptor sites is observed. In this situation, one speaks about a single class of receptor sites. Under these conditions, equation (10) simplifies to:

$$N_x = \frac{N[X]}{K_D + [X]} \tag{12}$$

where $K_D$ is the ligand–receptor dissociation constant. The fractional saturation $\bar{Y}_x$ in this case is given by:

$$\bar{Y}_x = \frac{N_x}{N} = \frac{[X]}{K_D + [X]} \tag{13}$$

Equation (13) is a typical rectangular hyperbola and typical for the Michaelis–Menten mode of binding.

### (iii) The less simple cases

When ligand binding cannot be described by a rectangular hyperbola, equation (13), one has to use the general expression, equations (5), (10), or (11). To a first approximation one can, however, use another equation:

$$N_x = \frac{N[X]^n}{K + [X]^n} \tag{14}$$

or

$$\bar{Y}_x = \frac{N_x}{N} = \frac{[X]^n}{K + [X]^n} \tag{15}$$

where $n \leqslant N$. K is not a dissociation constant and has no simple physicochemical meaning. One can show,[1] however, that K is related to the ligand saturation which yields 50% saturation by $K = [X_{0.5}]^n$.

For positively cooperative binding, $n$ assumes values higher than 1.0. Under such conditions, the binding curve is sigmoidal. In cases where the binding is negatively cooperative, $n$ is smaller than 1.0 and the binding curve is a flattened hyperbola. It should be emphasized, however, that when a heterogeneous set of binding sites binds the ligand, a flattened hyperbola is also obtained. Thus when an apparent negative cooperativity is observed, one cannot—on the basis of equilibrium data alone—exclude the existence of a heterogeneous population of sites. In the latter case the data should be fitted to the sum of terms similar to equation (13):

$$\bar{Y}_x = \frac{[X]}{K_1 + [X]} + \cdots + \frac{[X]}{K_i + [X]} + \cdots + \frac{[X]}{K_N + [X]} \tag{16}$$

or

$$\overline{Y}_x = \sum_{i=1}^{N} \frac{[X]}{K_i + [X]} \tag{17}$$

In order to delineate which of the two situations occurs, negative cooperativity or heterogeneous population of sites, other experiments must be performed.[2,3]

## III. GRAPHICAL PRESENTATION OF BINDING DATA

Five principal graphical methods can be used to describe the binding data of a ligand (Figure 2.1).

1. The direct plot. In this plot $\overline{Y}_x$ (or $N_x$) is plotted against the concentration of free ligand [X].
2. The logarithmic plot. In this plot $\overline{Y}_x$ (or $N_x$) is plotted against log[X]. The advantage in this plot is that all non-cooperative binding curves have identical slopes but are shifted, one with respect to the other, on the abscissa. On can take the derivative $d\overline{Y}_x./\log[X]$ according to equations (13) or (14) and investigate its value at 50% saturation ($[X] = K_D$). For example, from equation (13) it follows that:

$$\frac{d\overline{Y}_x}{d\log[X]} = [X]\frac{d\overline{Y}_x}{dX} = \frac{K_D[X]}{(K_D + [X])^2} = \frac{K_D{}^2}{4K_D^2} = \frac{1}{4} \tag{18}$$

3. The double reciprocal plot. In this plot $1/\overline{Y}_x$ (or $1/N_x$) is plotted against $1/[X]$. The extrapolated intercept on the ordinate yields the maximal binding: $1/\overline{Y}_x = 1$ or $1/N_x = 1/N$, whereas the intercept on the abscissa yields the negative value of the association constant, $K_A$ (where $K_A = K_D^{-1}$). When the binding is non-cooperative, the double reciprocal plot is described by the rearrangement of equation (12) to the Lineweaver–Burk equation:

$$\frac{1}{N_x} = \frac{1}{N} = \frac{K_D}{N} \times \frac{1}{[X]} \tag{19}$$

A similar rearrangement of equation (14) yields:

$$\frac{1}{N_x} = \frac{1}{N} + \frac{K}{N} \times \frac{1}{[X]^n} \tag{20}$$

When $n > 1.0$, the double reciprocal plot will be convex (Figure 2.1(c)) and the extrapolation on the ordinate will still yield $1/N$. When $n < 1.0$, the double reciprocal plot will be concave and, similarly, the extrapolation on the ordinate will yield $1/N$. In the two latter cases obviously the extrapolated value read on the abscissa has a complex meaning.

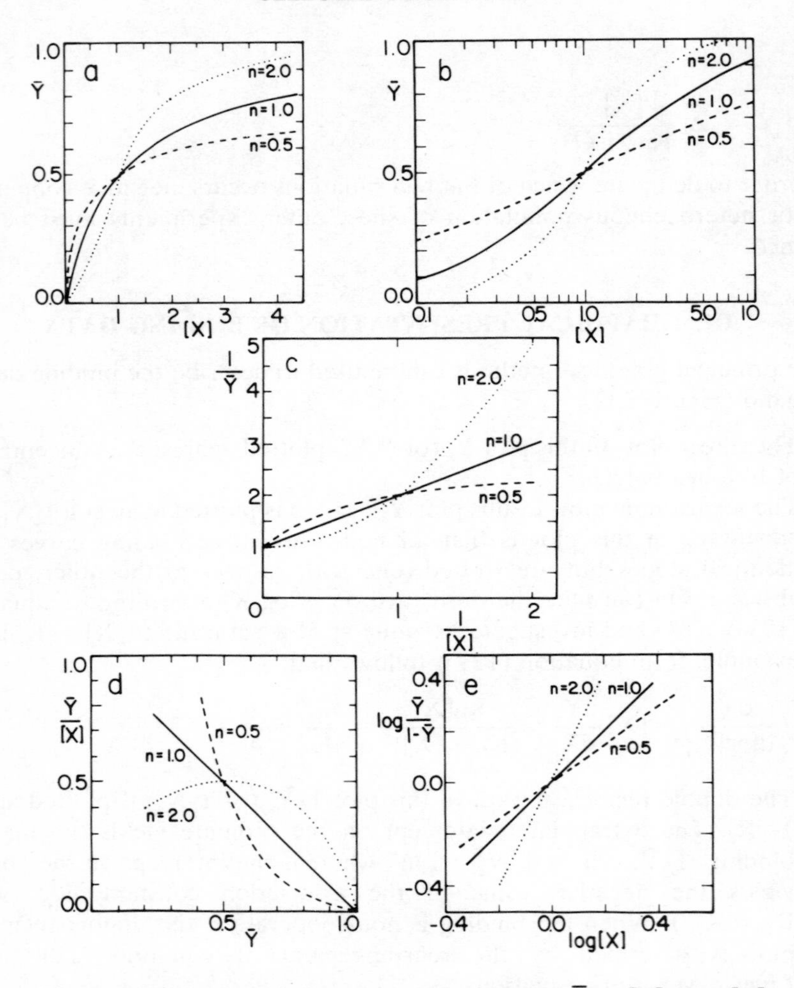


Figure 2.1. Plots of binding data. The equation $\bar{Y}_x = [X]^n/K + [X]^n$ is used to construct the graphs presented where K = 1 for simplicity. For positively cooperative binding, $n > 1$; for negatively cooperative binding $n < 1$; and for non-cooperativity $n = 1$.

(a) The direct plot: bound fraction versus free ligand concentration.
(b) The logarithmic plot: bound fraction versus log free ligand concentration.
(c) The double reciprocal plot: 1/bound fraction versus 1/free ligand concentration.
(d) The Scatchard plot: bound fraction/free ligand concentration versus bound fraction
(e) The Hill plot: log (bound fraction/1 − bound fraction) versus log(free ligand concentration)

4. The Scatchard plot. Upon rearrangement of equation (12), one obtains:

$$\frac{N_x}{[X]} = \frac{N}{K_D} - \frac{N_x}{K_D} \tag{21}$$

The plot of $N_x/[X]$ versus $N_x$ is known as the Scatchard plot.[4] The slope of this curve yields the ligand–receptor association constant ($K_A = K_D^{-1}$) and the extrapolation on the abscissa yields the maximal binding $N$. The Scatchard plot was designed for cases where non-cooperative binding is observed. In cases where the binding of ligand is positively cooperative, a plot of $N_x/[X]$ versus [X] yields a convex curve (Figure 2.1(d)). In a case where the binding is negatively cooperative or when the process of binding is to a heterogeneous population of sites, the Scatchard plot is concave (Figure 2.1(d)). In all cases, however, the extrapolation on the abscissa yields the maximal amount of binding, although in the non-linear cases the meaning of the slope is complex.[5,6]

5. The Hill plot. In this plot one plots log $\bar{Y}_x/1 - \bar{Y}_x$ versus log[X]. Thus equations (13) and (15) yield the Hill equations:

$$\log \frac{\bar{Y}_x}{1 - \bar{Y}_x} = \log \frac{1}{K_D} + \log[X] \tag{22}$$

$$\log \frac{\bar{Y}_x}{1 - \bar{Y}_x} = \log \frac{1}{K} + n\log[X] \tag{23}$$

The coefficient $n$ is known as the Hill coefficient and is, in fact, the slope of the Hill plot. The coefficient is usually measured at 50% ligand saturation and is defined as $n_H$ at this point. In the general case it is not expected that equation (15) holds, but rather the general expression given in equation (11). For the general case it is found that $n$ is not a constant and changes with [X]. One can show that the $n_H$ is related to the thermodynamic dissociation constant by a rather complex mathematical relationship.[3,5] The $n_H$ values, however, remain useful quantities as they reflect the degree of cooperativity that the system exhibits. When $n_H = 1.0$, the system is non-cooperative; when $n_H > 1.0$, the system is positively cooperative; and when $n_H < 1.0$, the system is either negatively cooperative or represents a heterogeneous set of binding sites (Figure 2.1(e)).

## IV. THE ASSAY OF RECEPTORS USING RADIOACTIVELY LABELLED LIGANDS

In order to establish the concentration of a receptor in a given preparation, the investigator must be equipped with a reliable assay. The method used for measuring receptor concentration requires a radioactively labelled compound that binds to the receptor with measurable affinity. Either the labelled agonist or hormone, or the labelled antagonist can fulfil this task. The success of a binding experiment requires that the concentration of the receptor–radioligand complex can be monitored with high precision. Two basic approaches can be, and have been, used to measure the binding of ligands to receptor: equilibrium dialysis and filtration techniques that measure directly the concentration of the receptor–ligand complex after its separation from the free radioligand by filtration or centrifugation. Let us consider the requirements for a successful binding experiment using these two approaches.

### (i) Equilibrium dialysis

In equilibrium dialysis the receptor containing preparation is allowed to equilibrate with a radioactively labelled ligand. The receptor containing material is confined to one of two compartments separated by a dialysis membrane, thus enabling the free flow of the radioactive ligand. After equilibrium is reached, the concentration of free radioligand $L^*$ is identical in the two compartments. In the receptor containing compartment the total concentration of radioligand is the sum of the free ligand $L^*$ and the bound ligand $L^*R$. Therefore the concentration of receptor bound radioligand is calculated by taking the difference in radioligand concentration between the two compartments. Thus, only when this difference is significant, can one perform the equilibrium dialysis experiment reliably.

When the receptor concentration is in the range of the receptor–ligand dissociation constant, good binding data can be obtained. When the receptor concentration is much higher than the ligand–receptor dissociation constant, most of the added ligand will be bound to the receptor and the free ligand concentration will be low. When the receptor concentration is much below the receptor–ligand dissociation constant, only a small fraction of the radioactive ligand in the receptor containing compartment will be in the bound form where most of it represents free ligand concentration. Under the latter conditions, the estimation of bound ligand relies on the small difference in ligand concentration between the two compartments and is, therefore, very inaccurate.

### (ii) Separation techniques

In the filtration technique the receptor is allowed to equilibrate with the radioactively labelled ligand in a test tube. After equilibration has occurred, the mixture is filtered and the receptor–ligand complex is collected on a filter, washed and counted for radioactivity. This technique is reliable only if the lifetime of the complex is much longer ($k_{off}$ very small) than the rate of filtration and washing. In order to slow down the rate of receptor–ligand dissociation, the washing step of the receptor–ligand is sometimes conducted at much lower temperatures (0–4 °C) as compared to the temperature of the binding experiment (25–37 °C). Separation techniques are usually found to be very effective when the receptor–ligand dissociation constant is in the range of $10^{-9}$ to $10^{-11}$ M. The very low dissociation constant usually reflects a very low $k_{off}$ value, namely a long lifetime for the receptor–ligand complex. Therefore, it is worthwhile for the investigator to put in some effort in the search for such a ligand, if the native agonist exhibits much lower affinity or cannot be obtained in a radioactively labelled form.

Another problem is the specific radioactivity of the ligand used. If one wishes to monitor a receptor at a concentration range of $10^{-11}$ M to $10^{-9}$ M, the maximal specific radioactivity for tritium $^3H$ attainable for the ligand is in most cases too low to allow for an accurate determination of ligand binding. Therefore one should seek a ligand that can be iodinated by $^{125}I_2$ and so obtain a high specific radioactivity. It is only then that the radioactive counts bound will be sufficient for an accurate determination.

Both techniques have been used extensively, but the separation techniques—mainly by filtration—have gained higher popularity among investigators.

## V. DISPLACEMENT EXPERIMENTS

In many cases the direct measurement of ligand binding to a receptor is not feasible because the ligand–receptor dissociation constant is high, and thus the receptor concentration accessible experimentally is too low to allow a meaningful binding experiment. For example, the affinity of catecholamines to the adrenergic receptor is, in most cases, in the range 1–100 $\mu$M, whereas the maximal receptor concentration accessible *in vitro* is 0.01 $\mu$M. For this reason high affinity antagonists, which bind to these receptors with dissociation constants of 0.001–0.01 $\mu$M, have been used. However, if one would still like to measure the affinity of a low affinity ligand to the receptor, it can be done by measuring its ability to displace the high affinity radioactively labelled antagonist. Let us analyse the procedure of a displacement experiment and examine how the dissociation constant of a low affinity ligand can be measured from a displacement curve. If the receptor binds the high affinity radioactively labelled antagonist I with a dissociation constant $K_I$ and the low affinity agonist H with a dissociation constant $K_H$, one can write:

$$[R_T] = [R] + [RI_1] = [R] + \frac{[R][I_1]}{K_I} \tag{24}$$

where $[I_1]$ is the concentration of free radioactively labelled antagonist and $[R_T]$ the total receptor concentration which equals the sum of free receptor concentration [R] and the antagonist bound receptor $[RI_1]$ when H is absent. Therefore, the fraction of occupied receptor is given by:

$$f_1 = \frac{[RI_1]}{[R_T]} = \frac{[I_1]/K_I}{1 + [I_1]/K_I} \tag{25}$$

When the low affinity ligand H is present, equation (24) is replaced by:

$$[R_T] = [R] + [RI_2] + [RH] = [R] + \frac{[R][I_2]}{K_I} + \frac{[R][H]}{K_H} \tag{26}$$

where $[I_2]$ and $[RI_2]$ are the concentrations of the free radioactive ligand and of the receptor–ligand complex respectively, and [RH] the concentration of the receptor complexed with the low affinity ligand. Under these conditions, the fraction of receptor occupied by the radioactive antagonist I is given by:

$$f_2 = \frac{[RI_2]}{[R_T]} = \frac{[I_2]/K_I}{1 + [I_2]/K_I + [H]/K_H} \tag{27}$$

Dividing equation (27) by equation (25) will give use the fraction of receptor occupied by the antagonist I in the presence of H, as compared to receptor occupancy in its absence:

$$F = \frac{f_2}{f_1} = \frac{1 + [I_1]/K_I}{1 + \dfrac{[I_2]}{K_I} + \dfrac{[H]}{K_H}} \times \frac{[I_2]}{[I_1]} \tag{28}$$

This equation is known as the displacement equation, where $[I_1]$ and $[I_2]$ are the free antagonist concentration in the absence and in the presence of H respectively. If one plots F as a function of [H] *or log*[H], one obtains a *displacement curve* from which the value of $K_H$ can be computed according to the procedure below (Figure 2.2). When 50% of the bound radioactive ligand I is displaced by H, $F = \frac{1}{2}$ and, after rearrangement, equation (28), takes the form:

$$K_H = \frac{[H]_{0.5}}{2\dfrac{[I_2]}{[I_1]} - 1 + \dfrac{[I_2]}{K_I}} \tag{29}$$

where $[H]_{0.5}$ is the concentration of free H which yields 50% displacement of I. In most experimental situations the receptor concentration is very low, as compared to the concentration I or H used:

$$[R_T] \ll [I],[H] \tag{30}$$

Under these conditions $[I_1]$ is approximately equal to $[I_2]$, namely:

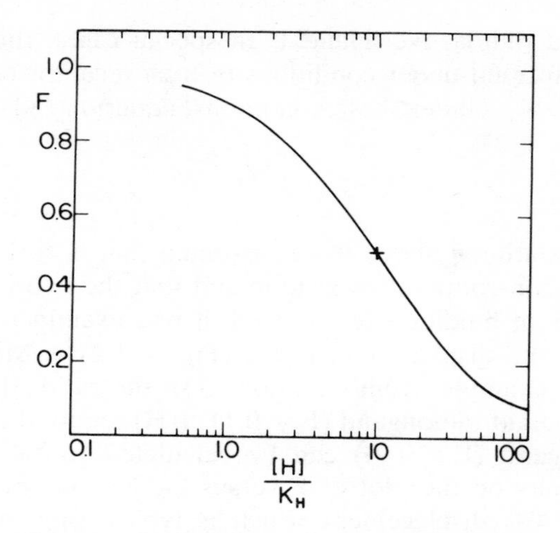


Figure 2.2. The displacement curve. Equation (33) was used to construct the displacement curve shown. In this particular case the ratio$[I]/K_I$ was taken to equal 90, and $[H]/K_H$ was varied according to the graph shown. F is the fraction of receptor occupied by the radioactive antagonist (see equation (28)), $K_I$ and $K_H$ the dissociation constants for the antagonist I and the low affinity agonist H respectively

$$[I_1] \sim [I_2] = [I] \qquad \text{or} \qquad [I]_{TOTAL} = [I]_{FREE} = [I] \tag{31}$$

and

$$[H]_{TOTAL} = [H]_{FREE} = [H] \tag{32}$$

Therefore equations (28) and (29) obtain the form:

$$F = \frac{1 + [I]/K_I}{1 + \frac{[I]}{K_I} + \frac{[H]}{K_H}} = \frac{1}{1 + \frac{[H]}{K_H\left(1 + \frac{[I]}{K_I}\right)}} \tag{33}$$

and

$$K_H = \frac{[H]_{0.5}}{\frac{[I]}{K_I} + 1} \tag{34}$$

where $[H]_{0.5}$ is the total (= free) ligand concentration required to displace

50% of the bound radioactive ligand I. In special cases, the displacement experiment is performed under conditions of high receptor occupancy by I, namely when $[I] \gg K_I$. Under these conditions, equation (34) simplifies to:

$$K_H = \frac{[H]_{0.5}}{[I]} K_I \tag{35}$$

In the analysis outlined above it was assumed that both I and H bind to the receptor in a non-cooperative fashion and that the two ligands compete for the same class of binding sites. Indeed, if one examines equation (33), one can see that the displacement curve (Figure 2.2) is Michaelian (non-cooperative). For example, from equation (33), the ratio: ([H] required to displace 90% of bound radioligand ($F = 0.1$)):([H] required to displace 10% of bound radioligand ($F = 0.9$)) can be calculated to be 81. Thus over almost two log units on the plot of F versus log[H], one moves from 10% displacement to 90% displacement which is typical of a non-cooperative (Michaelian) mode of binding for H (Figure 2.2). However, one can visualize a situation where the binding of I is non-cooperative, whereas the binding of H is cooperative. Under these conditions, of course, one must consider another set of equations where this situation is taken into account. More complex situations will arise if I and H compete for the same sites but bind in different modes (different cooperativities). Such situations have been encountered in regulatory enzymes[3] but so far have not been recognized in receptors.

## VI. NON-SPECIFIC BINDING

In most cases (except for intracellular steroid and thyroid hormone receptors) one measures the binding of a ligand to an impure receptor embedded in a membrane matrix. The ligands used for the assay of receptors may bind with low affinity to non-receptor components of the membrane. This process of binding is known as 'non-specific' binding. Since the non-specific binding is characterized by a low affinity process, the amount of radioactive ligand bound to the low affinity sites increases linearly with increasing free ligand concentration (Figure 2.3). In order to determine the extent of non-specific binding in a binding experiment, the binding of the radioactive ligand must be measured under conditions where it cannot bind to the specific receptor sites. Such conditions are met when the binding of radioactive ligand is measured in the presence of high concentrations (ten-fold or more the dissociation constant) of a non-radioactive ligand or in the presence of high concentrations of another competitive ligand. Thus, in order to obtain the binding isotherm to the specific receptor sites, the non-specific binding must be subtracted (Figure 2.3) from the total amount bound.

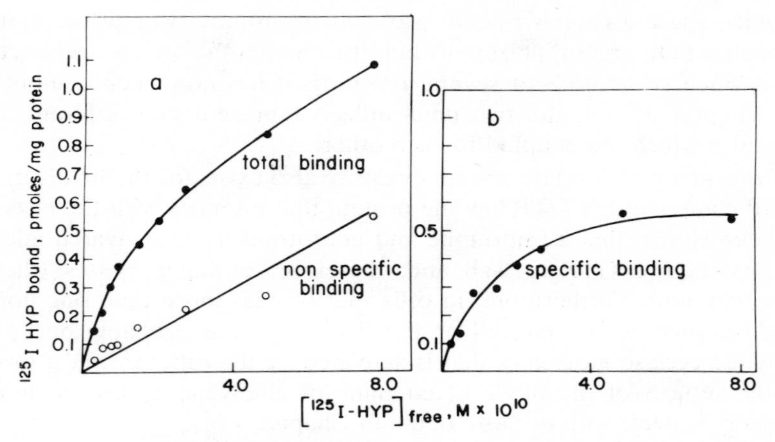


Figure 2.3. The non-specific binding. In this figure the actual binding data of [$^{125}$]hydroxybenzylpindolol ($^{125}$I-HYP) binding to turkey erythrocyte $\beta$-adrenergic receptors is shown.

(a) •–•–•. total binding; o–o–o, binding in the presence of $5.0 \times 10^{-6}$ dl-propranolol representing the non-specific binding.

(b) The specific binding obtained by subtracting the non-specific binding (–o–o–) from the total binding (–•–•–) in (a)

## VII. RECEPTOR OCCUPANCY AND RECEPTOR TO SIGNAL COUPLING

### (i) General remarks

When an agonist binds to a receptor, a biochemical signal is elicited. The signal elicited involves either the activation of enzyme or a change in permeability to specific ions. In the case of receptors coupled to an enzyme, only one enzyme has been so far identified to be coupled to receptors. This enzyme is the adenylate cyclase, first discovered by Sutherland and his co-workers. A variety of receptors to hormones and neurotransmitters have been found to be coupled to adenylate cyclase. Among these receptors one finds receptors to hormones such as the catecholamines, epinephrine and norepinephrine, (also known as adrenaline and noradrenaline), glucagon, ACTH, secretin, FSH, LH, TSH, and prostaglandin $E_1$, and receptors to neurotransmitters such as catecholamines (epinephrine, norepinephrine, and dopamine), histamine, and serotonin. It is reasonable to assume that in these cases the system is constructed basically as a regulatory enzyme, namely the receptor constitutes the regulatory unit and the adenylate cyclase the catalytic unit. These two basic units probably represent two different macromolecules that are organized within the membrane in an architecturally well-defined structure. For several receptors it has now been demonstrated (see Chapter 4) that the two units indeed represent two different macromolecules which are coupled to each other.

In the adenylate cyclase system evidence also exists for the involvement of a third component, a GTP binding protein that interacts with the enzyme. It has been shown that all hormone and neurotransmitter activated adenylate cyclases require GTP as well and, therefore, probably possess the GTP regulatory unit. Furthermore, in cells that possess more than one hormone receptor, such as the fat cell or the liver cell, it is probably one pool of adenylate cyclase molecules that is activated by the different receptors. The specific subject of the mode of coupling of adenylate cyclase to hormone receptors is dealt with in more detail in Chapter 4.

Many of the hormone or neurotransmitter elicited responses, however, are not mediated by cyclic AMP, and therefore do not involve the activation of the enzyme adenylate cyclase. In some cases, such as the nicotinic acetylcholine receptor, it has been demonstrated that a change in ion permeability occurs subsequent to the occupancy of the receptor by an agonist. In this case it has been postulated that the acetylcholine receptor is linked to an ionophore (or a 'gate') which allows the flow of $Na^+$ across the membrane. The nature of this ionophore is not known as yet, and attempts to identify the ionophore entity occupies the time and energy of many laboratories around the world at present.

The ultimate experiment to identify the catalytic moiety coupled to a receptor is obviously to reconstitute the system from its purified components. So far the only receptor that has been purified to homogeneity is the nicotinic acetylcholine receptor. In the latter case, however, it has not been possible to reconstitute the acetylcholine receptor with the ionophore and hence obtain the acetylcholine induced change in ion permeability upon exposure of the system to the agonist. In view of the rather primary stage at which the field of receptor to signal coupling is at present, we shall limit our discussions to general principles.

When one investigates a receptor mediated signal, one can in principle conduct three types of measurements.

### *(a) Direct ligand binding to the receptor*

This measurement allows determination of whether the population of receptors represents a single class of receptor sites, or whether the receptor consists of a heterogeneous population of ligand binding sites. Furthermore, the binding experiments reveal the cooperative nature of the binding and allow the determination of the ligand affinity to the receptor as well as the total number of receptors per milligram membrane protein (its density). It is also sometimes possible to measure ligand binding to intact cells using radioactively labelled ligands.

*(b) The measurement of the primary biochemical response as a function of ligand concentration*

This measurement yields the biochemical dose–response curve. For example, in the case of a hormone activated adenylate cyclase, one can determine the activation of the enzyme as a function of agonist (or partial agonist) concentration. From competitive experiments one can determine the inhibition constant of an antagonist. In the case where receptor occupancy results in changes in ion permeability, one can measure the change in ion permeability as a function of agonist concentration.

*(c) The measurement of a late biochemical or a physiological response as a function of ligand concentration*

In this case the signal measured is an event occurring subsequent to and as a result of the primary biochemical response. For example, the secretion of enzymes from the parotid gland is a $\beta$-adrenergic response elicited by the accumulation of cyclic AMP in the acinar cell of the gland. Thus, one can examine the extent of enzyme secretion as a function of the agonist norepinephrine. Similarly, the secretion of $K^+$ and water by the parotid gland is an $\alpha$-adrenergic response occurring subsequent to the influx of $Ca^{2+}$ (a change in permeability to the extracellular $Ca^{2+}$ ions) which is the primary $\alpha$-adrenergic response in the parotid gland.

The relation between these three types of ligand dose–response curves can teach the investigator a great deal about the quantitative relationship between receptor occupancy with the primary biochemical response and with the final response.

**(ii) Basic concepts**

The structure of the cell membrane is discussed in Chapter 3. Although little is known about the structure and organization of receptors within the membrane, even less is known about the molecular events that occur subsequent to ligand binding and lead to generation of the response. One can generally write the sequence of events af follows:

$$H + R \rightleftharpoons H{\cdot}R \rightarrow \rightarrow \ldots \rightarrow \rightarrow \text{Response} \qquad (36)$$

where the arrows indicate the events occurring subsequent to ligand binding but which precede the final response. H, indicates the ligand and R, the receptor.

When one measures a physiological or pharmacological response elicited by the binding of a ligand receptor, the results are usually expressed in the form of a dose–response curve. Very often the response is hyperbolic and represents a familiar situation encountered in the non-cooperative enzyme ligand interaction (Figure 2.1). Fortunately these simple cases are quite frequent and therefore we shall devote some space to their quantitative analysis. However, even in these simple cases little is known about the specific molecular events that take place subsequent to the binding of ligand to the receptor.

### *(a) The simple case*

Consider a situation where the hormone H binds to a receptor R coupled to a catalytic unit C (for example, adenylate cyclase or an ionophore). When the hormone binds to the receptor, a conformational change is induced and the catalytic unit is activated to C′:

$$H + R \cdot C \underset{K_H}{\rightleftharpoons} H \cdot R \cdot C \underset{L}{\rightleftharpoons} H \cdot R \cdot C' \tag{37}$$

where

$$K_H = \frac{[H][RC]}{[HRC]} \quad \text{and} \quad L = \frac{[HRC']}{[HRC]} \tag{38}$$

When one studies the *binding* of H to the receptor, both HRC and HRC′ are monitored. When the response is studied, only the state HRC′ is monitored.
Since:

$$[RC_T] = [RC] + [HRC] + [HRC'] \tag{39}$$

it is easy to show that the binding curve will be described by the equation:

$$[HRC] + [HRC'] = \frac{[RC_T][H]}{\frac{K_H}{1 + L} + [H]} \tag{40}$$

whereas the response curve will be given by:

$$[HRC'] = \frac{\frac{L}{1 + L}[RC_T][H]}{\frac{K_H}{1 + L} + [H]} \tag{41}$$

Both equations (40) and (41) describe a non-cooperative process where the measured apparent dissociation constant is:

$$K_{D_{app}} = \frac{K_H}{1 + L} \tag{42}$$

Thus in the simplest case, the response curve and the binding curve yield the same dissociation constant. If the activation of the catalytic moiety depends also on the presence of another effector such as an ion or a nucleotide, it is possible to study the binding with and without the effector. Without the effector the transition from HRC to HRC′ does not take place; hence, the dissociation constant monitored under these conditions will be $K_H$.

$$K_{D_{app}} = K_H \tag{43}$$

In the presence of the effector, the analysis outlined above will be applicable.

In many systems the final biochemical response depends on a chemical produced by the activated form of the catalytic unit C′ such as cyclic AMP. In such cases the characteristics of the response are determined by the mode of activation of the biochemical process by this chemical. This is also true if C is an ionophore and, in its activated form, allows the selective insertion of an ion through the cell membrane. For example, the secretion of enzymes by the parotid gland is mediated by cyclicAMP. It has been observed[7] that the response of adenylate cyclase to 1-epinephrine is a non-cooperative process with an apparent dissociation constant of $K_{D_{app}} = 1.0 \times 10^{-5}$ M. The secretion of $\alpha$-amylase in response to 1-epinephrine is also a non-cooperative process with an apparent dissociation constant $K_{D_{app}} = 1.0 \times 10^{-7}$ M. This finding means (1) that 1% occupancy of the $\beta$-adrenergic receptor by the $\beta$-agonist is required to generate the maximal final biochemical response and (2) that the process induced by cyclic AMP and which is responsible for the enzyme secretion occurs by a simple non-cooperative mechanism. Thus we have a situation where one can speak about ‘receptor reserve’ or ‘spare receptors’.

A number of workers have observed in a number of systems that the full pharmacological response of a system can be elicited by the occupancy of less than 1% of the receptors by the agonist. (These cases are discussed in detail in Reference 8.) In the latter cases, however, the distance between the primary event of ligand binding and the final pharmacological response is much larger than the cases discussed above where the full response curve can be obtained for all intermediate stages. Therefore we have restricted our discussion to the specific case of enzyme secretion from the parotid gland. It should be emphasized that the notion ‘receptor reserve’ or ‘spare receptors’ may be rather misleading. Let us assume that the extent of the signal is proportional to the concentration of the ligand occupied receptor [H·R].

This concentration is given by:

$$[H \cdot R] = \frac{[H]\{[R_T] - [H \cdot R]\}}{K_D} \quad (44)$$

where [H] is the free ligand concentration, $K_D$ the receptor ligand dissociation constant, and $[R_T]$ the total concentration of receptors. When $[H] \ll K_D$, equation (44) obtains the form:

$$[H \cdot R] = \frac{[R_T]}{K_D}[H] \quad (45)$$

If the physiological response saturates at hormone concentrations much below its dissociation constant to the receptor, the extent of the response, generated by H · R, will be *linear* with the total receptor concentration. This situation seems to apply, for example, for the 1-epinephrine induced enzyme secretion from the parotid gland.

This is only one of the many examples where it has been recognized that the hormone or neurotransmitter mediated process operates by a mechanism that involves the interaction of a ligand possessing low affinity to the receptor, with a large excess of the receptor. When such a situation is combined with the fact that the maximal response is attained at very low ligand occupancy, the response becomes a linear function of the number of receptors on the target cell. This state of affairs allows the system to adjust its response by modulation of the number of receptors on the cell surface. Cells, as we know, can adjust the total number of receptors according to physiological needs. Another possible advantage is that different tissues, possessing the same receptor, will respond to a different degree when exposed to a certain level of hormone, if they differ in receptor concentration and if the physiological concentration of the hormone is much below its dissociation constant to the receptor. Thus, the 'spare receptor' mechanism may allow for a discriminatory action of the same hormone on different tissues.

Catecholamines (when functioning as hormones), glucagon, and histamine are responsible for a multitude of biochemical processes in different tissues, and in all three cases there is evidence that the physiological level of the ligand is well below its dissociation constant to the receptor. Two approaches have been used to analyse the existence of 'spare receptors'. One is a detailed comparison between receptor occupancy and the dose response curve. Another approach is the one used originally by Nickersen, which is to block irreversibly a fraction of the receptors and determine whether full response can still be attained. Indeed, Nickersen[9] was able to block 99% of the histamine receptors by an irreversible blocker and obtain a full response to the histamine agonist. The dose response curve was, however, shifted to higher agonist concentrations. Equation (45) reveals that, in fact, one requires a large number of receptors ($[R_T]$) in order to generate a certain required response, although the occupied receptor under these conditions is only a small fraction of the total receptor concentration. Thus, the role of 'spare receptors' is to provide the necessary total number of receptors, such that, at certain levels of hormone, a certain absolute amount of receptors will be occupied. If this is indeed the role of 'spare receptors', the terminology 'spare' or 'reserve' may be inadequate as all the receptors participate in the generation of the signal.

*(b) The more complex case*

One can imagine a situation in which the receptor is an oligomeric structure that binds a number of ligand molecules. If the response occurs only when more than one receptor site is occupied, the dependence of the response on ligand concentration will be positively cooperative. This is true, even if the binding process itself is non-cooperative.[10] The maximal degree of cooperativity expected in such a case, as measured by the Hill coefficient, equals the number of ligand binding sites on the receptor.[10] One can also imagine a situation in which the oligomeric receptors are arranged in clusters, as in the case of the nicotinic acetylcholine receptor (see Chapter 18). In the latter situation, the degree of cooperativity of the system may even increase.[10]

The binding of hormone or of a neurotransmitter may be negatively cooperative,[2] as was found in the case of insulin binding to insulin receptors.[11] In this case it is important, again, to study the correlation between the binding of insulin, the dependence of the primary biochemical response on insulin concentration, and the dependence of the final biochemical response on the concentration of the hormone. Such correlative studies are not yet available in full detail and, therefore, it is difficult to assess the physiological significance of the negative cooperativity in binding found for insulin. In principle, however, a negatively cooperative response can be an efficient mechanism to sensitize[1] the responsiveness of the system to hormone concentration, if the fluctuations in hormone concentration occur at concentrations well below the 50% saturation point. This can be easily seen from Figure 2.1(a). The response at very low hormone concentrations rises steeply as a function of hormone concentration in this range. However, if one operates in the neighbourhood of 50% saturation, it is clear from Fgure 2.1(a) that positive cooperativity is the mechanism of choice, if one requires large changes in response as a result of small fluctuations in hormone concentration. It should be emphasized that both types of cooperativity, negative as well as positive cooperativity, require interaction between receptor sites.

## VIII. REFERENCES

1. Levitzki, A. *J. Theor. Biol.,* **44**, 367–372, (1974).
2. Levitzki, A., and Koshland, D. E., Jr. in *Current Topics in Cell Regulation* (Horecker, B. L., and Stadman, E. R., eds.), Academic Press, New York, Vol. 10, pp. 1–40, (1976).
3. Henis, Y. I., and Levitzki, A. *Eur. J. Biochem.,* (in press).
4. Scatchard, G. *Ann. N.Y. Acad. Sci.,* **51**, 660–666, (1949).
5. Levitzki, A. Quantitative Aspects of Allosteric Mechanisms, in *Molecular Biology, Biochemistry and Biophysics*, Vol. 28, Springer Verlag, Heidelberg, (1978), 105 pages.
6. Henis, Y. I., and Levitzki, A. *Euro. J. Biochem.,* **71**, 529–532, (1976).
7. Batzri, S., Selinger, Z., Schramm, M., and Robinovitch, R. *J. Biol. Chem.,* **248**, 361–368, (1973).
8. Triggle, D. J., and Triggle, C. R. in *Chemical Pharmacology of the Synapse*, Academic Press, New York, pp. 164–186, (1976).
9. Nickersen, M. *Nature (London),* **178**, 697–699, (1956).
10. Levitzki, A., Steer, M. L., and Segel, L. E. *J. Mol. Biol.,* **91**, 125–130, (1975).
11. De-Meyts, P., Roth, J., Neville, D. M., Gavin, J. R., Jr., and Lesinate, M. A. *Biochem. Biophys. Res. Commun.,* **55**, 154–158, (1973).

Cellular Receptors for Hormones and Neurotransmitters
Edited by D. Schulster and A. Levitzki


CHAPTER 3

# Membrane structure and receptor organization

Graham Warren
*European Molecular Biology Laboratory, Postfach 10 2209, 6900 Heidelberg, W. Germany*

Miles Houslay
*Department of Biochemistry, University of Manchester Institute of Science and Technology, Manchester, U.K.*

## I. INTRODUCTION

There are exceedingly few hormone receptors in almost all biological membranes. In the membrane of the turkey erythrocyte, for example, there are about 1,000,000 membrane proteins, but only 500 of these (or 0.05%) are $\beta$-receptors able to couple to adenylate cyclase. The scarcity of hormone receptors has seriously hindered attempts to isolate them. They are also difficult to purify because they are membrane proteins. In consequence we know very little about the disposition of hormone receptors in biological membranes. We cannot relate structure to function at the level of the isolated protein. In contrast, a great deal is now known about the organization of several major membrane proteins. In many instances the over-all structure can be related to function and their disposition in the membrane is known to be dictated by the mechanisms that are available for membrane protein biosynthesis. The rules which govern the biosynthesis and hence the disposition of proteins in the membrane are very simple and there is every reason to believe that they will also be applicable to minor membrane components such as hormone receptors. With this in mind the chapter has been divided into several sections. In earlier sections the structure and functioning of biological membranes is discussed. The rules governing the biosynthesis and functioning of proteins is detailed with reference to some of the major proteins of biological membranes. In the last part of the chapter a few receptors have been selected and examined in the light of the results obtained for the major membrane proteins. It is hoped that this discussion will emphasize the similarities between the major and minor components of biological membranes and indicate those areas that would be most fruitfully explored if an insight is to be gained into the functioning of hormone receptors at the molecular level.

All eukaryotic cells are bound by a plasma membrane composed of lipids, proteins, and carbohydrates. The lipids and proteins are usually present in roughly equal amounts but the carbohydrate comprises no more than 10% of the total mass. The membrane is a thin, flexible sheet of lipid studded with membrane proteins. It is permeable only to certain ions and small molecules and only certain information can be passed across the membrane. It is the membrane proteins that determine these specific functions. Each protein has a single function and the characteristic features of a membrane are determined by the number and type of each protein found there. The lipids provide the flexible matrix of the membrane that is impermeable to all ions and almost all neutral molecules.

Much of what we have to say concerns the plasma membrane of eukaryotic cells. It is here that many hormone and neurotransmitter receptors are to be found. We shall discuss in some detail the human erythrocyte membrane since this has been the object of intensive study.[1,2]

## II. LIPIDS

All membrane lipids are elongated, amphipathic molecules; they are hydrophilic at one end and hydrophobic at the other. In aqueous media they spontaneously form continuous bimolecular sheets; the hydrophobic portion of the molecule buries itself in the middle of the bilayer leaving the hydrophilic part to interact with the aqueous medium. The formation of the bilayer is largely dictated by entropic forces and the structure has been amply confirmed by a variety of physical techniques.[3] Membrane lipids usually account for about 40% of the mass of biological membranes. There are three main classes.[4]

1. The phospholipids comprise the majority and there are two subclasses. The phosphoglycerides possess a hydrophilic headgroup (phosphate bound to any one of choline, ethanolamine, serine, or inositol) attached to two long fatty acid chains by a glycerol backbone. The sphingomyelins have a choline headgroup but the backbone is sphingosine. The over-all conformation of the two subclasses is very similar (see Figure 3.1). The fatty acid chains usually comprise a linear array of 16 or 18 methylene groups and, in most phospholipid molecules, one of the two chains has an unsaturated double bond.
2. The neutral lipids. Cholesterol is the main neutral lipid to be found in eukaryotic cells. It is found mainly in the plasma membrane and always in association with phospholipids. The molecule is oriented so that the hydroxyl group interacts with the aqueous media and the plate-like steroid rings interact with those halves of the phospholipid fatty acid chains that are nearest to the headgroup (Figure 3.1).[5] The effect of this

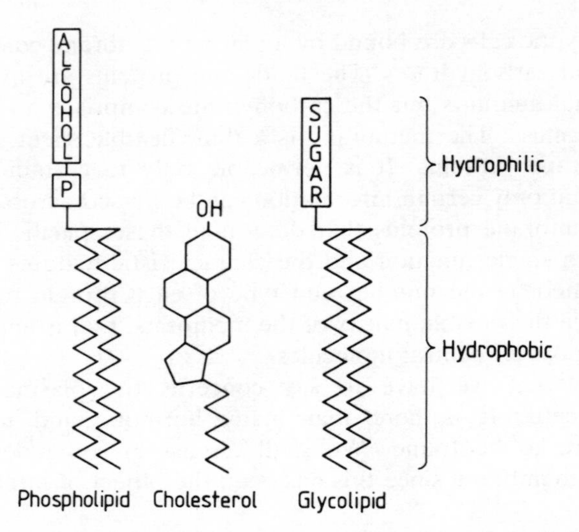


Figure 3.1. Schematic structures of the three main lipid classes in eukaryotic plasma membranes

interaction with phospholipid molecules is to decrease the motion of the fatty acid chains and hence increase the viscosity of the bilayer. The permeability of the bilayer is also decreased.

3. The glycolipids are minor components of most biological membranes. They resemble phospholipids in their over-all shape but the headgroup comprises one or more sugar molecules (Figure 3.1). The number of sugar molecules can vary from one (in cerebrosides) to as many as seven (in gangliosides).

**(i) The lipid matrix of biological membranes is a bilayer**

Membrane lipids alone form bilayers, but is this their organization in biological membranes where approximately half of the total mass is made up of protein? In 1925 Gorter and Grendel[6] extracted the lipids from red blood cell membranes and spread them out as a monomolecular sheet. The surface area of this sheet was twice that of the red cell membranes that were extracted. They concluded that the lipids in this membrane were arranged in the form of a bilayer. This point was later emphasized by Danielli and Davson[7] who suggested that the proteins in the membrane must therefore be layered on both sides of the lipid bilayer. This arrangement for the membrane proteins became increasingly untenable and doubt was cast on the original proposal of the lipid bilayer by Gorter and Grendel. Alternative arrangements of lipids in the membrane were proposed.[1] The doubt

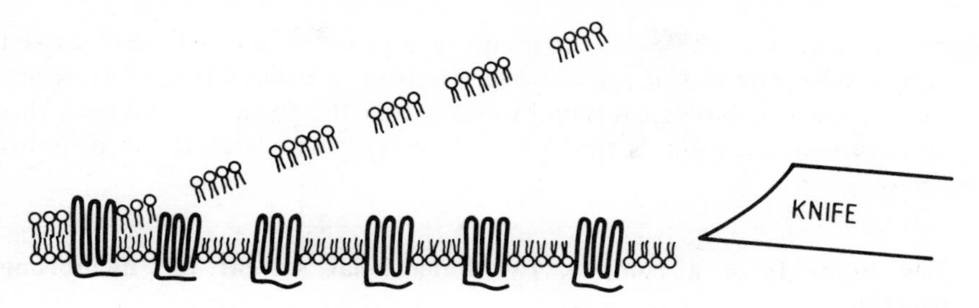


Figure 3.2. A schematic view of the freeze–fracture process

concerning the bilayer structure of lipids in biological membranes has now been cleared. X-ray techniques have been used to show extensive areas of lipid bilayer in several biological membranes.[3] Other techniques have confirmed and extended these observations. The freeze–fracture technique is perhaps the best.[8] If a frozen membrane is fractured, the plane of fracture will pass preferentially through the middle of the bilayer (Figure 3.2) separating the bilayer into two monolayer halves. No biological membrane is resistant to freeze–fracture so that all biological membranes must contain extensive areas of lipid bilayer. The fracture face of a human red cell membrane (Figure 3.3) reveals a large number of small particles that are

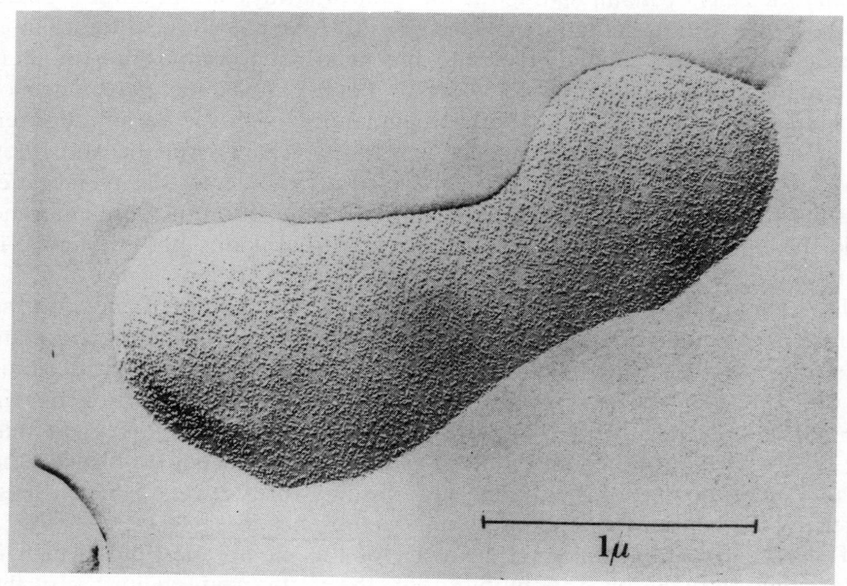


Figure 3.3. The surface structure of a freeze-fractured red blood cell membrane (kindly provided by Dr D. Branton, Biological Laboratories, Harvard, USA)

resistant to fracture. These are the membrane proteins that will be discussed in detail below. There is at present every reason to believe that most, if not all, of the lipids in biological membranes are in the form of a bilayer; this applies even to those lipids that interact directly with membrane proteins (see below).

### (ii) The lipids form a complex two-dimensional solvent for membrane proteins

The human red cell membrane contains approximately 200 species of lipid.[9] This diversity is important for proper cell functioning but the detailed reasons for the diversity are still unclear. What is clear is that two bulk parameters affected by changes in lipid composition markedly affect the functioning of biological membranes. These are the over-all charge of the membrane lipid headgroups and the viscosity of the bilayer.

The over-all charge of a eukaryotic plasma membrane is determined largely by the ratio of zwitterionic phospholipids (phosphatidylcholine, sphingomyelin, and phosphatidylethanolamine) to anionic phospholipids (phosphatidylserine, phosphatidylinositol, and phosphatidic acid). This ratio can vary widely from organism to organism but for a particular cell it is usually constant. Attempts to manipulate the headgroup composition of eukaryotic cells by manipulating the diet almost always end in failure. There are however, certain mutants of *Neurospora crassa* that do incorporate lipid precursors supplied in their diet and the headgroup composition of their membrane lipids can be varied widely.[10] For example, the percentage of phosphatidylcholine or phosphatidylethanolamine could be varied between 1% and 50%. Yet these mutant cells grow remarkably well because they have not lost the ability to control the over-all charge on the membrane. Despite the wide differences in the chemical composition of the membrane lipid headgroups the ratio of zwitterionic to anionic phospholipids is maintained.

The charge on a phospholipid headgroup can have a marked effect on the activity of membrane proteins. Mitochondrial $\beta$-hydroxybutyrate dehydrogenase is only active in the presence of the zwitterionic phosphatidylcholines[11] while the $Na^+$, $K^+$-ATPase is particularly active in the presence of negatively-charged lipids.[12] It is reasonable to suggest that the diverse headgroup composition of membranes is needed to provide the correct environment for membrane proteins and this need explains, at least partially, the constant charge on plasma membranes.

The second parameter is the viscosity of the membrane lipids which is determined largely by the mobility of the fatty acid chains.[13] In the crystalline state the phospholipid molecules are packed closely together; the fatty acid chains are elongated and rigid. If the lipid is pure it will melt at a precise temperature; during melting solid clusters of lipid will coexist with lipid in the liquid state. Above the melting point the fatty acid chains are very mobile and the viscosity of the hydrophobic part of the bilayer is equivalent to that of a light oil.

Biological membranes contain a complex mixture of lipids and do not crystallize at a precise temperature.[14] Instead, as the temperature is lowered a series of phase separations occur with the higher melting point lipids separating out first as solid clusters in a fluid lipid matrix. In biological membranes there is no convincing evidence that solid clusters of lipid exist at the growth temperature. In fact many organisms react to a lowering of the environmental temperature by incorporating more unsaturated (and hence lower melting point) lipids into their membranes, thereby maintaining them in a fluid state.[15] There are at least two reasons for this. First, many membrane proteins become inactive when the lipid in their immediate environment becomes more viscous and eventually freezes. These proteins can only function in a fluid lipid environment.[15] Secondly, the permeability properties of a lipid bilayer depend on the physical state of the lipid and it is important for a functional membrane that the lipid matrix be essentially impermeable to ions and almost all small molecules. When the bilayer is in a crystalline state the permeability is essentially zero but a crystalline bilayer is inflexible and could easily crack, spilling the cell contents into the environment. At intermediate temperatures where solid clusters of lipid coexist with lipid in the liquid state, the permeability to small ions and molecules is increased dramatically because of the discontinuity at the interface between the liquid and crystalline lipid.[16] It is only at temperatures where all of the lipid is just in the fluid state that the membrane is both flexible and impermeable. At higher temperatures there is an increasing permeability to ions and small molecules. Cell membranes are neither too fluid nor too viscous; they support the activity of membrane proteins and remain impermeable to ions and small molecules.

### (iii) The lipid classes are unequally distributed between the two halves of the bilayer

Phospholipid molecules move rapidly in the plane of the membrane but they do not move spontaneously from one side of the bilayer to the other. It takes about 1 $\mu$s for a lipid to move 40 Å in the plane of the membrane; it takes 1,000,000,000 times longer for it to traverse the 40 Å thickness of the bilayer. If a phospholipid is inserted on one side of a membrane it will be restricted to lateral motion in that monolayer. If lipid biosynthesis results in an unequal insertion of different lipid classes into both sides of the membrane, the result will be a stable yet asymmetric bilayer.[17]

The lipids in the human erythrocyte membrane are asymmetrically distributed. The choline-containing phospholipids (phosphatidylcholine and sphingomyelin) are found mainly in the non-cytoplasmic half of the bilayer while the amino phospholipids (phosphatidylethanolamine and phosphatidylserine) are almost entirely located in the cytoplasmic half of the bilayer.[18] This distribution holds for a number of eukaryotic plasma membranes but not for intracellular membranes; these do not appear to show any marked asymmetry. The asymmetry of lipids in prokaryotic membranes varies widely. The biological function of lipid asymmetry is unknown and it may simply reflect limitations in the biosynthetic pathway. This conclusion is suggested by experiments showing that the enzymes of phospholipid synthesis are located on the cytoplasmic side of the membrane so they presumably insert phospholipid into the cytoplasmic half of the bilayer. How then is the phospholipid transported to the non-cytoplasmic monolayer? It seems that there are proteins in the membrane, called flippases, which specifically transport certain classes of lipid from one side of the membrane to the other. They recognize the phospholipid headgroups but not the fatty acid chains and the distribution of a lipid class between the two monolayers then simply depends on the equilibrium constant for the flippase. These proteins are abundant in growing cells; they are sparsely distributed in adult cells such as the human erythrocyte.

### (iv) Membrane proteins can determine the lateral organization of lipid molecules

Between 20% and 30% of the lipids in biological membranes interact directly with the penetrant surfaces of transmembrane proteins. These lipids can be visualized as a single bilayer shell of lipid that coats the hydrophobic protein surface and is termed the lipid annulus.[19] This structure has been described for several transmembrane proteins. The properties of the lipids in the annulus are distinct from those in the rest of the bilayer. They are more immobile because they interact with a protein surface that is rigid relative to that offered by other lipid molecules. They also exchange more slowly with the lipids in the bilayer than do lipids in the bilayer with each other. They solvate the membrane proteins preventing contact between the penetrant, hydrophobic protein surfaces. The interaction between annulus lipids and protein has been examined in most detail for the $Ca^{2+}$-ATPase from sarcoplasmic reticulum.[20] The ATPase activity of this protein is dependent on the viscosity of the fatty acid chains and the charge on the headgroup of the lipids in the annulus. The activity is not determined by those lipids outside the annulus. When the protein is embedded in a mixture of defined phospholipids, the lipids redistribute so that the annulus comprises those lipids that support highest ATPase activity. The lipids that support low activity are in the bilayer outside the annulus where they have no effect on activity. As an example, an annulus of cholesterol molecules supports negligible ATPase activity but if a phospholipid such as dioleoylphosphatidylcholine (DOPC) is added in quantities sufficient to form an annulus of DOPC molecules, nearly all the cholesterol is displaced from the protein surface and the high ATPase activity then reflects the interaction of the protein with DOPC.[19]

Many other membrane proteins exhibit a specificity for certain phospholipids and many membranes will contain proteins with opposing needs. If the annulus of each membrane protein comprises those lipids that support function and lipids which do not support function are excluded, then proteins with differing lipid specificities could function optimally in the same membrane.

## III. PROTEINS

Although the concept of the lipid as a bimolecular layer in biological membranes is more than 50 years old, the over-all disposition of proteins in biological membranes has only become clear during the past five years. However, the three-dimensional structure of at least one membrane protein is now known at the 7 Å level and others should follow rapidly. Much of the initial work on the topology of membrane proteins was carried out on the major proteins of the human red cell membrane.[1] This has proved to be an ideal membrane for study for several reasons: it is the only membrane in the mature human erythrocyte so that there is no contamination by intracellular membranes; it can be prepared in large quantities thereby facilitating biochemical studies; it can be broken and resealed either the right-side out or inside-out. This last point is particularly important since it allows us to probe independently those parts of a protein that are exposed on each side of the membrane. In this section we shall discuss three of the major proteins of the human erythrocyte membrane: the anion transport protein, which has an apparent molecular weight of about 100,000; glycophorin (molecular weight of about 30,000); and one of the cytoskeletal components of the erythrocyte membrane known as spectrin (a dimer with a molecular weight in excess of 400,000). Other membrane proteins will be referred to where they serve to emphasize a particular point.

### (i) Many membrane proteins span the bilayer

There has been a curious reluctance to accept the possibility that the polypeptide chain of a membrane protein might repeatedly span the bilayer. This reluctance was based in part on the experiments that led to the concept of a lipid bilayer. The experiments that first demonstrated the spanning nature of membrane proteins are central to our present understanding of the disposition and functioning of proteins in membranes and so they will be discussed in some detail below. Nowadays it is difficult to conceive how a membrane could function if the proteins did not span the bilayer.

The first clear indications that membrane proteins might span the membrane bilayer came from the appearance of membranes that had been frozen and fractured. The particles of protein seen in the fracture plane of human erythrocyte membranes (Figure 3.3) would seem capable of spanning the bilayer in the native membrane. Definitive evidence had to await, however, the development of several new techniques. One was electrophoresis in polyacrylamide gels in the presence of sodium dodecyl sulphate (SDS–PAGE):[21] membranes are disrupted by the strong, anionic detergent SDS which displaces lipid from the hydrophobic protein surfaces and swamps the intrinsic charge of the protein. When sieved in a polyacrylamide gel the polypeptide chains migrate according to their molecular weight. This technique is simple and has a high resolving power. When applied to the plasma membrane of the human erythrocyte it reveals seven major proteins one of which (glycophorin) bears most of the membrane carbohydrate. The second technique involved the development of probes that would modify membrane proteins but would not pass across an intact membrane vesicle. It was then possible to determine which parts of a membrane protein were exposed on each side of a membrane. The probes that have been used most widely are proteases and chemical labelling reagents, particularly [$^{35}$S]diazosulphanilic acid.[17] All of the major proteins of the red cell membrane are available for digestion by proteases or chemical labelling if the membrane is fragmented. This immediately shows that all the major proteins are exposed on one or both surfaces and that no protein is inaccessible because it is located wholly within the hydrophobic phase of the bilayer. In intact red blood cells only two of the seven major proteins can be labelled; most of the protein is therefore located on the cytoplasmic side of the bilayer. By showing that these two proteins could be differentially labelled in intact and fragmented membranes, Bretscher was able to show that both spanned the bilayer.[22,23] The two proteins are the anion transport protein and glycophorin; they have entirely different functions in the membrane.

That part of a membrane protein which spans the bilayer and interacts directly with membrane lipids is comprised of hydrophobic amino acids. When the membrane lipids are displaced, this hydrophobic surface binds detergent instead. As a general rule each molecule of a transmembrane protein will bind one micelle (approximately 100 molecules) of Triton X-100. Membrane proteins that do not span the bilayer and do not even penetrate the hydrophobic phase of the lipid bilayer do not in general bind detergent. By determing the degree to which a membrane protein binds detergent, it is possible to estimate the extent to which it penetrates the lipid bilayer.[24]

### (ii) Membrane proteins that do not span the bilayer are bound to those that do

If a membrane protein does not span the bilayer it can only be labelled from one side of the membrane. The extent to which they penetrate the bilayer can be estimated from the amount of detergent that can be bound[24] and there are many membrane proteins that bind no detergent at all. One example is spectrin, an elongated molecule which is bound indirectly to the anion transport protein on the cytoplasmic side of the membrane. This will be discussed later. The most important point to note here is that membrane proteins which do not span the membrane must be anchored, however indirectly, to those that do. This applies even to those membrane proteins that penetrate the lipid bilayer quite deeply even though they do not span. The reason lies in the specialized nature of eukaryotic cell membranes. Each organelle is bound by a membrane uniquely designed to serve the needs of that organelle. The number and types of protein molecules, lipids, and carbohydrates found there will differ from organelle to organelle. One face of all organelle membranes share the same environment, the cytoplasm. It is obviously important that the proteins on the cytoplasmic face of one organelle membrane do not migrate to the membrane of another organelle. If a membrane protein is attached to the lipid bilayer either by interacting with the lipid headgroups or with the fatty acid chains, there is no reason why this protein should not partition between the various membrane surfaces available to it.[1] Transmembrane proteins cannot partition because they comprise a hydrophobic segment separating two hydrophilic segments. The only way to prevent migration of a non-spanning protein is to anchor it to one that does span.

### (iii) Transmembrane proteins

#### *(a) All transmembrane proteins are structurally asymmetric and all copies face the same way*

The cytoplasmic face of the anion transport protein can be labelled more extensively than the non-cytoplasmic face.[22] This clearly shows that most of the protein is exposed on the side of the membrane facing the cytoplasm and that the molecule is structurally asymmetric. In intact cells, treatment with pronase[25] reduces the molecular weight of the anion transport protein by about one-third; all copies are digested by pronase; none are resistant. Hence these asymmetric molecules must all face the same way in the membrane. These results are corroborated by the appearance of the anion transport protein in freeze–fracture replicas (Figure 3.3). All the particles are associated with the cytoplasmic monolayer and few, if any, with the non-cytoplasmic monolayer. The bulk of the protein on the cytoplasmic face causes the particles to go with the cytoplasmic monolayer.

### (b) *The transmembrane polypeptides do not wander in the hydrophobic bilayer*

The use of chemical and enzymic probes reveals many important features of membrane protein structure, but to understand the disposition of the protein at the atomic level it is essential to use X-ray techniques. The difficulties in applying these techniques to membrane proteins are technical; it is difficult to crystallize membrane proteins. Some proteins exist as two-dimensional crystals in the native membrane. An example is bacteriorhodopsin, a prokaryotic membrane protein that pumps protons upon absorption of light. The three-dimensional structure at the 7 Å level has been elucidated by Henderson and Unwin[26] and is shown in Figure 3.4. There is one striking feature that we are concerned with here. Each polypeptide chain spans the bilayer and does so more or less at right angles to the plane of the bilayer, there is no extensive wandering in the hydrophobic phase. This, together with the fact that 85% of the protein is within the membrane makes it difficult to see how the protein pumps $H^+$. The complete amino acid sequence has still to be elucidated.

On the other hand the three-dimensional structure of glycophorin is unknown but the amino acid sequence has been elucidated.[27] The transmembrane polypeptide is a stretch of 23 hydrophobic amino acids which is just sufficient to span the bilayer as an $\alpha$-helix at right angles to the plane of the membrane. Again there is no extensive wandering of polypeptide in the hydrophobic phase.

This feature, if general, is important and we feel it underscores the limitations of the biosynthetic machinery discussed below.

### (c) *Transmembrane proteins are globular or fibrous*

Bretscher has distinguished two types of membrane protein which are exemplified by glycophorin and the anion transport protein and are drawn schematically in Figure 3.5. The anion transport protein is a globular transmembrane protein because its polypeptide chain spans the bilayer many times. Because of this globular membrane proteins are visible in membranes that have been frozen and fractured. That part of the protein that is not within the membrane is found mainly on the cytoplasmic side of the bilayer.

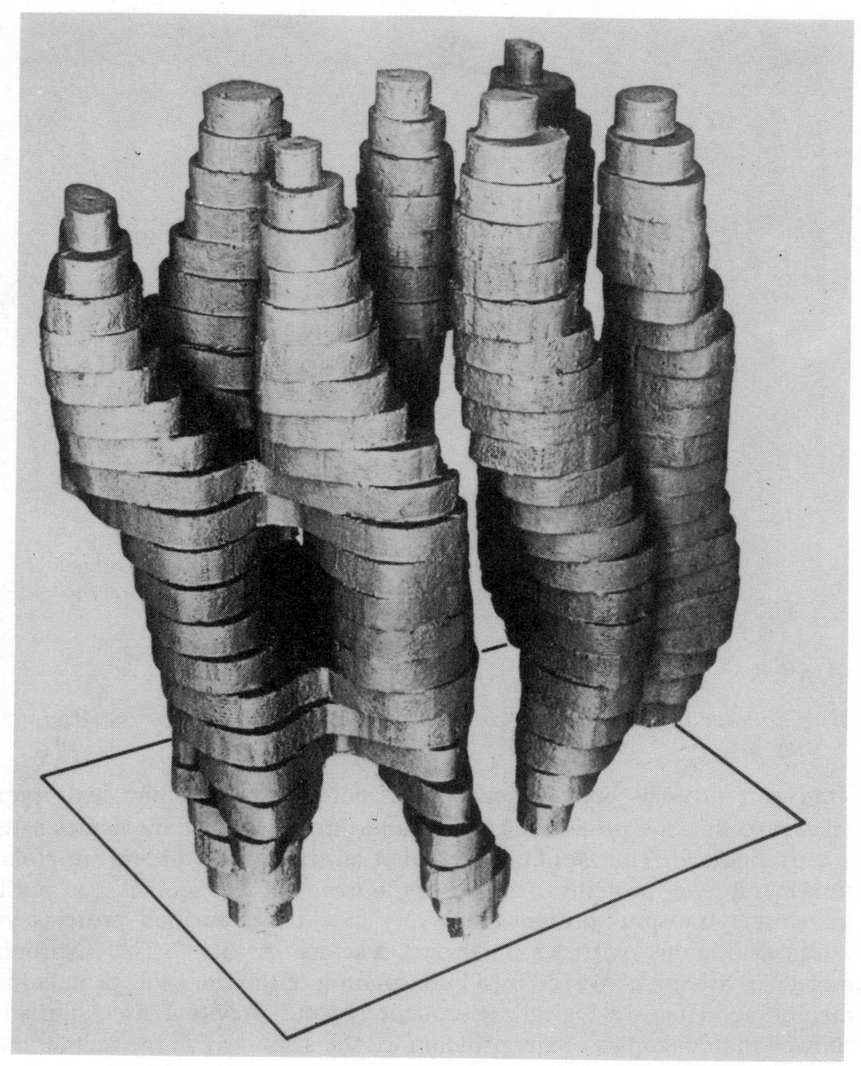

Figure 3.4. A model of bacteriorhodopsin at the 7 Å level. (Reproduced with permission from Reference 26)

Glycophorin is a fibrous transmembrane protein because it passes only once a cross the lipid bilayer and hence is not readily visible in freeze–fractured membranes. The particles seen in fractured erythrocyte membranes (Figure 3.3) are the anion transport proteins though they could represent a complex of this protein with glycophorin. These two types of membrane protein have different functions.

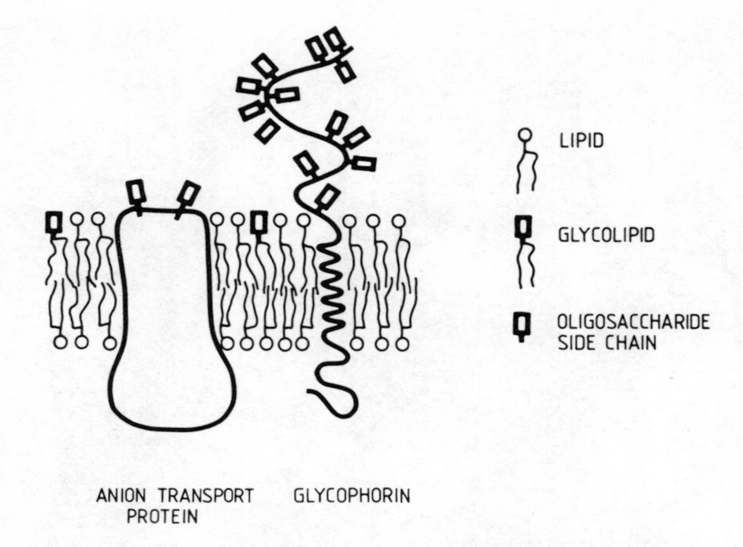


Figure 3.5. A schematic cross-section of a red cell membrane showing the disposition of the lipids, carbohydrate, and the two major transmembrane proteins

*(d) Globular transmembrane proteins convey information or material across the bilayer*

Transport proteins convey hydrophilic molecules across the hydrophobic bilayer so that the protein must surround the hydrophilic molecules as they are transported. This would suggest that all transport proteins are globular transmembrane proteins and this is borne out by studies on purified membrane transport proteins. In every case these purified proteins yield intramembranous particles in freeze–fracture replicas. The hydrophilic molecules are not conveyed by a bulk rotation of the transport protein in the membrane. This is a logical consequence of our previous discussion which showed that all copies of any protein face the same way in the membrane; if the membrane lipids move across the bilayer only once in three months the proteins would not be expected to traverse the bilayer at all. This conclusion underscores a fundamental feature of membrane function and organization: the passage of material and information across the membrane is not accompanied by a movement of membrane components from one side of the bilayer to the other. The transport of small ions and molecules across membranes must occur via a channel in the core of the transport protein. Information is conveyed across the bilayer by globular transmembrane proteins such as hormone receptors. These are discussed below.

*(e) Fibrous transmembrane proteins can anchor globular enzymes or act as a recognition marker*

The aminopeptidase from the intestinal brush border is a globular enzyme that is bound to the membrane by a short hydrophobic segment which spans the bilayer.[28] The enzyme can be released from the membrane by cleavage with specific proteases and this has no significant effect on the enzymic activity of the protein. This globular enzyme is bound to the membrane surface so that its digestion of proteins in the intestine releases amino acids near to the amino acid transport proteins of the brush border membrane.

Fibrous transmembrane proteins such as glycophorin serve no clear function. There are even individuals who lack glycophorin in their red cell membranes and they seem to suffer no ill-effects.[29] There is however, evidence that glycophorin serves to identify the red cell to neighbouring cells and that the carbohydrate side chains are involved in this process. It is interesting that those individuals lacking glycophorin have more carbohydrate residues attached to the anion transport protein. In normal red cells for example the removal of terminal sialic acid results in their removal from the blood stream by the liver. Liver cells can clearly sense the carbohydrate fingerprint of red cell membranes.

**(iv) The lateral mobility of proteins**

Frye and Edidin[30] were the first to demonstrate that membrane proteins could diffuse in the plane of the membrane. They took two cell populations and labelled the plasma membrane proteins with different fluorescent markers. Fusion between a cell of each type was followed by a fairly rapid intermixing of the labelled membrane proteins. The rate of diffusion did, however, suggest that the lateral mobility was being restricted perhaps because the proteins were being anchored to underlying cytoskeletal elements. Such attachments have been invoked for macromolecular processes such as the capping of surface antigens in lymphocytes after they have been cross-linked by antibodies or lectins. There has recently been some direct evidence for this[31] but the only clear example is the tethering of spectrin molecules to the anion transport protein. This tethering restricts the lateral mobility of the anion transport protein. The attachment is however, indirect; spectrin is bound to the transport protein via a protein with a molecular weight of 70,000.[32]

Many membrane proteins are free to move in the plane of the bilayer. Rhodopsin is almost the only membrane protein in the retinal discs of the rod photoreceptor and these molecules collide 100,000 times every second. It has been suggested that this high lateral mobility is fundamental to the functioning of the protein;[33] when a quantum of light bleaches one of the rhodopsin molecules, it binds to an unbleached molecule and this forms a pore which allows $Ca^{2+}$ to escape. The release of $Ca^{2+}$ into the cytoplasm hyperpolarizes and triggers the rod cell because it blocks the $Na^{+}$ channel in the plasma membrane.

Several microsomal proteins are known to derive a large kinetic advantage by being anchored to the membrane and their functioning depends on their lateral mobility in the membrane which allows them to interact with each other.[38] The activity of any single microsomal protein is not affected by the fluidity of the bilayer because the fibrous transmembrane portion is simply an anchor. In this they differ from most globular transmembrane proteins which are affected by the viscosity of the bilayer.

Perhaps the most elegant example of the need for lateral mobility is the coupling of hormone receptors to adenylate cyclase. This is discussed below.

### (v) Oligomers of membrane proteins

The functional unit of membrane proteins need not be the monomeric form: the anion transport protein for example is a dimer in the red cell membrane. The evidence for other membrane proteins is more equivocal. If the oligomer consists of many subunits its function can be to locate specific proteins in specific areas of the membrane. This is the case for the acetylcholine receptor at the motor end plate (see below).

### (vi) Membrane protein biosynthesis

The biosynthetic pathway of membrane proteins is dictated by their disposition in the membrane. For plasma membrane proteins, three routes of biosynthesis can be distinguished.

1. The transmembrane proteins are synthesized at, and inserted into the membrane of the endoplasmic reticulum. Here they are free to intermix with other membrane proteins. Those proteins destined for the plasma membrane must then be segregated from other proteins and they are then transported via the Golgi complex to the plasma membrane. The synthesis of the anion transport protein and glycophorin are examples of proteins that are synthesized by this route which is depicted schematically in Figure 3.6.
2. The membrane proteins which are found on the cytoplasmic side of the plasma membrane and do not span the bilayer are synthesized in the same way as cytoplasmic proteins. As an example, spectrin is synthesized on cytoplasmic ribosomes and then diffuses to the plasma membrane where it interacts, indirectly, with the anion transport protein (see Figure 3.6).

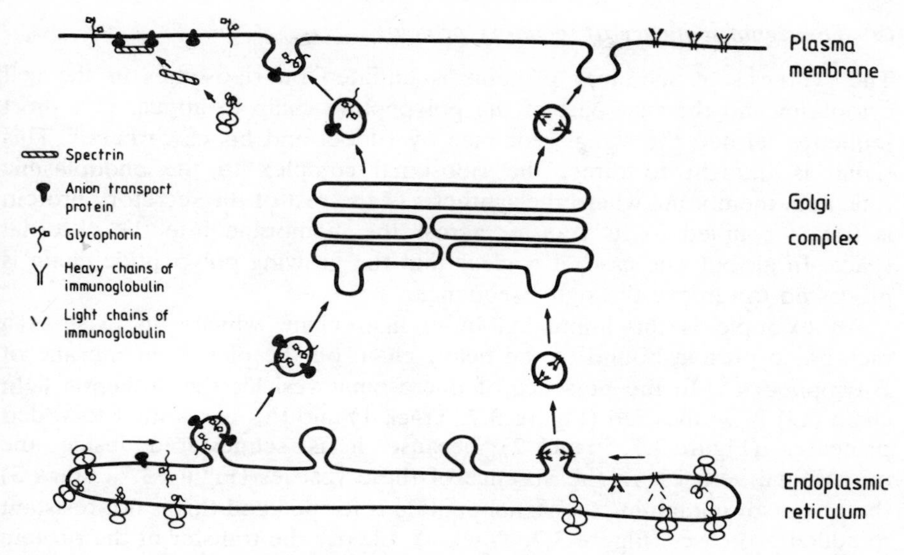


Figure 3.6. A scheme for the biosynthesis of different classes of plasma membrane protein

3. The membrane proteins which are located on the non-cytoplasmic side of the plasma membrane and do not span the bilayer are synthesized in the same way as secretory proteins. Membrane-bound ribosomes transfer the nascent polypeptide chain into the lumen of the endoplasmic reticulum where it is then free to interact with the appropriate extracellular sites on the newly-synthesized transmembrane proteins. An example, depicted in Figure 3.6, is the light chain of immunoglobulin which together with the heavy chain forms antibody molecules. In B-lymphocytes the heavy chain is bound to the membrane (it either spans the membrane or is bound to another protein which does span) and the secreted light chain can bind to this.

All three classes of plasma membrane protein are synthesized on cytoplasmic ribosomes *in vivo*. If they are synthesized *in vitro* in the absence of membrane vesicles, the transmembrane and secretory proteins (classes 1 and 3) cannot then be inserted into a membrane. They must be inserted into the membrane during synthesis. The cytoplasmic ribosomes must therefore be directed to the endoplasmic reticulum membrane if they are synthesizing a transmembrane or secretory protein but not if they are synthesizing a cytoplasmic protein. What then is the nature of the information that causes ribosomes to insert a growing polypeptide chain into the endoplasmic reticulum membrane?

*(a) The signal sequence of secretory proteins*

The synthesis of secretory proteins is initiated on ribosomes in the cell cytoplasm and the first part of the polypeptide chain to appear is a short sequence termed the signal sequence by Blobel and his coworkers.[35] This signal is thought to direct the ribosomal complex to the endoplasmic reticulum membrane where the synthesis of the rest of the secretory protein is tightly coupled to its transfer across the membrane into the cisternal space. In all but one case, it is clear that the growing polypeptide chain is processed to remove the signal sequence.

An example is the immunoglobulin light chain which can exist as a membrane protein bound to the heavy chain on the plasma membrane of B-lymphocytes. In the presence of microsomal vesicles the authentic light chain (Li) is synthesized (Figure 3.7, Track 1) and this is resistant to added proteases (Figure 3.7, Track 2) because it is sequestered inside the microsomal vesicles. In the absence of these vesicles (Figure 3.7, Track 3) the precursor containing the signal peptide is formed and this is not resistant to added proteases (Figure 3.7, Track 4). Clearly the transfer of the protein into the vesicles is tightly coupled to the removal of the signal peptide. The signal sequences of many secretory proteins are now known. They are 15–30 amino acids in length, hydrophobic, and the sequence varies from protein to protein. They must recognize certain proteins on the endoplasmic reticulum membrane. It is this feature that distinguishes the synthesis of transmembrane and secretory proteins from cytoplasmic proteins. If microsomal vesicles are washed with 0.5 M KCl, membrane proteins are

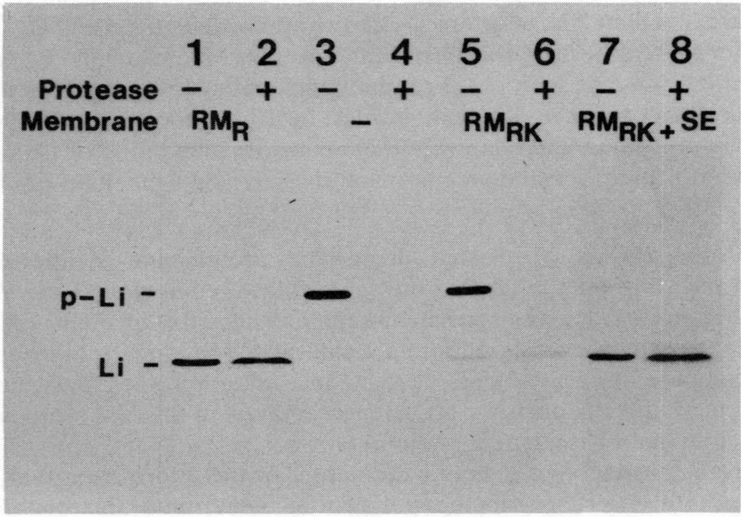


Figure 3.7. The *in vitro* synthesis of immunoglobulin light chain

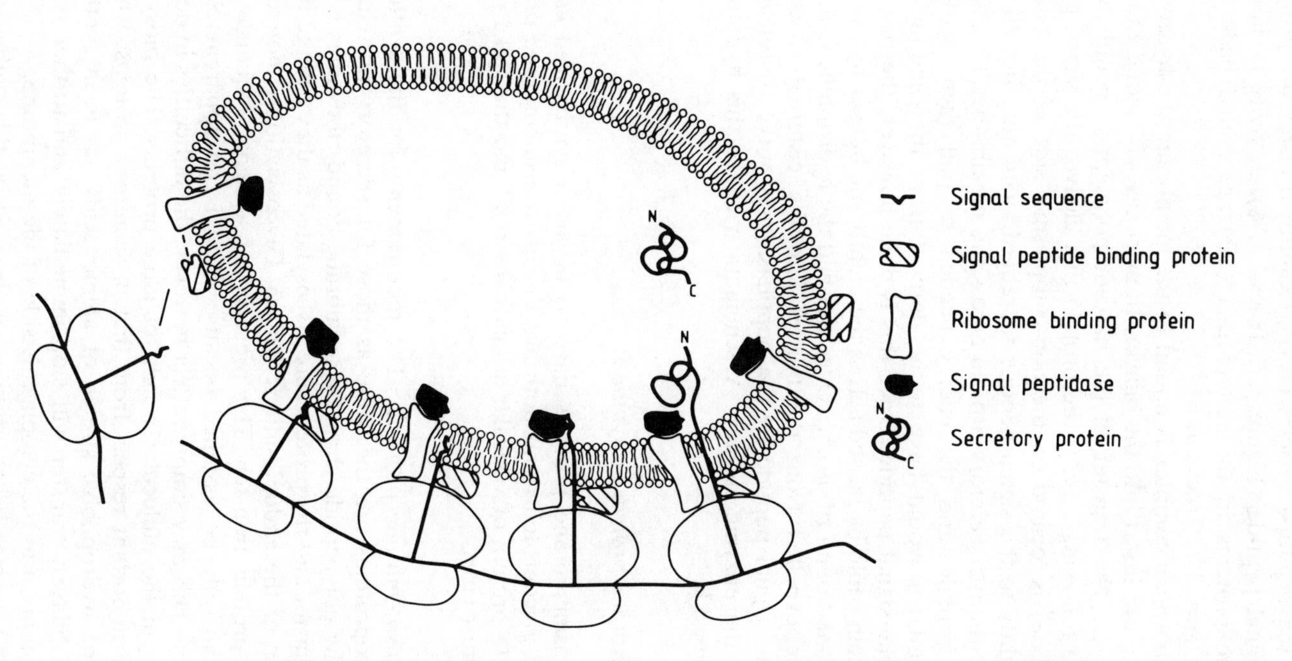


Figure 3.8. A model for the transfer of secretory proteins across the endoplasmic reticulum membrane

removed and the resulting vesicles ($RM_{RK}$) cannot transfer the protein or remove the signal peptide (Figure 3.7, Tracks 5, 6). Activity is, however, restored if the proteins in the salt extract (SE) are added back to the membranes (Figure 3.7, Tracks 7, 8).[36]

Once the ribosomal complex is bound to the membrane the hydrophobic signal sequence can insert into the bilayer. In most cases it is long enough to span the bilayer. The transfer of the nascent polypeptide could then be effected in the following way. If the protein that cleaves the signal peptide (signal peptidase) is located on the non-cytoplasmic side of the bilayer it could bind tightly to the signal peptide thereby drawing the first 20 amino acids of the authentic secretory protein across the membrane. These amino acids are hydrophilic and the energy needed to pull them across the hydrophobic bilayer would have to be derived from the binding of the peptidase to the signal peptide. Once this is done however, the rest of the secretory protein could be passed across the membrane bilayer by the action of the membrane-bound ribosome. Once finished the hydrophilic C-terminal end would be expelled from the bilayer into the cisternal space. The cleavage of the signal peptide by the peptidase may serve to release this peptide from the enzyme surface. A schematic view of this hypothesis is shown in Figure 3.8.

### *(b) Transmembrane protein biosynthesis*

Only two transmembrane proteins have so far been synthesized *in vitro*; both are viral glycoproteins and both are fibrous transmembrane proteins. They are the G protein of vesicular stomatitis virus[37] and the spike protein of Semliki Forest virus.[38]

*Fibrous transmembrane proteins* The requirements for the synthesis of the viral glycoproteins are the same as those for secretory proteins. The synthesis of the polypeptide chain on membrane-bound ribosomes is tightly coupled to transfer and the proteins are glycosylated as they emerge into the cisternal space of the endoplasmic reticulum. Glycosylation is not required however for protein transfer. The presence of a signal sequence at the N-terminus can only be assumed because there is no apparent cleavage during transfer. This may simply reflect however, the difficulties in detecting small changes in the molecular weight of large proteins. The biosynthesis differs in one important respect from that of secretory proteins: when the transmembrane hydrophobic stretch of amino acids has been synthesized and span the bilayer, transfer will stop immediately and further synthesis will expel protein on to the cytoplasmic side of the membrane.

The fibrous transmembrane proteins synthesized by this pathway would have the N-terminal amino acid on the non-cytoplasmic side of the membrane. This is true for both viral glycoproteins. There is also no limitation to the amount of protein on each side of the membrane. Globular enzymes on either side of the membrane could be anchored to fibrous transmembrane segments. Increasingly, however, there is evidence for globular enzymes on the non-cytoplasmic side of the bilayer that are attached to the membrane by N-terminal extensions distinct from the signal sequence. One example is the bacterial penicillinase from *Bacillus licheniformis*. These may represent proteins that are secreted and then bind to membrane.

*Globular transmembrane proteins* A major part of these proteins is embedded in the hydrophobic part of the bilayer and it is reasonable to conclude that the polypeptide chain of these proteins traverse the bilayer a number of times. An example that has already been discussed is bacteriorhodopsin. Although attempts to synthesize this type of protein *in vitro* are now under way there are no conclusive results. How do we envisage the pathway of biosynthesis of this type of protein? Bretscher has argued that the protein could be synthesized in the cytoplasm and then migrate to the membrane. It would eventually be located in the endoplasmic reticulum membrane because it is here that sugars are added to those parts of the protein appearing on the non-cytoplasmic side of the membrane. The addition of these sugar residues would prevent further partitioning between membranes.[1] An alternative view and one that we favour for plasma membrane proteins is that they are synthesized by a modification of the pathway used to synthesize fibrous transmembrane proteins. After the insertion of the first transmembrane segment of the protein the synthesis of a sequence of hydrophilic amino acids will result in its expulsion on the cytoplasmic side of the membrane. If this is followed by a sequence containing only neutral and hydrophobic amino acids this could insert into and across the membrane. It would be anchored by glycosylation. This process could be repeated until the globular transmembrane protein is completed. This mechanism relies mainly on the presence of amino acid sequences that would insert into the hydrophobic phase of a lipid bilayer. To minimize the need for such sequences we would predict that the transmembrane segments would not wander extensively in the bilayer and that most of the protein would be present on the cytoplasmic face of the membrane; protein exposed on the non-cytoplasmic face would have needed to cross the hydrophobic phase. In so far as there are examples, they bear out these predictions. The transmembrane segments of bacteriorhodopsin have already been discussed and the anion transport protein has most of the protein on the cytoplasmic face of the membrane. This is also true for the $Ca^{2+}$-ATPase of sarcoplasmic reticulum. Furthermore, both of these proteins have few sites available for labelling on the non-cytoplasmic side of the membrane.

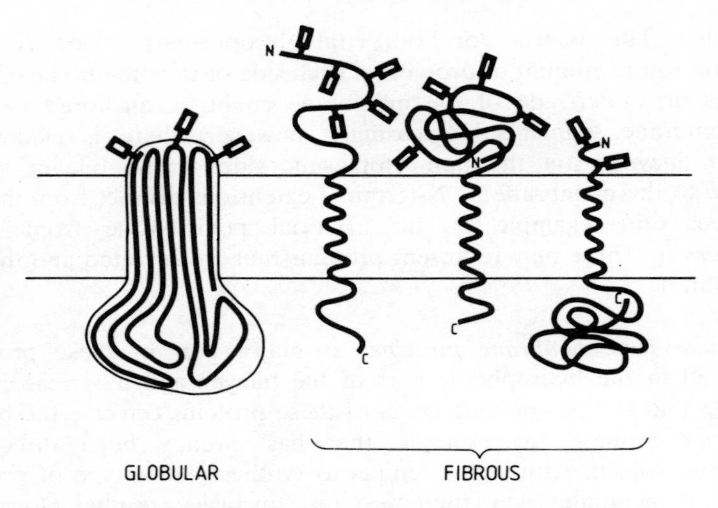


3.9. The disposition of the transmembrane proteins that can be synthesized in eukaryotic cells by ribosomes bound to the endoplasmic reticulum membrane

The biosynthetic pathways that have just been described clearly limit the disposition of proteins in the membrane (see Figure 3.9) but their most important feature is that they cause proteins to span the bilayer. Without this, information or material could not be easily conveyed across biological membranes. So far we have only discussed the signal that causes plasma membrane proteins to be inserted into the endoplasmic reticulum membrane. Additional information is also needed to transfer these proteins to the plasma membrane. The means by which this is accomplished is still obscure though one hypothesis will be explored in the next section.

## IV. CARBOHYDRATES

The sugar molecules bound to membrane lipids and proteins are only minor components of biological membranes. The enzymes that construct and transfer the oligosaccharide chains are located on the non-cytoplasmic side of the endoplasmic reticulum membrane so that only those lipids and parts of proteins on this side are covalently attached to sugar residues. As a general rule, if a protein is known to be associated with the cytoplasmic face of the membrane and it carries one or more oligosaccharide chains we can conclude that this membrane protein spans the bilayer. The type of oligosaccharide chain attached to a plasma membrane protein depends on the position it has reached on its way to the plasma membrane. This suggests that oligosaccharides might help to direct membrane proteins to their correct location in the cell. As an example we can take the addition of oligosaccharides to the G protein of vesicular stomatitis virus.[39] In the endoplasmic reticulum, two oligosaccharides, rich in mannose, are covalently attached to asparagine residues in the protein. As soon as these oligosaccharides are attached several glucose residues are removed. The protein is then transported to the Golgi complex. Here the oligosaccharides are trimmed by removal of all but three of the mannose residues. *N*-acetylglucosamine, galactose, fucose, and *N*-acetylneuraminic acid are then added. The protein is then transported to the plasma membrane.

If these sugar residues do function as signals for transfer to specific locations in the cell, it would be important to have components in the cell that recognize specific sugar residues and sugar sequences. Plant lectins have been widely used to probe the cell surface because they bind specific sugar residues. It is of interest that recent work has demonstrated the existence of lectins as components of mammalian cell membranes.[40]

## V. HORMONE RECEPTORS

The specific binding of a hormone to a receptor on the plasma membrane leads to an event that affects the cell interior. It is important to understand the nature of these events at the molecular level, although for many receptors this remains a remote goal. Studies that have been carried out on the disposition and functioning of the glucagon receptor, insulin receptor, and nicotinic acetylcholine receptor yield results for these minor membrane components that are compatible with the work done on major proteins that we have already discussed.

### (i) Glucagon receptor

This receptor is an intrinsic glycoprotein with all copies having their binding site for glucagon facing the cell exterior, where it is susceptible to attack by agarose-immobilized trypsin.[41] Glucagon immobilized on agarose beads will yield a biological response in intact cells[41] giving further support to the notion that functional glucagon receptors are orientated with their hormone binding sites to the outside. The stimulation of a cellular response by an immobilized hormone rules out the possibility that a hormone–receptor complex rotates bodily in the plane of the membrane to trigger this effect. It is still not known, however, if the glucagon receptor spans the bilayer. Acidic phospholipids appear to be important in the binding of hormone to the receptor[41] (see Chapter 10), and the fluidity of the cell membrane can modulate the biological response[42] (see Chapter 4). The binding of glucagon to its receptor achieves an increase in intracellular cyclic AMP levels by activating adenylate cyclase with its active site exclusively exposed at the cytosol surface of the membrane. This asymmetric organization of the two proteins achieves a vectorial flow of information across the bilayer membrane by a mechanism implying that both components are free to diffuse laterally in the plane of the bilayer and is discussed in Chapter.4.

**(ii) Insulin receptor**

Treatment of intact adipocytes with non-lytic concentrations of trypsin or agarose-immobilized trypsin destroys both the response of the cells to insulin, and their ability to bind hormone.[41] This demonstrates that the functional receptor site for insulin is exposed on the non-cytoplasmic surface of the plasma membrane of cells. This can be confirmed by freeze–fracture studies with ferritin–insulin molecules, and autoradiographic studies with radiolabelled insulin.[41,43] Agarose-immobilized insulin can also react with intact cells and stimulate a response, which eliminates mechanistic models suggesting a bodily rotation of the hormone–receptor complex in the plane of the membrane.

Carbohydrate associated with the receptor appears to be located at the extracytoplasmic surface and be involved with binding.[41] The molecule itself is an intrinsic protein (see Chapter 10) all copies of which are similarly asymmetrically disposed in the bilayer membrane. At present the mechanism is not appreciated by which insulin binding to receptors achieves its well-characterized physiological response (see Chapter 10).

**(iii) Nicotinic acetylcholine receptor**

The nicotinic receptor from *Torpedo* electroplax tissue is an intrinsic glycoprotein consisting of several polypeptide chains (see Chapter 18). The molecule is asymmetrically disposed in the bilayer with the carbohydrate moieties exclusively attached to those subunits located on the extracytoplasmic side of the membrane.[43] The different subunits may represent components of both the receptor unit and ion-translocation unit which are assembled as an oligomeric complex in the membrane. This receptor complex itself spans the bilayer membrane as ferritin conjugated antibodies against the purified soluble receptor binds to both sides of fragmented microsacs, and to the outer surface of intact microsacs.[44] This binding is not due to copies of receptor orientated at both sides of the bilayer, as the total receptor number as indicated by $\alpha$-bungarotoxin binding (see Chapter 18) is not increased upon fragmenting intact microsacs.

Although the density of receptors in subsynaptic membranes is so high that the receptor protein exists in an almost crystalline array, the lipid micro-environment of the receptor forms a fluid structure.[45] Indeed the relative fluidity of the lipid environment can determine to some extent the degree of opening of the channel.[46]

In the neuromuscular junction, nicotinic receptors attain a density some 100–1,000 times higher than in extrasynaptic regions forming a stable structure in which the receptors show no tendency for free-lateral diffusion even when the motor end plate is 'uncovered'. The stability of this structure may be related to the oligomeric nature of the receptor which allows the formation of such an assembly when initiated perhaps by some modification of their nature or by interaction with cytoskeletal components underlying the membrane in a similar fashion to the tethering of the anion transport protein to spectrin.[32] Thus the extra-junctional receptors undergoing free-lateral diffusion may be modified and incorporated into a tight lattice of clustered receptors.[47]

In common with the studies carried out on major membrane proteins, we can say for these hormone receptors that the molecules are structurally asymmetric with all copies facing the same way in the bilayer. They are all intrinsic membrane proteins with a surface exposed at the extracytoplasmic side of the bilayer where the carbohydrate moiety is exclusively located. It is likely that they all span the bilayer membrane and are able to undergo free-lateral diffusion in the plane of the membrane unless tethered by specific interactions to other components. They do not achieve their response by a gross rotation of a hormone–receptor complex in the plane of the bilayer, but achieve it by conformational changes or interaction with other components that may be modulated by the lipid environment.

## VI. REFERENCES


1. Bretscher, M. S. *Science,* **181**, 622–629, (1973).
2. Bretscher, M .S., and Raff, M. C. *Nature,* **258**, 43–49, (1975).
3. Oseroff, A. R., Robbins, P. W., and Burger, M. M. *Ann. Rev. Biochem.,* **42**, 647–682, (1973).
4. Stryer, L. *Biochemistry* Freeman & Co., San Francisco, pp. 227–252, (1975).
5. Franks, N. P. *J. Mol. Biol.,* **100**, 345–358, (1976).
6. Gorter, M. D., and Grendel, F. *J. Exp. Med.,* **41**, 439–443, (1925).
7. Danielli, J. F., and Davson, H. *J. Cell. Physiol.,* **5**, 495, (1956).
8. Branton, D. *Proc. Natl. Acad. Sci. U.S.A.,* **55**, 1048–1056.(1966).
9. Van Deenen, L. L. M. *Pure Appl. Chem.,* **25**, 25–56, (1971).
10. Hubbard, S. C., and Brody, S. *J. Biol. Chem.,* **250**, 7173–7181, (1975).
11. Fleischer, S., Bock, H. G., and Gazzotti, P. in *Membrane Proteins in Transport and Phosphorylation* Azzone, G. F., Klingenberg, M. E., Quagliarello, E., and Siliprandi, N., eds.) North Holland Publishing Co., Amsterdam, pp. 124–135, (1974).
12. De Pont, J. J. H. H. M., Van-Prooijen-Van-Eeden, A., and Bonting, S. L. *Biochem. Biophys. Acta,* **508**, 464–477, (1978).
13. Lee, A. G. *Biochem. Biophys. Acta,* **472**, 237–281, (1977).
14. Lee, A. G. *Biochim. Biophys. Acta,* **472**, 285–344, (1977).
15. Kimelberg, H. K. *Dynamic Aspects of Cell Surface Organisation* (Poste, G., and Nicolson, G. L., eds.) pp. 205–293, Elsevier/North Holland Biomedical Press, Amsterdam, pp. 205–293, (1977).

16. Marsh, D., Watts, A., and Knowles, P. F. *Biochemistry,* **15**, 3570–3578, (1976).
17. Rothman, J. E., and Lenard, J. *Science,* **195**, 743–753, (1977).
18. Bretscher, M. S. *Nature New Biol.,* **236**, 11–12, (1972).
19. Warren, G. B., Houslay, M. D., Metcalfe, J. C., and Birdsall, N. J. M. *Nature (London),* **255**, 684–687, (1975).
20. Metcalfe, J. C., and Warren, G. B. *Proceedings of the 1st International Congress of Cell Biology* (Brinkley, B. R., and Porter, J. R., eds.) Rockefeller University Press, New York, pp. 15–23, (1976).
21. Weber, K., and Osborn, M. *J. Biol. Chem.,* **244**, 4406–4412, (1969).
22. Bretscher, M. S. *J. Mol. Biol.,* **59**, 351–357, (1971).
23. Bretscher, M. S. *Nature New Biol.,* **231**, 229–232, (1971).
24. Helenius, A., and Simons, K. *Proc. Natl. Acad. Sci. U.S.A.,* **74**, 529–532, (1977).
25. Bender, W. W., Garan, H., and Berg, H. C. *J. Mol. Biol.,* **58**, 783–797, (1971).
26. Henderson, R., and Unwin, N. *Nature (London),* **257**, 28–31, (1975).
27. Tomita, M., and Marchesi, V. T. *Proc. Natl. Acad. Sci. U.S.A.,* **72**, 2964–2968, (1975).
28. Louvard, D., Semeriva, M., and Maroux, S. *J. Mol. Biol.,* **106**, 1023–1035, (1976).
29. Gahmberg, C. G., Myllyla, G., Leikola, J., Pirkola, A., and Nordling, S. *J. Biol. Chem.,* **251**, 6108–6116, (1976).
30. Frye, L. D., and Edidin, M. *J. Cell. Sci.,* **7**, 319–335, (1970).
31. Flanagan, J., and Koch, G. L. E. *Nature (London),* **273**, 278–281, (1978).
32. Bennett, V. *J. Biol. Chem.,* **253**, 2292–2299, (1978).
33. Montal., M., Darszon, A., and Tissl, H. W. *Nature (London),* **267**, 221–225, (1977).
34. Strittmatter, P., and Rogers, M. J. *Proc. Natl. Acad. Sci. U.S.A.,* **72**, 2658–2661, (1975).
35. Blobel, G. *Proceedings of the 1st International Congress of Cell Biology* (Brinkley, B. R., and Porter, K. R., eds.) Rockefeller University Press, New York pp. 318–325, (1976).
36. Warren, G., and Dobberstein, B. *Nature (London),* **273**, 569–571, (1978).
37. Katz, F. N., Rothman, J. E., Knipe, D. M., and Lodish, H. F. *J. Supramol. Struc.,* **7**, 353–370, (1977).
38. Garoff, H., Simons, K., and Dobberstein, B. *J. Mol. Biol*., **124**, 587–600, (1978).
39. Hunt, L. A., Etchison, J. R., and Summers, D. F. *Proc. Natl. Acad. Sci. U.S.A.,* **75**, 754–758, (1978).
40. Kaeasaki, T., and Ashwell, G. *J. Biol.Chem.,* **252**, 6536–6543, (1977).
41. Cuatrecasas, P. *Ann. Rev. Biochem.,* **43**, 169–215, (1974).
42. Dipple, I., and Houslay, M. D. *Biochem. J.,* **174**, 179–190, (1978).
43. Kahn, R. C. *J. Cell Biol.,* **70**, 261–286, (1976).
44. Tarrab-Hazdai, R., Geiger, B., Fuchs, S., and Amsterdam, A. *Proc. Natl. Acad. Sci. U.S.A.,* **75**, 2497–2501, (1978).
45. Bienvue, A., Rousslet, A., Kato, G., and Devaux, P. F. *Biochemistry,* **16**, 841–848, (1977).
46. Lass, Y., and Fisbach, G. D. in *Synapses* (Cottrell, G. A., and Usherwood, P. N. R., eds.) Blackie, Glasgow and London, pp. 338–339, (1976).
47. Anderson, M. J., and Cohen, M. W. *J. Physiol. (London),* **268**, 757–773, (1976).

Cellular Receptors for Hormones and Neurotransmitters
Edited by D. Schulster and A. Levitzki


CHAPTER 4

# Modes of membrane receptor–signal coupling

Ravi Iyengar, Lutz Birnbaumer
*Department of Cell Biology, Baylor College Med., Houston, Texas 77030, U.S.A.*
Dennis Schulster
*Hormones Division, National Institute for Biological Standards and Control, Hampstead, London, U.K.*
Miles Houslay
*Department of Biochemistry, University of Manchester Institute of Science and Technology, Manchester, U.K.*
Robert H. Michell
*Department of Biochemistry, University of Birmingham, Birmingham, U.K.*

## I. INTRODUCTION

A wide variety of regulatory ligands, including the peptide and glycoprotein hormones, some neurotransmitters, and the prostaglandins are now recognized as utilizing cyclic AMP to transmit information to the intracellular machinery.[1,2] In recent years, considerable efforts have been made to extract from the membrane the different molecular components of the adenylate cyclase system responsible for regulating cyclic AMP. Although our understanding of how these components interact in the intact cell remains limited, information is accruing on the characteristics of the proteins constituting the hormone receptor, the catalytic subunit, and the components involved in the GTP effect.

For the characterization of receptors associated with the cell membrane many different radioactively labelled ligands have been used. Because of the remarkably small number of receptors per cell, it is necessary to use a radioligand of very high specific activity with high affinity for the receptor. This is necessary to determine the specific binding as opposed to the binding of ligand to other membrane components with low affinity, contributing to non-specific binding. Analogues (agonists) and competitive inhibitors (antagonists) of the active ligand that have been radiolabelled with $I^{125}$ or $^{3}H$ to high specific activity, have been used for these studies.[3]

For actual analysis of the receptor–adenylate cyclase relationship, one prerequisite is the use of a labelled agonist with all the biological properties of the physiological ligand. Because of the heterogeneity of the labelled population of molecules the general use of iodinated ligands for this purpose is of questionable merit. This is particularly so for the small peptide hormones, in which the bulky iodine atom is more likely to induce alterations in the conformation and biological properties of the hormone. In these instances tritiated hormones (e.g. [$^{3}$H]ACTH; [$^{3}$H]vasopressin) have been utilized, but then only relatively low specific activities are attainable. A second methodological requirement relates to the measurement of adenylate cyclase activity. In order to evaluate the instantaneous rate of adenylate cyclase activity, and thereby determine the relationship between receptor occupancy and adenylate cyclase activation, highly sensitive methods for this evaluation are required. The use of [$\alpha$-$^{32}$P]ATP provides the required sensitivity, allowing trace doses of the substrate to be added to the hormonal response system at any time. [$^{32}$P]Cyclic AMP accumulating within a minute or two can thus be determined, while other constituents (e.g. substrate concentration) remain essentially unchanged. Furthermore, recent work has demonstrated that a variety of effectors, such as divalent cations and guanine nucleotides, influence the receptor–adenylate cyclase relationship. Thus a third practical requirement is that hormone binding and adenylate cyclase activation must be evaluated under identical experimental conditions.[4]

In this section we discuss the components of the cyclic nucleotide system responsible for modulating cellular machinery. This focuses on adenylate cyclase as the site of synthesis of cyclic AMP and its modulation by intracellularly acting effectors such as guanine nucleotides, adenosine, and divalent cations.

Also detailed is the current convincing evidence that the catalytic unit of adenylate cyclase and the hormone receptor are separate proteins able to undergo free lateral diffusion in the plane of the bilayer. They then interact when hormone binds to the receptor, achieving a vectorial transfer of information across the cell membrane by virtue of their asymmetric disposition in the bilayer.

In the final section another mechanism is discussed by which a number of hormones transmit messages across the bilayer membrane: the use of $Ca^{2+}$-dependent systems. For example, calcium can be demonstrated to be intimately involved in the action of $\alpha$-adrenergic and muscarinic receptors and to be the 'trigger' signal in the explosive degranulation of mast cells that leads to histamine release. A link between the $Ca^{2+}$-dependent hormonal systems and the cyclic AMP-dependent systems may operate via a $Ca^{2+}$-dependent modulator protein, also known as calmodulin. Removal of cyclic AMP from inside cells is regulated by phosphodiesterases, whose activity can be stimulated by $Ca^{2+}$ in a calmodulin-mediated process.

## II. COMPONENTS OF CYCLIC NUCLEOTIDE-LINKED SYSTEMS

### (i) Characteristics of adenylate cyclases and hormone receptors

Adenylate cyclase [ATP, pyrophosphate-lyase (cyclizing); E.C.4.6.1.1.] is predominantly a membrane-bound enzyme. It catalyses the formation of cyclic AMP and $PP_i$ from ATP. Hormonally sensitive adenylate cyclase is believed to be a multi-enzyme complex consisting of at least three physically separable components, (1) hormone receptor, bearing one specific binding site for hormone and possibly a second for nucleotide; (2) catalytic subunit of adenylate cyclase (C-component), bearing the catalytic site responsible for the cyclization reaction; and (3) regulatory subunit (G-component), bearing a guanine nucleotide binding site. Recent data indicate that the G-component is $\alpha$ GTPase.

The basic regulatory features of hormone-sensitive adenylate cyclases can be summarized as follows: (1) GTP stimulates adenylate cyclases in the absence of hormone; (2) non-hydrolysable analogues of GTP such as p(NH)ppG and GTP$\gamma$S are very potent and convert adenylate cyclase to a persistently active state; (3) the guanine nucleotide regulatory site appears to have a phosphatase activity (GTPase) which controls the 'turn off' process of the activated adenylate cyclase; (4) guanine nucleotides, in addition to their regulation of the catalytic activity of adenylate cyclase, modulate both the hormone–receptor interaction and the coupling of hormone–receptor complex to adenylate cyclase; (5) guanine nucleotides may regulate hormone binding and coupling through a distinctly different site from the one that is involved in regulation of cyclizing activity and is responsible for activation of basal activity by p(NH)ppG in the absence of hormone; (6) divalent cations, allosterically, affect the catalytic activity and are essential for guanine nucleotide regulation of the system; (7) divalent cations also modulate hormone–receptor interactions as well as coupling of the hormone–receptor to adenylate cyclase; (8) treatment of adenylate cyclase with cholera toxin (subunit $A_2$ in the presence of $NAD^+$) results in a stimulated state of activity in the presence of GTP; and (9) adenylate cyclases that can be activated by GTP are also stimulated by fluoride ions.

Most eukaryotic adenylate cyclases are intrinsic membrane proteins and require rather drastic treatments with substances such as non-ionic detergents to be solubilized from their membrane environment. The first serious study on the physical characteristics of the membrane-bound adenylate cyclase was done by Neer[5] with the solubilized renal enzyme. She found that the enzyme had an apparent molecular weight of 160,000 and a partial specific volume of 0.71 ml $g^{-1}$, the latter being a value similar to that obtained with naturally soluble globular proteins. She concluded that the adenylate cyclase moiety did not bind substantial amounts of the detergents and hence had only a small hydrophobic surface. Subsequent studies by Neer and others have shown that the adenylate cyclase from bovine cerebral cortex and mouse lymphoma S49 cells have much higher partial specific volumes (0.78 ml $g^{-1}$) and bind up to 0.3 mg detergent per mg protein. The bovine enzyme has a sedimentation coefficient of 8.1 S, a Stokes radius of 7 nm and a molecular weight of 220,000. The enzyme from lymphoma S49 cells has a sedimentation coefficient of 7.5 S, a Stokes radius of 7.1 nm and a molecular weight of 270,000. These size estimates for adenylate cyclase include both the catalytic subunit and the guanine nucleotide binding subunit since the enzymes studied were stimulatable by guanine nucleotides. In addition to the guanine nucleotide-sensitive, detergent-soluble adenylate cyclases, there are at least two naturally soluble adenylate cyclases that are insensitive to guanine nucleotide regulation: (1) a bacterial adenylate cyclase from *Brevibacterium liquefaciens* using MgATP as substrate and (2) a mammalian adenylate cyclase found in rat testis and bovine sperm, both catalytically inactive with MgATP as substrate, but active with MnATP as substrate.

That β-adrenergic receptors are physically separable from adenylate cyclase has been demonstrated[6] by subjecting the p(NH)ppG-stimulated enzyme system of frog erythrocyte membranes to gel filtration over Sepharose-6B and separating receptor from cyclase activity. A physical study[7] of the β-adrenergic receptors of the S49 cell revealed the receptor to be a very hydrophobic protein capable of binding 0.8 mg detergent per mg of protein with a sedimentation coefficient of 3.1 S, a partial specific volume of 0.83 ml $g^{-1}$, a Stokes radius of 6.4 nm, and an estimated molecular weight of 130,000 (or 75,000 when allowance was made for bound detergent).

Studies on the Triton X-100 solubilized gonadotropin receptor[8] showed this to have a sedimentation coefficient of 6.5 S, a Stokes radius of 6.4 nm, and a molecular weight of about 170,000. The existence of a naturally soluble lipid-associated gonadotropin receptor in luteinized ovaries has also been reported. The physicochemical properties of this naturally soluble receptor are reportedly similar to Triton X-100 extracted receptors.

### (ii) Guanine nucleotide regulation of adenylate cyclase in the absence of hormones

An important role for GTP in regulating the activities of adenylate cyclases was found in the glucagon-stimulated system of rat liver plasma membranes.[9,1]. It was subsequently found that the 'non-hydrolysable' analogues of GTP, such as p(NH)ppG and GTPγS (i.e. analogues with non-hydrolysable linkages between their β- and γ-phosphates, or with a modified phosphate) are potent activators of membrane-bound adenylate cyclases. Activation by analogues of GTP has been shown to be very persistent, although p(NH)ppG activation of adenylate cyclase is time-dependent, requiring up to 20 min to reach steady state. In the article first describing the actions of p(NH)ppG,[10] the reason for its potency was proposed as the non-hydrolysable nature of its NH link between the β- and γ-phosphates, and it was suggested that the lesser effects of GTP might be due to hydrolysis of its γ-phosphate at the regulatory site. Rodbell and coworkers proposed that slow activation of adenylate cyclase by p(NH)ppG was a two step process involving three states wherein p(NH)ppG binds rapidly to enzyme, yielding a non-active complex which then slowly isomerizes to its final active state.[11] Studies on the activation of mouse neuroblastoma cell membrane adenylate cyclase[12] showed that activation by p(NH)ppG proceeded without any detectable lag if the enzyme was subjected to preincubation at 30 °C for 5–7 min. Based on this and on the fact that GDP is a potent inhibitor of p(NH)ppG activation, it was proposed that adenylate cyclase activation proceeded slowly due to slow displacement of a tightly bound inactive guanine nucleotide (GDP) rather than by slow isomerization. Current research indicates that both explanations apply and that in different systems the ratios of the rates of isomerization to rates of nucleotide dissociation vary. Since these initial observations, guanine nucleotide activation of membrane bound adenylate cyclase has been shown to be a universal feature and qualitatively similar in widely differing species such as flukes, turkeys, and rats.

### (iii) GTPase activity

Cassel and Selinger found a catecholamine-stimulated, cholera toxin-inhibited GTPase activity in the turkey erythrocyte membrane.[13,14] This GTPase has an apparent $K_m$ for GTP of 0.1 $\mu$M, similar to the concentration of GTP needed to half maximally activate adenylate cyclase in a number of systems. Inhibition of the GTPase activity occurred concomitantly with activation of adenylate cyclase by cholera toxin and these authors proposed that adenylate cyclase was an enzyme in constant turnover and regulated by an intrinsic GTPase. In their model the active form is the adenylate cyclase-GTP complex. In this complex form, GTP is the substrate for the GTPase leading to release of free adenylate cyclase (inactive), GDP and $P_i$. Hence, inhibition of GTPase by cholera toxin leads to activation of the adenylate cyclase activity of the system. Steady-state activities of the system depend on both rate of formation of GTP-enzyme complex and on its rate of degradation. Hormonal stimulation of cyclizing activity was proposed[14] to be

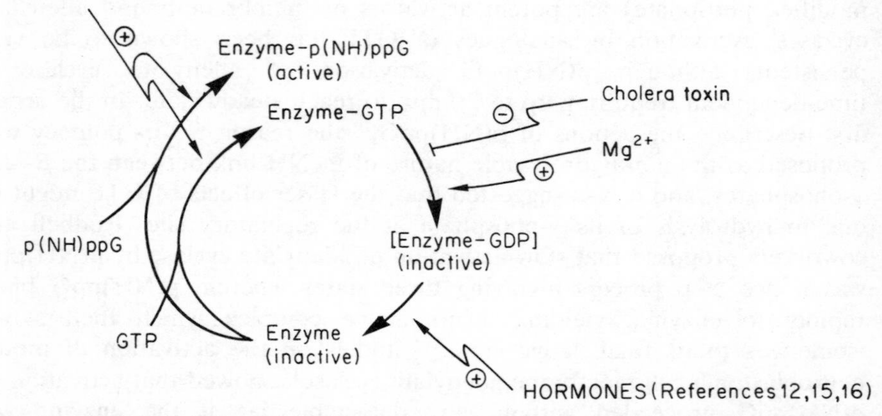


Figure 4.1. GTPase based turnover models for adenylate cyclases. This scheme is based on the observations of Cassel and Selinger[13,14,16] and Blume and coworkers.[12,15] For further details see text, (Section II. iii) Superior numbers in the figure indicate references

due to increased rates of formation of active GTP-enzyme complex which also resulted in an increased supply of substrate for the GTPase activity. Thus, GTPase activity represents a measure of the turnover of the enzyme system between its inactive (free) and active (GTP-occupied) states. The views of Blume and coworkers would include the intermediary formation of GDP from this complex as the site of action of hormonal regulation.[15] Indeed a recent report by Cassel and Selinger[16] shows that hormonal stimulation of membranes to which [$^3$H]GDP was bound results in release of this nucleotide. While the system needs further work to describe its details, it is clear that adenylate cyclases are regulated by turnover rather than by simple equilibrium reactions. This scheme is represented in Figure 4.1. However, it still remains to be shown that there is in mammalian systems, a GTPase similar to the one discovered in the turkey erythrocyte system.

Two lines of evidence led to the conclusion that the basic adenylate cyclase system is composed of C (catalytic) and G (GTP-binding) components. One, developed by Pfeuffer,[17] is based on the finding that affinity chromatography of detergent extracts of pigeon erythrocyte membranes over a GTP-Sepharose matrix results in resolution of an unretarded fraction with adenylate cyclase activity unresponsive to GTP (or p(NH)ppG) and a p(NH)ppG-dependent eluate that confers p(NH)ppG sensitivity back to the resolved enzyme.

The other line of evidence indicating the existence of C and G components stems from the discovery by Ross and Gilman,[18,19] using mutants of lymphoma S49 cells, that the so-called cat$^-$ S49 cells (i.e. cells that contain no measurable adenylate cyclase activity) contain a very heat-sensitive (10 min at 30 °C) component that is common to basal, p(NH)ppG and fluoride-stimulated activities, and that can restore activity to heat treated extracts from wild type S49 cells. After prolonged heating at 52 °C, extracts from wild type cells can not normally have their activity reconstituted, but this can be partially overcome by protection with p(NH)ppG.

From this it can be concluded that wild type cells contain two components. One is relatively heat stable, affected by guanine nucleotide and is in all probability the GTP-binding component; the other is very heat labile and common to both basal and nucleotide-stimulated activity: probably the catalytic component. The phenotypically cat$^-$ cell line is actually C-component$^+$ and G-component$^-$. It follows that both components are required for optimal activity. These results are summarized in Table 4.1.

Studies with [$^{32}$P]NAD$^+$ and cholera toxin revealed that concomitant with adenylate cyclase activation, a component with properties of the guanine nucleotide binding component is covalently modified, i.e. ADP-ribosylated. This was shown by SDS-polyacrylamide gel electrophoresis in both the pigeon erythrocyte system[20] and in the S49 cell system.[21] An apparent

Table 4.1. Some of the cell-free complementation assays carried out on Lubrol-extracts of S49 cell membranes. Reproduced with permission from Ross and Gilman[18] and Ross *et al.*[19]

| Source of extract(s) | Characteristics of components | Cyclizing activity |
|---|---|---|
| Wild type | $C^{+}$–$G^{+}$ | Active |
| Wild type heated at 37 °C for 10 min | $C^{-}$–$G^{+}$ | Inactive |
| $cat^{-}$ | $C^{+}$–$G^{-}$ | Inactive |
| $cat^{-}$ plus heated wild type | $C^{+}_{(cat^{-})}$–$G^{+}_{(WT)}$ | Active |

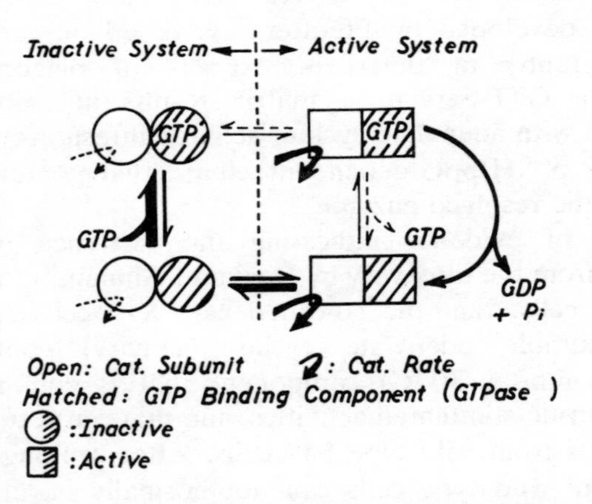


Figure 4.2. Two-state model of the adenylate cyclase system. The guanine nucleotide regulated, GTPase-dependent regulation of a basic adenylate cyclase system in the absence of hormone is represented. The system, composed of C- and G-components, is allowed to exist in only two states: an inactive one (round forms) and an active one (square forms). In the absence of nucleotide (GTP), the inactive conformation of both subunits is the preferred one; in the presence of nucleotide, the active conformation is the preferred one (concerted activation). However, since activation of the G-component leads to removal of GTP, the resulting free, active state soon isomerizes back to the inactive state, thus completing a cycle.

The degree of activation of cyclizing activity obtained with a given nucleotide depends on: (1) rates of nucleotide binding to free inactive form (rapid for all nucleotides tested); (2) rate of isomerization of the nucleotide-occupied inactive form to its active form (slow with p(NH)ppG and more rapid with GTP); (3) the sum of rates of dissociation of nucleotide from the active form (slow for all nucleotides tested) plus rate of active removal via GTPase activity; and (4) the rate of back-isomerization of the unoccupied active form of the system to its inactive form (rapid). Thickness of arrows is intended to represent relative rate values

molecular weight of 42,000–44,000 has been ascribed to both the guanine nucleotide regulatory component resolved by affinity chromatography[17] and the ADP-ribosylated component mediated by cholera toxin.[20] That cholera toxin does indeed lead to covalent modification of the G-component of the adenylate cyclasc system has been shown by functional reconstitution analysis.[22] It was demonstrated that while cholera toxin-treated wild type S49 cells yielded a G-component capable of conferring toxin-treated properties to the $cat^-$ adenylate cyclase system, toxin treatment of $cat^-$ cells led to reconstitution of adenylate cyclase activities that displayed no characteristics of the toxin-treated enzyme. In addition, G-component isolated by affinity chromatography from extracts of toxin-treated pigeon erythrocytes conferred toxin-treated characteristics to a resolved adenylate cyclase preparation derived from control pigeon erythrocytes.[20]

Detailed kinetic analysis of the mode of stimulation by p(NH)ppG, GTP and GTP$\gamma$S using the rat liver system indicates that adenylate cyclase is in a state of turnover. This is consistent with the existence of an intrinsic GTPase activity associated with cyclizing activity which is induced by guanine nucleotide binding.[23] A model of adenylate cyclase regulation by nucleotide in which the basic system consists of two subunits existing in only two states of activity (inactive and active) and in which GTP binding results in activation of both cyclizing activity and GTPase activity is shown in Figure 4.2.

### (iv) Divalent cation regulation of adenylate cyclase

The existence of an allosteric site for divalent cations, whose affinity is modified by hormone, was originally proposed by Birnbaumer *et al.*[24] Similar observations were made with brain adenylate cyclase,[1,2] and these studies were extended by Rodbell and coworkers to liver, adrenal and fat cell adenylate cyclases.[11] Early work suggested that adenylate cyclase, rather than being regulated by an allosteric site for divalent cations, was regulated through the catalytic site which had very high affinity for inhibitory uncomplexed $HATP^{3-}$, and that large concentrations of divalent cations (in excess over total ATP added) were required to release this inhibition. Activation of adenylate cyclase by various stimulants was thus viewed as being due to alteration of the affinity of the catalytic site for $HATP^{3-}$. However, extensive kinetic analysis of detergent-dispersed brain adenylate cyclase,[25] has provided strong evidence that cations are required as activators, irrespective of any effect they may have on free ATP concentration. Reassessment of the effects of divalent cations by Rodbell and coworkers led to the conclusion that $HATP^{3-}$ was not a potent inhibitor of adenylate cyclase and that divalent cations did stimulate adenylate cyclase through an allosteric site.

$Mg^{2+}$ itself, in the absence of any direct interactions with exogenously added guanine nucleotide, was found to be an activator of adrenocortical adenylate cyclase.[26] Moreover in guinea-pig heart and rat adrenal adenylate cyclase it was shown[27,26] that there was an absolute requirement for divalent cations to obtain persistent activation with p(NH)ppG, and it has been suggested that guanine nucleotide regulation of adenylate cyclase is modulated by cations.

### (v) Regulation of hormone binding and hormonal stimulation of adenylate cyclase

In the liver system, Rodbell and coworkers found that guanine nucleotides stimulated the rate of dissociation of [$^{125}$I]glucagon from its receptor and decreased the affinity of the receptor for its hormone.[9] There was also an obligatory guanine nucleotide requirement for glucagon stimulation of adenylate cyclase.[9] It has been confirmed using highly purified assay components,[28] that the receptor coupling is dependent on guanine nucleotide addition. In a GTP-sensitive LH-stimulatable adenylate cyclase from rabbit corpus luteum, on the other hand, ATP and not GTP is required for optimal coupling of hormone receptor to adenylate cyclase.[29] This strongly suggests that more than one type of nucleotide site must exist in adenylate cyclase systems: one GTP-specific regulating basal activity, and another site, specific for guanine nucleotides in liver and for ATP in the corpus luteum, regulating receptor activities and coupling. Recent work[30] has indeed shown that the effects of guanine nucleotides on glucagon binding are separable from those on adenylate cyclase, and Iyengar *et al.*[31] have shown that effects of guanine nucleotides on coupling can be isolated from effects on the catalytic complex. These and other studies confirm the existence of two separate nucleotide sites. Interestingly, Kimura and Nagata[28] and Iyengar and Birnbaumer (unpublished) have shown that coupling of the glucagon receptor to adenylate cyclase can be achieved, independently of the GTP-mediated regulation of adenylate cyclase turnover, by using GDP as the nucleotide. This indicates clearly that while the dissociation rate of GDP from certain systems can be affected by hormones, it is not a necessary feature for hormonal stimulation.

The role of nucleotides in coupling is not clear. One interesting and not yet explained finding of Maguire *et al.*,[32] who studied the binding of [$^{125}$I]hydroxybenzylpindolol (a $\beta$-adrenergic antagonist) to S49 cell membranes, is that guanine nucleotides alter the binding of agonists but not of antagonists. Further studies[33] on the modulation of binding of agonists and activation of adenylate cyclase, found five- to 100-fold discrepancies between the $K_d$ (apparent binding constant) and $K_{act}$ (apparent activation constant) with which the agonist isoproterenol interacts with S49 cell

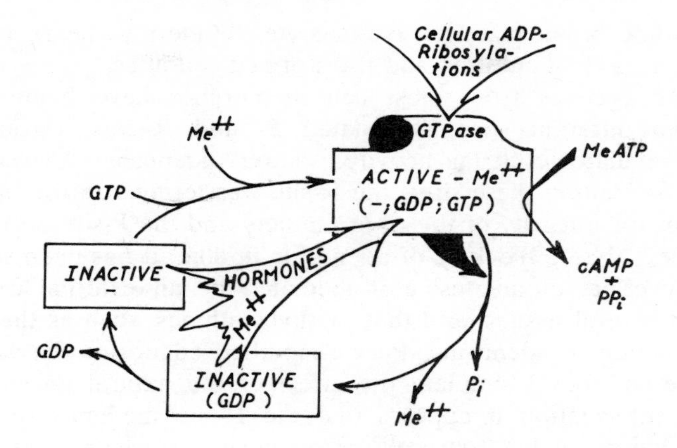


Figure 4.3. Metal ion-dependent, two-level regulation of adenylate cyclase by hormones and GTPase. Note that the active state of the system formed under the influence of hormone has a divalent cation ligand ($Mg^{2+}$ or $Mn^{2+}$) and does not depend on having a nucleotide bound to it. Activation of the system in the absence of hormone is dependent on presence of both GTP and divalent cation

membranes in the presence of the various nucleotides. The $K_d : K_{act}$ ratio can be as high as 1,000–5,000 in some cells.[34] The mechanism for this high efficiency of coupling between the $\beta$-adrenergic receptors and adenylate cyclase is poorly understood. However, it is clear the coupling efficiency is greatly diminished upon cell disruption.

In summary, with respect to receptor coupling, much remains to be discovered. In those systems where it has been looked at, both the hormone–receptor interaction and receptor-dependent stimulation of cyclizing activity are regulated by a nucleotide via a site that is distinct from the GTPase site involved in the regulation of basal activity. Indeed, coupling can be obtained with a nucleotide that blocks the turnover cycle of the basic adenylate cyclase system, indicating that two pathways of regulation must be at work: one mediating hormonal effects and dependent on extracellular stimuli, and the other mediating inputs stemming most probably from inside of the cell. Figure 4.3 presents a view of how both modes of regulation may act to stimulate adenylate cyclase systems.

### (vi) Adenosine

The role of adenosine in activating adenylate cyclase has recently been reviewed.[35] Adenosine is a potent activator of adenylate cyclase in many cell types including brain, platelets, lymphocytes, epidermis, heart, bone cells, and a wide variety of cultured and transformed cell lines.

Adenylate cyclases from most cell membranes have been shown to contain two different sites, designated P- and R-sites, through which adenosine regulates cyclizing activity. Various adenosine derivatives have been used to identify these sites; the R-site was termed on the basis of the requirement for integrity of the ribose moiety and the P-site on that of the purine moiety. Using blockers of adenosine uptake, it has been shown that the activity effect of adenosine is mediated by an external cell surface receptor in several tissues, and that methylxanthines, such as theophylline, block this action of adenosine. Purine-modified adenosine derivatives (i.e. with unaltered ribose) simulate the effect of the natural nucleoside, and from such information it appears that the R-site mediates some of the biological actions of low extracellular adenosine concentrations. So many cells have been found to contain an R-type receptor for adenylate cyclase activation, that this receptor may be as ubiquitous as the $\beta$-adrenergic receptor.[35]

No biological actions of adenosine can yet be ascribed to the P-site effects. However, the P-site is widely distributed, and adenylate cyclase can be inhibited by micromolar concentrations of adenosine at this intracellularly located site. A role has been suggested for the P-site in mediating feedback inhibition of adenylate cyclase activity.

The properties of the P- and R-sites clearly distinguish them from each other, and these are consistent with the arrangement shown in Table 4.2. The basal adenylate cyclase activity of membrane preparations, in the absence of hormone, has been of continuing interest in the field. It now appears likely that a substantial portion of basal activity in many membranes results from adenosine stimulation.

In turkey erythrocyte membranes it has been found[36] that whereas all of the adenylate cyclase pool could be activated by the $\beta$-adrenergic receptor,

Table 4.2. Adenylate cyclase adenosine receptor sites. Reproduced with permission from Londos *et al.*[35]

| Characteristics | P-site | R-site |
|---|---|---|
| Analogues bound | Ribose modified— | Purine modified–adenosines |
| Methylxanthines | No effect | Antagonize |
| $Mg^{2+}$; $Mn^{2+}$ | Enhance | No effect |
| Effect | Inhibition | Activation (except fat cell: inhibition) |
| Location | Internal membrane surface | External membrane surface |

only 60–70% of the enzyme could be activated by the adenosine receptor. The remaining 30–40% could not be activated by adenosine, and it was suggested that the adenosine receptor is precoupled to the enzyme. This has been confirmed recently using an affinity label directed towards the adenosine receptor.[37] It was also concluded that, in this system, adenosine and catecholamine activate adenylate cyclase via a common guanine nucleotide regulatory site.

### (vii) Cyclic GMP

The observations of Goldberg *et al.*[38] stimulated interest in the role of cyclic GMP as an alternative second messenger. It was found that most of the receptors which stimulate adenylate cyclase do not markedly alter cellular cyclic GMP concentration, and also that there is a separate group of receptors that raise tissue cyclic GMP concentrations without affecting cyclic AMP concentration. These receptors appear to be associated with those designated as members of the calcium-mobilizing group (see later; Table 4.3). Goldberg suggested that two opposing influences controlled the balance in most cells. Such cells were said to show bidirectional control, one mediated by cyclic GMP and the other by cyclic AMP. Some cells, however, were said to show unidirectional control, and in these the effects of the receptors were complementary in raising the levels of both cyclic nucleotides. From these observations it was suggested that cyclic GMP could function as a second messenger, but usually influenced cells in a manner opposite to that of cyclic AMP.

This 'Yin–Yang hypothesis' as it was known, has, however, received little further support. Addition of cyclic GMP or its analogues to cells can reproduce few of the biological responses initiated via the appropriate receptors. Moreover, the stimulation of exocytosis, contractility, and many other effects of these receptors are established responses to increased intracellular $Ca^{2+}$ concentration, making cyclic GMP appear somewhat redundant.

In many cells, these receptor-linked sites in cyclic GMP concentrations are abolished in the absence of extracellular calcium, and it thus appears that, like most other cell responses, increased cyclic GMP is a consequence of raised intracellular $Ca^{2+}$ concentrations. Although this is consistent with the observation of $Ca^{2+}$-stimulated guanylate cyclase in many cells, the role of cyclic GMP is still unclear and its biological function remains to be elucidated.[39] It has recently been suggested[40] that some hormones may be linked to guanylate cyclase through specific redox systems and that cellular events involving oxidation–reduction may represent a general mechanism for regulating cyclic GMP and guanylate cyclase.

## III. COUPLING BETWEEN RECEPTOR, GTP-BINDING COMPONENT, AND ADENYLATE CYCLASE

Several cells have a number of different receptors on the cell surface that are each specific for different hormones. This is most amply demonstrated in isolated adipocytes where no less than eight different hormones can elevate intracellular cyclic AMP levels through activation of adenylate cyclase.[41] These hormones appear to compete to activate a common pool of adenylate cyclase catalytic units, as addition of saturating levels of each hormone does not yield simple additive increases in cyclic AMP production.[42] The unlikelihood of a supramolecular complex of eight receptors with individual molecular weights of 200,000–300,000 modulating an individual catalytic unit, together with the new knowledge of protein mobility in the plane of the bilayer, led to the formulation of the mobile receptor model for adenylate cyclase.[2,41,43] In its simplest definition, this envisages that in the absence of hormone both the receptor and catalytic unit of adenylate cyclase are free to migrate in the plane of the bilayer; upon hormone binding to the receptor this complex can couple with and activate the catalytic unit, forming a multicomponent system spanning the bilayer membrane.

### (i) Cell fusion transfer of receptor from one cell to couple with adenylate cyclase of another

As discussed in Chapter 3, Frye and Edidin demonstrated the lateral mobility of plasma membrane proteins by fusing two different cell types with their plasma membrane proteins labelled with different fluorescent markers. The fusion between two different cell types resulted in the rapid intermixing of labelled membrane proteins. Orly and Schramm[44,45] reasoned that by fusing different cell membranes together in a similar manner, it should be possible to activate the adenylate cyclase system of one membrane via the hormone–receptor system of the other, if the cyclase and receptor did indeed exist as separate entities. The experiments utilized intact turkey erythrocytes i.e. nucleated red blood cells possessing a $\beta$-receptor able to stimulate an endogenous adenylate cyclase. These were heat-treated or exposed to the alkylating reagent *N*-ethylmaleimide, to inactivate the adenylate cyclase activity (without affecting the maximal binding of hormone to the $\beta$-receptor). These cells were then fused with Friend erythroleukaemia cells that possessed no $\beta$-receptor, but did exhibit adenylate cyclase activity. The cell ghosts of such fusion systems exhibited an adenylate cyclase activity that was markedly stimulated by the addition of isoproterenol, a $\beta$-receptor agonist. It was concluded that fusion of the plasma membranes of these two cells allowed the mixing of both receptor and catalytic units to occur by free lateral diffusion in the plane of the bilayer. Such experiments have been extended to examine the interaction of

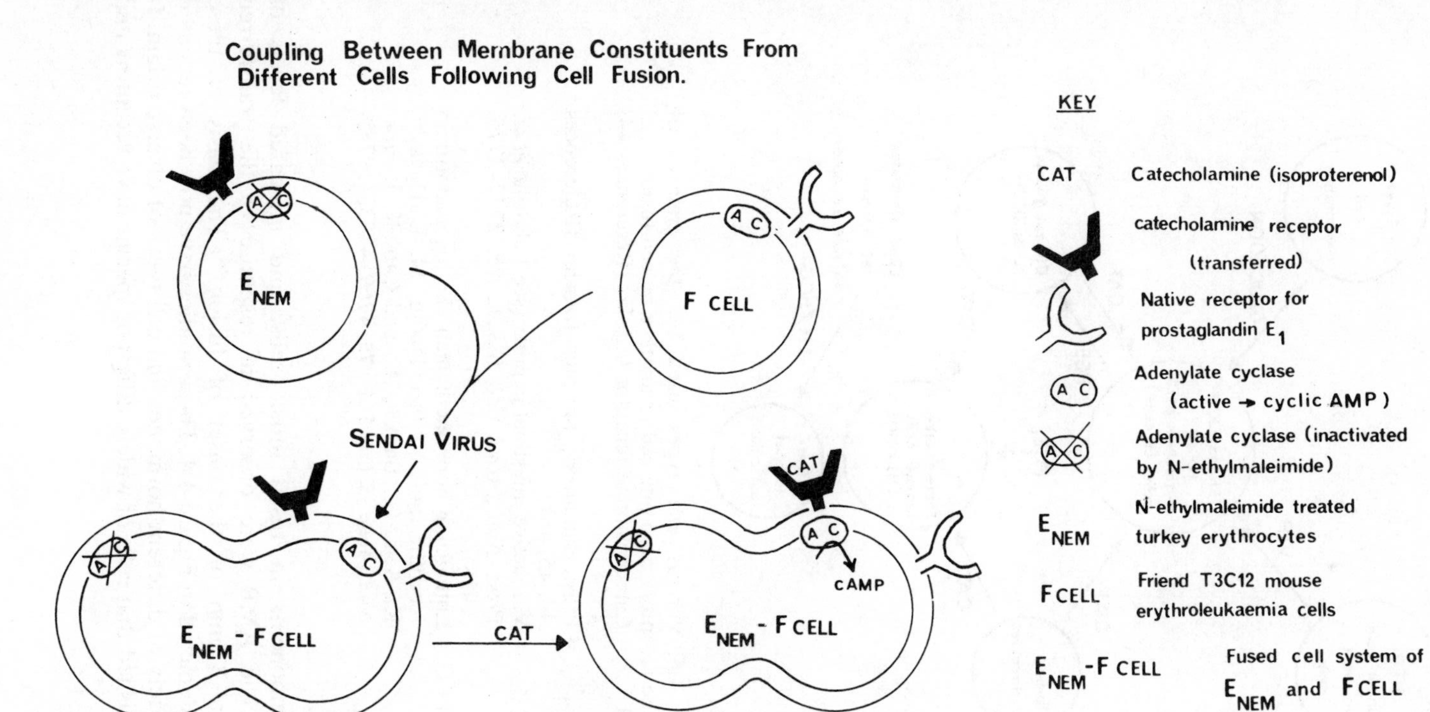


Figure 4.4. Coupling between membrane constituents from different cells following cell fusion using Sendai virus. (For further details see text and References 44, 45, 46)

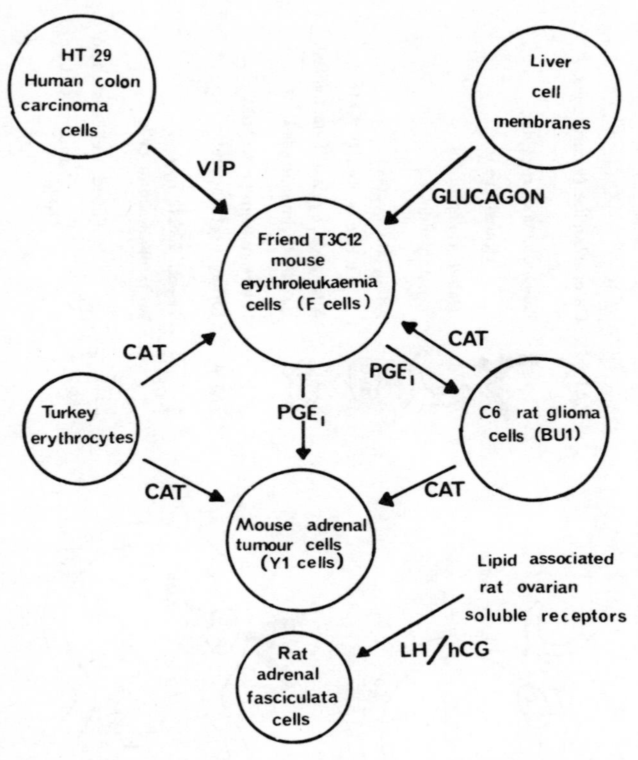


Figure 4.5. Various cell types used for the transfer of hormone receptors from one cell to another by cell fusion.

CAT = Catecholamine receptor transfer (References 44, 45, 46)

$PGE_1$ = Prostaglandin $E_1$ receptor transfer (References 44, 45, 46)

VIP = Vasoactive intestinal peptide (see Laburthe *et al.* *Proc. Natl. Acad. Sci. U.S.A.,* **75**, 2772–2775, (1978))

LH/hCG = Luteinizing hormone/human choriogonadotropin receptor transfer: (see Dufau, M. L. Hayashi, K., Sala, G., Baukal, A., and Catt, K. J. *Proc. Natl. Acad. Sci. U.S.A.*, **75**, 4769–4773, (1978))

membrane components in fused intact cells, and increased levels of intracellular cyclic AMP were observed in response to the transferred receptor within 3 min of the onset of fusion.[46] The basis of these experiments is depicted in Figure 4.4. These studies have now been repeated with a wide variety of different hormones and cell types of diverse origin. It has been established that cells of widely different species have receptors and adenylate cyclase units that are compatible with each other. These results are summarized in Figure 4.5. It is clear that receptors are membrane components, distinct and separable from the catalytic subunit.

The observation that coupling can occur between receptors and adenylate cyclase of widely different species, suggests that the mechanism for this coupling may be universal for all eukaryotic cells. Furthermore, it is probable that when a hormone binds to its specific receptor, the change that is induced and leads to activation of adenylate cyclase, is also identical for all the different receptors. From these considerations, it seems likely that receptors coupling to adenylate cyclase will all be found to have similar molecular structures, with similar highly conserved domains found on all receptors and catalytic units. These may be likened to the common domains exhibited by the immunoglobulins which, for individual classes, differ in essence only at the variable domains of the antigen binding site.

### (ii) Irradiation inactivation studies indicate that activation of adenylate cyclase is achieved by the interaction of independently migrating receptors and catalytic units

The apparent target sizes of the glucagon receptor and the catalytic unit of adenylate cyclase in rat liver plasma membranes have been measured by the technique of irradiation inactivation in a beam of 12 MeV electrons.[47] The freeze-dried membranes were exposed to varying doses of high energy electrons and the decay in activity then measured as a function of the dose. A single 'hit' with an electron, anywhere within a functional protein complex, inactivates the entire complex. The rate of decay therefore depends on the size of the functional protein complex which can be determined empirically using standard enzyme markers. When rat liver plasma membranes were irradiated in the uncoupled state (absence of glucagon) the activity of fluoride-stimulated adenylate cyclase (dependent on the catalytic unit only) decayed as a complex of molecular weight 160,000—in agreement with estimates of the molecular weight of adenylate cyclase obtained by other methods. The decay in glucagon-binding to these membranes gave a functional size for the glucagon receptor of molecular weight 217,000, and for glucagon-sensitive adenylate cyclase activity the functional size of molecular weight 389,000 was almost exactly the sum of the sizes of the two independent components (160,000 + 217,000 = 377,000). Such studies thus indicated that the glucagon receptor is functionally distinct from adenylate cyclase and only couples to it in the presence of glucagon. To show that the two complexes are physically as well as functionally separate, the membranes were pretreated with glucagon, to couple physically the two complexes, and then irradiated (Figure 4.6). For a dose which destroyed 50% of the functional complexes, 50% of the original glucagon-stimulated

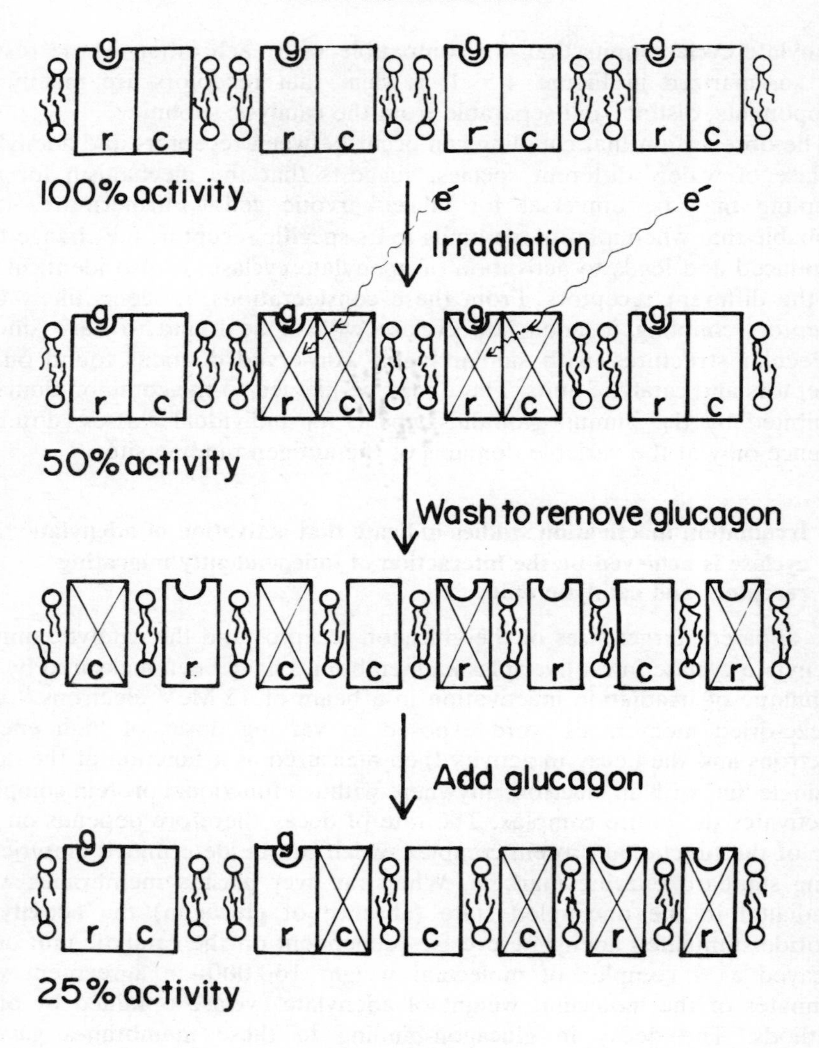


Figure 4.6. Radiation inactivation studies indicate receptors and catalytic units can migrate independently

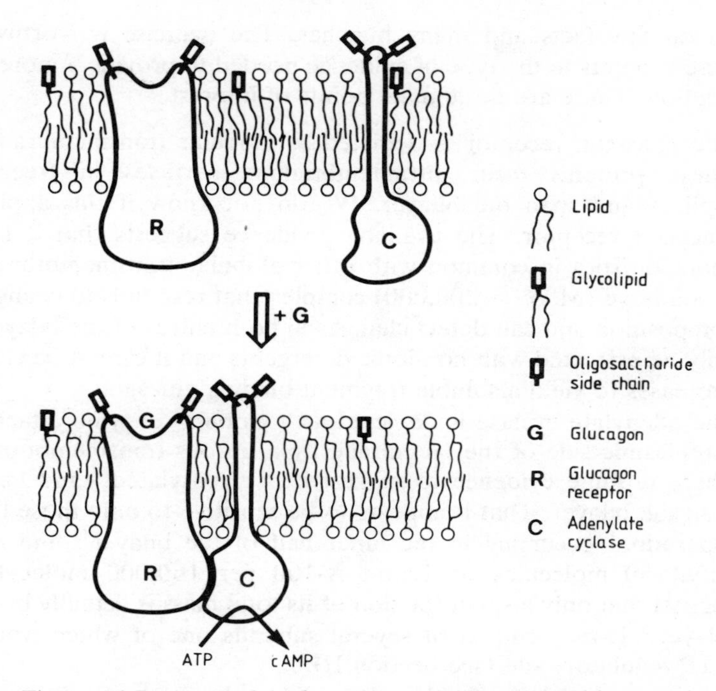


Figure 4.7. A model for the possible disposition and functioning of the glucagon receptor and adenylate cyclase in the plasma membrane

adenylate cyclase activity was recovered providing the glucagon was not removed before the assay. If glucagon is now removed the complexes are uncoupled and if the receptor and adenylate cyclase are physically as well as functionally separate they will migrate independently in the bilayer (Figure 4.6). Adding back glucagon at this stage then allows formation of inactive complexes between functional receptors and inactivated catalytic units, reducing by half the number of active complexes that can be formed between functional receptors and functional catalytic units. Those functional catalytic units unable to find an active partner exhibit negligible activity. As shown in Figure 4.6 recovery of only 25% of the original glucagon-stimulated adenylate cyclase activity may be predicted. That this was actually the result obtained lends strong support to the independent nature of the glucagon receptor and adenylate cyclase in biological membranes.

Although there is no definitive evidence or weight of evidence that allows us to decide the disposition of the glucagon receptor and adenylate cyclase in the membrane, a model for this is depicted in Figure 4.7. This model is based on few facts and many hunches. The exercise is worthwhile only because it points to the type of evidence needed to provide a more definitive description. There are three main points of interest.

1. The glucagon receptor is depicted as globular transmembrane protein. These proteins form intramembranous particles in freeze–fracture replicas and span the bilayer. We do not know if this applies to the glucagon receptor. The available evidence suggests that it has certain characteristics in common with other globular transmembrane proteins. It is a large (M.W. $> 200,000$) complex that responds to changes in lipid composition and can detect changes in both halves of the bilayer.[48] It can only be extracted with non-ionic detergents and it cannot be cleaved with proteases to yield a soluble fragment binding glucagon.
2. The adenylate cyclase is depicted as a globular enzyme attached to the cytoplasmic side of the membrane by a fibrous transmembrane protein. There is some evidence to suggest it is glycosylated[49] and hence might span the bilayer. That it appears to be sensitive to only those lipid phase separations occurring in the inner half of the bilayer[48] and only binds about 50 molecules of Triton X-100 per 160,000 molecular weight suggest that only a small fraction of its total mass is actually buried in the bilayer.[5] It may consist of several subunits one of which would be the GTP regulatory site (see Section II).
3. The interaction between the receptor and the catalytic unit is depicted as occurring in the hydrophilic phase of the cytoplasm and not, as is usually depicted, in the hydrophobic phase of the bilayer. The annular lipids (see Chapter 3) solvate the membrane proteins keeping their hydrophobic surfaces apart, otherwise a non-specific aggregation of membrane proteins would occur. Specific interactions between proteins are optimal in the aqueous and not in the hydrophobic phase.

## IV. CALCIUM IONS AS INTRACELLULAR MESSENGERS

The concept that many hormonal and neurotransmitter stimuli induce an increase in the level of ionized calcium in the cytosol compartment of cells ($[Ca^{2+}]_{cyt}$), and that these changes control at least some of the intracellular biochemical responses to these hormones and neurotransmitters, has evolved from the convergence of many lines of evidence from different disciplines. Thus there is an emerging picture which indicates that a rise in $[Ca^{2+}]_{cyt}$ is a common denominator in the actions of a wide variety of stimuli in many tissues (Table 4.3).[50,51]

However, it is not yet known whether all of these receptors achieve the mobilization of $Ca^{2+}$, which leads to a rise in $[Ca^{2+}]_{cyt}$, by the same basic mechanism. In general, the recognition of detailed differences in the behaviour of the various systems (e.g. differences in sensitivity to removal

Table 4.3. Some ligands which seem to exert some or all of their cellular effects through the mobilization of $Ca^{2+}$

| Receptor | Examples of $Ca^{2+}$-mediated responses in appropriate tissues | Stimulation of phosphatidylinositol metabolism? |
|---|---|---|
| Acetylcholine (muscarinic) | $K^+$ efflux, contraction, protein secretion, Ⓖ | Yes |
| Epinephrine* (adrenaline) ($\alpha$-) | $K^+$ efflux, contraction, protein secretion, glycogenolysis, Ⓖ | Yes |
| Histamine* ($H_I$-) | $K^+$ efflux, contraction, Ⓖ | Yes |
| 5-Hydroxytryptamine* | fluid secretion, contraction, Ⓖ | Yes |
| Pancreozymin | protein secretion, contraction, Ⓖ | Yes |
| Bombesin | protein secretion | Yes |
| Substance P | $K^+$ efflux, protein secretion | Yes |
| Angiotensin II* | glycogenolysis, contraction, Ⓖ | Yes |
| Vasopressin* | glycogenolysis, contraction, Ⓖ | Yes |
| Antigen | histamine secretion | Yes |
| *f*Met–Leu–Phe | chemotaxis, protein secretion | Yes |
| Thyrotropin* | glucose oxidation? | Yes |
| Oxytocin | contraction | ? |
| Parathyrin* | bone resorption? | Yes |

*All of the hormones indicated by asterisks have also been claimed with excellent evidence, to exert some or all of their effects through activation of adenylate cyclase. In two cases, namely epinephrine (adrenaline) and histamine, it is clear that different types of receptors control adenylate cyclase and $Ca^{2+}$ mobilization. In the other situations where one hormone controls both adenylate cyclase and mobilization of calcium the situation remains open: there may be two discrete families of receptors, each coupled to a different effector, or one family of receptors which exerts control on the two different effector systems

Ⓖ elevation of tissue cyclic GMP

of extracellular $Ca^{2+}$) has until now been used to emphasize disparities between systems; common features have often been underplayed. It is notable, though, that the differences between various tissues in their responses to activation of any single type of receptor (e.g. the effects of muscarinic cholinergic stimuli on ileum smooth muscle and on exocrine pancreas) are often as large as the differences in behaviour between more disparate agonist–tissue combinations. This may argue in favour of a widely distributed fundamental design of the $Ca^{2+}$-mobilizing mechanism which is subject to 'minor' variations in various tissues and receptors.

The conclusion that mobilization of $Ca^{2+}$ is likely to be the key response evoked by a ligand–receptor interaction is usually based on a number of experimental observations: (1) reduction or loss of response upon withdrawal of extracellular $Ca^{2+}$; (2) inhibition by $La^{3+}$, $Mn^{2+}$, local anaesthetics or other 'calcium-antagonistic' agents; (3) evocation of the same response when $Ca^{2+}$ is admitted to cells either with an electrode or an ionophore; and (4) marked enhancement of $Ca^{2+}$ fluxes across the plasma membrane of the stimulated cells.

Often only some of these diagnostic tests give clear results, and a decision as to the importance or otherwise of $Ca^{2+}$ in cellular responsiveness tends to be a finely balanced judgment based on somewhat inconclusive evidence. As a consequence, there are only a limited number of situations in which most workers would concur that changes in $[Ca^{2+}]_{cyt}$ do constitute the key factor by which hormones control cell activity. There are, however, many more situations in which this incomplete evidence is adequate to convince many, but not all, workers. Among the more important cellular responses brought about by receptor-controlled $Ca^{2+}$ mobilization in various tissues are contraction (many smooth muscles), secretion (a host of exocytotic secretory tissues), efflux of $K^+$ ions (smooth muscles, liver, and salivary glands) and glycogenolysis (salivary gland and the liver of some species: see Table 4.3).

### (i) Calcium ion mobilization and phosphatidylinositol metabolism

If the idea that different calcium-mobilizing hormones bring about their effects on target tissues through variations upon some common mechanistic theme is correct, then one should be able to identify events, such as enzymic coupling reactions or conformational changes by macromolecules involved in stimulus–response coupling, which occur in response to, or in conjunction with, all $Ca^{2+}$-mobilizing stimuli. At present the only event which appears to be a plausible candidate for such a role is the hormone-stimulated breakdown of phosphatidylinositol. Many years ago it was discovered that various stimuli evoke an increased turnover of phosphatidylinositol, a quantitatively minor membrane phospholipid, in target secretory tissues. Only recently was it demonstrated that this sequence of events is initiated by phosphatidylinositol breakdown and that it seems to be evoked by all hormones which belong to the proposed '$Ca^{2+}$-mobilizing family' (Table 4.3).[39,52] Despite this association between phosphatidylinositol breakdown and $Ca^{2+}$ mobilization, there is good evidence that phosphatidylinositol breakdown, unlike other responses such as secretion, contraction, glycogenolysis, and accumulation of cyclic GMP, is not a consequence of the increase in $[Ca^{2+}]_{cyt}$ which is brought about by stimulation. Its function is still not understood, but it has been suggested[39,52] that its known characteristics make it an excellent candidate for some essential role in the mobilization of $Ca^{2+}$ which is brought about by hormones (see Chapter 19).

### (ii) Control of cellular activity by cytosol calcium ion concentration

In the case of cyclic AMP, it seems likely that the regulatory subunit of a protein kinase is the only intracellular constituent with which the nucleotide has to interact for it to bring about diverse cellular responses: the variety of different responses are achieved as a result of the cyclic AMP-stimulated protein kinase phosphorylating, and thus modulating the activity of, a variety of different intracellular proteins.

No single target enzyme activity of this type has been found for cytosol $Ca^{2+}$ ions. Instead, it appears that the cytosol of cells contains several structurally related $Ca^{2+}$-binding proteins, with at least some of these responsible for conferring $Ca^{2+}$-sensitivity upon appropriate enzymes.[53] The first to be identified was troponin C of skeletal and cardiac muscle, and it is now clear that this protein acts as the $Ca^{2+}$-recognition site through which the ATPase activity and contractility of actomyosin are very rapidly switched on in stimulated muscle.

Troponin C does not, however, seem to be responsible for mediating the majority of cellular responses to a rise in $[Ca^{2+}]_{cyt}$: in many non-muscle cells it is not even present. There is, though, a very similar $Ca^{2+}$-binding protein which appears to be a ubiquitous component of eukaryotic cells and which may play a role in many $Ca^{2+}$-triggered cell responses. This is calmodulin (also known as calcium-dependent modulator (or regulator) protein), a small (M.W. = 17,000) and acidic (pI = 4.2) protein which possesses four $Ca^{2+}$-binding sites. Like troponin C, this protein progressively binds $Ca^{2+}$ in the concentration range 0.1–10 $\mu$M, which is the general range of concentrations over which changes in $[Ca^{2+}]_{cyt}$ are thought to exert their major effects.[54] Further, calmodulin has the remarkable property that its complex with $Ca^{2+}$ is capable of interacting with, and activating, a substantial variety of enzymes. Among those described to date are cyclic nucleotide phosphodiesterase, brain adenylate cyclase, a protein kinase of plasma membranes, phosphorylase kinase, the $Ca^{2+}$-transport ATPase of erythrocytes, myosin light chain kinase, and an NAD kinase of plants, with the list rapidly growing.[55] These findings raise the intriguing possibility that the cellular enzymes responsible for these diverse $Ca^{2+}$-regulated processes all possess similar binding domains through which they recognize and respond to the $Ca^{2+}$-calmodulin complex. Most tissues, however, contain far more calmodulin than would be needed simply for the activation of these enzymes: the reason for this is not yet clear.

Moreover, it is notable that the above list of calmodulin-modulated processes includes nothing which is a clear candidate for the key $Ca^{2+}$-regulated step(s) which must be involved in exocytotic secretion, even though such secretory responses obviously depend critically upon some rate-limiting $Ca^{2+}$-controlled step.[56] Although troponin C and calmodulin are the only cytoplasmic $Ca^{2+}$-binding proteins to which any clear control functions have yet been ascribed, in some cells there are other closely-related $Ca^{2+}$-binding proteins whose functions are still not clearly understood (e.g. the parvalbumins).[53]

Although the first reported effects of calmodulin were on cyclic nucleotide

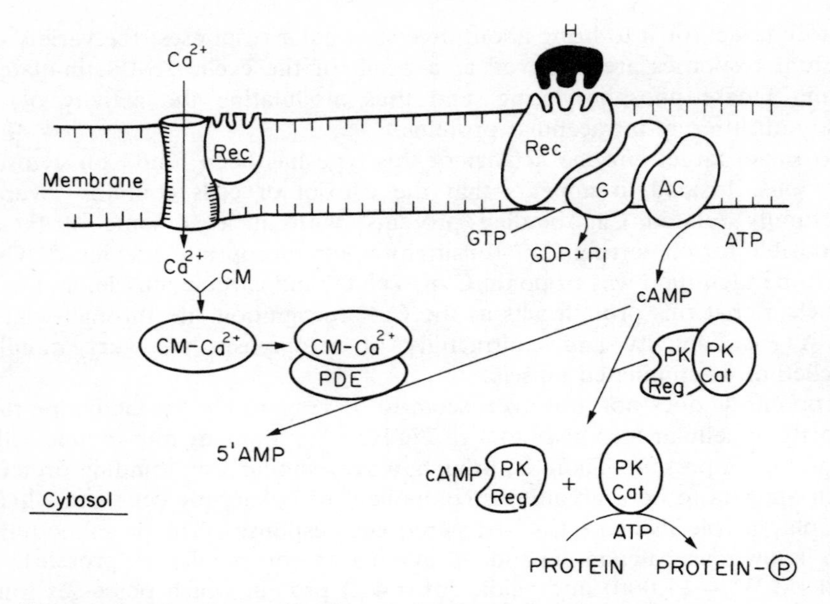


Figure 4.8. A scheme for the mechanisms of cell membrane receptors linked to $Ca^{2+}$ mobilization and to cyclic AMP. Interrelationships between hormone (H), receptor (Rec), GTPase (G), adenylate cyclase (AC), cyclic AMP (cAMP), protein kinase catalytic subunit (PK Cat) and regulatory subunit (PK Reg), phosphodiesterase (PDE), and calmodulin (CM)

phosphodiesterase and adenylate cyclase, the interrelationships between changes in $[Ca^{2+}]_{cyt}$ and cellular cyclic nucleotide levels are still not clearly defined. Stimulation of cyclic nucleotide phosphodiesterase by $Ca^{2+}$ appears to be a relatively general phenomenon, thus indicating that hormones which elevate $[Ca^{2+}]_{cyt}$ should tend to decrease [cyclic AMP]. This is often the case, as for example with $\alpha$-adrenergic or muscarinic cholinergic stimuli in a variety of cells. The situation with regard to adenylate cyclase is far more confused. Some brain adenylate cyclase preparations are activated by $Ca^{2+}$ in a calmodulin-mediated manner, some cells (e.g. avian erythrocytes) possess adenylate cyclases which are sensitive to inhibition by small concentrations of $Ca^{2+}$, and yet others show hormonal stimulation of adenylate cyclase only when extracellular $Ca^{2+}$ is available (the $Ca^{2+}$ may not, however, need to interact with the adenylate cyclase itself). Finally, it should be noted that one of the most common cellular responses to stimuli which bring about a rise in $[Ca^{2+}]_{cyt}$ is an increase, at least transiently, in cyclic GMP level.[57] An important factor in this response is probably activation of guanylate cyclase by $Ca^{2+}$ ions, with a rise in concentration occurring despite the fact that in such cells there would simultaneously be activation of cyclic nucleotide phosphodiesterase, an enzyme with a higher affinity for cyclic GMP than for cyclic AMP.

In summary, an attempt to interrelate the hormone receptors linked to adenylate cyclase with those linked to mobilization of calcium ions, is shown in Figure 4.8.

## V. REFERENCES

1. Perkins, J. P. *Adv. Cyclic. Nucl. Res.*, **4**, 1–64, (1978).
2. Birnbaumer, L. *Biochim. Biophys. Acta*, **300**, 129–158, (1973).
3. Tell, G. P., Haour, F., and Saez, J. M. *Metabolism*, **27**, 1566–1592, (1978).
4. Jard, S., Roy, C., Barth, T., Rajerison, R., and Bockaert, J. *Adv. Cyclic Nucl. Res.*, **5**, 31–52, (1975).
5. Neer, E. J. in *Receptors and Hormone Action*, (O'Malley, B. W., and Birnbaumer, L., eds.) Academic Press, New York, Vol. 1, pp. 463–483, (1977).
6. Limbird, L. E., and Lefkowitz, R. E. *J. Biol. Chem.*, **252**, 799–802, (1977).
7. Haga, T., Haga, K., and Gilman, A. G. *J. Biol. Chem.*, **252**, 5776–5782, (1977).
8. Dufau, M. L., Ryan, D. W., Baukal, A. J., and Catt, K. J. *J. Biol. Chem.*, **250**, 4822–4824, (1975).
9. Rodbell, M., Birnbaumer, L., Pohl, S. L., and Krans, H. M. J. in *Structure–Activity Relationships of Protein and Polypeptide Hormones*, (Margoulis, M., and Greenwood, F. C., eds.), Excerpta Medica International Congress Series No. 241, Excerpta Medica, Amsterdam, Vol. 1, 199–211, (1971).
10. Londos, C., Salomon, Y., Lin., M. C., Harwood, J. P., Schramm, M., Wolff, J., and Rodbell, M. *Proc. Natl. Acad. Sci. U.S.A.*, **71**, 3087–3090, (1974).
11. Rodbell, M., Lin, M. C., Salomon, Y., Harwood, J. P., Martin, B. R., Rendell, M., and Berman, M. *Adv. Cyclic Nucl. Res.*, **5**, 3–30, (1975).
12. Blume, A. J., and Foster, C. J. *J. Biol. Chem.*, **251**, 3399–3404, (1976).
13. Cassel, D., and Selinger, Z. *Biochim. Biophys. Acta*, **452**, 538–551, (1976).
14. Cassel, D. and Selinger, Z. *Proc. Natl. Acad. Sci. U.S.A.*, **74**, 3307–3311, (1977).
15. Levinson, S. L., and Blume, A. J. *J. Biol. Chem.*, **252**, 3766–3774, (1977).
16. Cassel, D. and Selinger, Z. *Proc. Natl. Acad. Sci. U.S.A.*, **75**, 4155–4159, (1978).
17. Pfeuffer, T. *J. Biol. Chem.*, **252**, 7224–7234, (1977).
18. Ross, E. M., and Gilman, A. G. *J. Biol. Chem.*, **252**, 6966–6969, (1977).
19. Ross, E. M., Howlett, A. C., Ferguson, K. M., and Gilman, A. G. *J. Biol. Chem.*, **253**, 6401–6412, (1978).
20. Cassel, D., and Pfeuffer, T. *Proc. Natl. Acad. Sci. U.S.A.*, **75**, 2669–2673, (1978).
21. Johnson, G. L., Kaslow, H. R., and Bourne, H. R. *J. Biol. Chem.*, **253**, 7120–7123, (1978).
22. Johnson, G. L., Kaslow, H. R., and Bourne, H. R. *Proc. Natl. Acad. Sci. U.S.A.*, **75**, 3113–3117, (1978).
23. Birnbaumer, L., and Kaumann, A. J. in *Medicinal Chemistry: Proceedings of the Vth International Symposium, University of Sussex, Brighton, 4–7 September 1978*, (Simkins, M. A., ed.), Cotswold Press Ltd., Oxford, Great Britain, in press, (1979).
24. Birnbaumer, L., Pohl, S. L., and Rodbell, M. *J. Biol. Chem.*, **244**, 3468–3476, (1969).

25. Johnson, R. A., and Garbers, D. L. in *Receptors and Hormone Action*, (O'Malley, B. W., and Birnbaumer, L., eds.), Academic Press, New York, Vol. 1, pp. 549–572, (1977).
26. Glynn, P., Cooper, D. M. F., and Schulster, D. *Biochim. Biophys. Acta.*, **524**, 474–483, (1978).
27. Alvarez, R., and Bruno, J. J. *Proc. Natl. Acad. Sci. U.S.A.*, **74**, 92–95, (1977).
28. Kimura, N., and Nagata, N. *J. Biol. Chem.*, **252**, 3829–3835, (1977).
29. Birnbaumer, L., and Yang, P.-C. *J. Biol. Chem.*, **249**, 7867–7873, (1974).
30. Welton, A. F., Lad, P. M., Newby, A. C., Yamamura, H., Nicosia, S., and Rodbell, M. *J. Biol. Chem.*, **252**, 5947–5950, (1977).
31. Iyengar, R., Swartz, T. L., and Birnbaumer, L. *J. Biol. Chem.*, **254**, 1119–1123, (1979).
32. Maguire, M. E., Van Arsdale, P. M., and Gilman, A. G. *Mol. Pharmacol.*, **12**, 335–339, (1976).
33. Ross, E. M., Maguire, M. E., Sturgill, T. W., Biltonen, R. L., and Gilman, A. G. *J. Biol. Chem.*, **252**, 5761–5775, (1977).
34. Terasaki, W. L., and Brooker, G. *J. Biol. Chem.*, **253**, 5418–5425, (1978).
35. Londos, C., Wolff, J., and Cooper, D. M. F. in *Physiological Regulatory Functions of Adenosine and Adenine Nucleotides* (Baer, H. P., and Drummond, G. I., eds.), Raven Press, New York, pp. 271–282, (1979).
36. Tolkovsky, A., and Levitzki, A. *Biochemistry*, **17**, 3811–3817, (1978).
37. Braun, S., and Levitzki, A. *Biochemistry,* **18**, 2134–2138 (1979).
38. Goldberg, N. D., Haddox, M. K., Nicol. S. E., Glass, D. B., Sanford, C. H., Kuehl, F. A., and Estensen, R. *Adv. Cyclic Nucl. Res.*, **5**, 307–330, (1975).
39. Michell, R. H. in *Companion to Biochemistry* (Bull, A. T., Lagnado, J., Tipton, K., and Thomas, J. O., eds.). Longmans, London, Vol. 2, pp. 205–228, (1979).
40. Goldberg, N. D., Graff, G., Haddox, M. K., Stephenson, J. H., Glass, D. B., and Moser, M. E. *Advances in Enzyme Regulation* (Webber, G., ed.) Pergamon Press, Oxford, Vol. 16, pp. 165–191, (1978).
41. Cuatrecasus, P., Hollenberg, M. D., Chang, K. J., and Bennett, V. *Rec. Progr. Hormone Res.*, **31**, 37–94, (1975).
42. Kahn, R. C. *J. Cell. Biol.*, **70**, 261–286, (1976).
43. De Haen, C. *J. Theor. Biol.*, **58**, 383–400, (1976).
44. Orly, J., and Schramm, M. *Proc. Natl. Acad. Sci. U.S.A.,* **73**, 4410–4414, (1976).
45. Schramm, M., Orly, J., Eimerl, S., and Korner, M. *Nature (London),* **268**, 310–313, (1977).
46. Schulster, D., Orly, J., Seidel, G., and Schramm, M. *J. Biol. Chem*, **253**, 1201–1206, (1978).
47. Houslay, M. D., Ellory, J. C., Smith, G. A., Hesketh, T. R., Stein, J. M., Warren, G. B., and Metcalfe, J. C. *Biochim. Biophys. Acta*, **467**, 208–219, (1977).
48. Houslay, M. D., and Palmer, R. W. *Biochem. J.,* **174**, 909–919, (1978).
49. Storm, D. R., and Chase, R. A. *J. Biol. Chem.,* **250**, 2539–2545, (1975).
50. Rasmussen, H., and Goodman, D. P. B. *Physiol. Revs.,* **57**, 421–509, (1977).
51. Berridge, M. J. *Adv. Cyclic Nucl. Res.*, **6**, 1–98, (1975).
52. Michell, R. H. *Trends in Biochemical Science*, **4**, 128–131, (1979).
53. Kretsinger, R. H. *Intern. Rev. Cytol.*, **46**, 323–393, (1976) and Wolff, D. J. and Brostrom, C. O. *Adv. Cyclic Nucl. Res.*, **11**, 27–88, (1979).
54. Wang, J. H. in *Cyclic 3',5'-Nucleotides: Mechanisms of Action* (Cramer, H., and Schultz, J. eds.) John Wiley & Sons, London, pp. 37–56, (1977).

55. Dedman, J. R., Welsh, M. J., and Means, A. R. *J. Biol. Chem.*, **253**, 7515–7521, (1978).
56. Rubin, R. P. *Calcium and the Secretory Process*, Academic Press, New York, (1974).
57. Goldberg, N. D., and Haddox, M. K. *Ann. Rev. Biochem.*, **46**, 823–896, (1977).

Cellular Receptors for Hormones and Neurotransmitters
Edited by D. Schulster and A. Levitzki


CHAPTER 5

# Determination of cyclic AMP and the assay of adenylate cyclase

Yoram Salomon
*Department of Hormone Research, The Weizmann Institute of Science, Rehovot, Israel*



Challenge of some cells or tissues with certain hormones results in an increase in intracellular levels of cyclic AMP. Receptor activity in these cases may therefore be studied by determining the changes in steady-state levels of this nucleotide following hormonal stimulation. Direct measurement of the adenylate cyclase reaction may be performed using broken cell preparations or purified plasma membranes. This primary reaction is stimulated by the hormone *in vitro* and often shows a relationship to hormone concentration similar to that seen in the intact parent target cell or tissue.

Listed below are some of the techniques currently used to determine the absolute or relative concentrations of cyclic AMP in biological fluids or in cell/tissue extracts. Also described is the method for determination of adenylate cyclase activity.

## I. METHODS FOR DETERMINATION OF ABSOLUTE AMOUNTS OF CYCLIC AMP

### (i) Radioimmunoassay

This analytical tool has been adapted to 3′ : 5′ cyclic AMP determination by Steiner and his colleagues in 1969.[1] Its principle is based upon the competition of 3′:5′ cyclic AMP (to be determined) and the tyrosine methyl ester derivative of [$^{125}$I]succinyl cyclic AMP for binding sites on a specific antibody. The antibody–cyclic AMP complex is separated from free cyclic AMP by precipitation with ammonium sulphate. The limit of sensitivity of this procedure is about 0.05 pmol of cyclic AMP. (For detailed description of the method see References 2 and 3.)

### (ii) Protein binding assay

The principle of this method introduced by Gilman[4] is similar to that described for the radioimmunoassay. Cyclic AMP in the unknown sample is determined by its ability to compete with isotopically labelled ($^{3}$H or $^{32}$P) cyclic AMP on binding sites of 3′:5′ cyclic AMP binding protein (see also equation (2)). The cyclic AMP protein complex is separated from unbound cyclic AMP by being retained on millipore filters according to Gilman[4] or by eliminating the unbound cyclic AMP by charcoal adsorption according to Brown *et al.*[5] The limit of sensitivity in this method is about 0.05 pmol of cyclic AMP. (For detailed description of the methods see References 6, 7, and 8.)

### (iii) Protein kinase activation

Cyclic AMP is determined by its ability to stimulate the activity of cyclic AMP-dependent protein-kinase. Using ATP, this enzyme catalyses the phosphorylation of protein substrates such as histones according to the equation (1).

$$\text{Histone} + [\gamma\text{-}^{32}\text{P}]\text{ATP} \xrightarrow[\text{cyclic AMP} \atop \text{Mg}^{2+}]{} [^{32}\text{P}]\text{Histone} + \text{ADP} \quad (1)$$

The labelled phosphorylated protein is isolated by precipitation with trichloroacetic acid. The limit of sensitivity of this method is 0.3 pmol of cyclic AMP. (For detailed description of this method see References 8 and 9.)

The preparation of calibration curves with known amounts of cyclic AMP is required for the execution of these three methods.

## II. DETERMINATION OF RELATIVE CYCLIC AMP CONCENTRATIONS

The relationship between hormonal stimulation and intracellular levels of cyclic AMP may be determined by measuring changes in cyclic AMP levels which take place in response to hormonal stimulation. Isotopically labelled adenine ($^{14}C$ or $^{3}H$) is efficiently incorporated into the intracellular pool of ATP when incubated with cells or tissue slices. The relative amounts of cyclic AMP newly synthesized from ATP may thus be determined by isolating the nucleotide from cell/tissue extracts as suggested by Humes *et al.*[10] This is efficiently achieved by chromatography on Dowex 50 cation exchanger followed by nascent barium sulphate precipitation according to Krishna *et al.*[11] or by following the double column procedure according to Salomon *et al.*[12] This methodology is rapid, accurate, and sensitive but does not yield absolute values owing to incomplete isotope equilibration and unknown sizes of internal pools of adenine, adenosine and adenosine nucleotides.

## III. DETERMINATION OF THE INTRACELLULAR STATE OF AGGREGATION OF CYCLIC AMP DEPENDENT PROTEIN KINASE

3′:5′ Cyclic AMP has been demonstrated to activate cyclic AMP-dependent protein kinase according to equation (2)[8]:

$$R_2C_2 + 2cAMP \longrightarrow R_2\text{-}cAMP_2 + 2C \qquad (2)$$

This enzyme is a tetramer[13,14,15] where $R_2C_2$ is the holoenzyme (bearing little or no catalytic activity). R is the regulatory subunit (bearing the specific binding site for cyclic AMP) and C is the free and fully active catalytic subunit. It can be seen that a physiologically relevant increase in cyclic AMP concentration within the cell will shift the equilibrium shown in equation (2) to the right, thereby decreasing the concentration of $R_2C_2$ and increasing the concentration of $R_2$–$cAMP_2$ and C. Under conditions which preserve the equilibrium state that prevailed *in situ* (equation (2)) throughout the extraction procedure one may determine cyclic AMP-induced changes using crude tissue extracts in one of three ways:

1. A direct determination of the relative concentrations of $R_2C_2$, $R_2$–$cAMP_2$ and C using the elegant and rapid technique reported by Rangel-Aldao and Rosen.[16] This technique makes use of $\omega$-aminohexyl-agarose[17] for chromatographic separation of the various molecular forms of the enzyme. The same goal may be achieved by gel filtration as suggested by Corbin *et al.*[18]
2. Determination of the change in protein kinase activity ratio:

$$\text{activity ratio} = \frac{\text{activity in the absence of added cAMP}}{\text{activity in the presence of added cAMP}}$$

Activity ratio reaches the value of one when the enzyme in the absence of added cyclic AMP is fully activated, Corbin *et al.*[18] The details of this reaction have been described earlier (see equation (1)).

3. Determination of the free and occupied cyclic AMP binding sites according to Do Khac *et al.*[19] Minute changes in cyclic AMP levels too small to be detected with any of the former methods have been determined using this approach. Binding of [$^3$H]cyclic AMP in cell/tissue extracts is determined as described earlier by Gilman[4] and represents unoccupied binding sites ($R_2C_2$). Occupied cyclic AMP binding sites ($R_2$–$cAMP_2$) in the tissue extract are determined as follows: $R_2$–$cAMP_2$ is quantitatively adsorbed to an ion exchanger. Cyclic AMP liberated from the bound protein by heat inactivation is determined by radioimmunoassay. By means of this methodology Dufau *et al.*[20] recently showed that a good correlation exists between cyclic AMP formation and testosterone production elicited in testicular Leydig cells by human chorionic gonadotropin.

## IV. ADENYLATE CYCLASE

Adenylate cyclase activity in broken cell preparations or purified plasma membranes may be determined by measuring the formation of labelled 3′:5′ cyclic AMP from labelled ATP according to the following reaction:

$$\text{Adenosine } 5'\text{-}^{32}\text{P-P-P} \xrightarrow[\text{GTP Mg}^{2+}]{\text{Hormone}} \text{Adenosine } 3'5'^{32}\text{P} + \text{P-Pi}$$

Enzyme activity is absolutely dependent on the presence of magnesium ions. Hormone stimulation is dependent in most instances on the presence of low concentrations of GTP ($10^{-6}$ M). In the absence of hormone, enzyme activity is low (basal activity), but rises upon addition of increasing concentrations of hormone. In the example depicted (Figure 5.1) a membrane preparation derived from the rat ovary was incubated in the presence of increasing concentrations of luteinizing hormone (LH) and adenylate cyclase activity was determined.[12]

Enzyme activity is determined in a small reaction volume (0.05–0.1 ml). The reaction mixture contains buffer to maintain constant pH (7.6), magnesium ions, substrate [$\alpha$-$^{32}$P] ATP and an ATP regeneration system (to maintain ATP levels). In addition, unlabelled 3′:5′ cyclic AMP is added to protect the newly formed [$^{32}$P] 3′:5′ cyclic AMP from being hydrolysed by endogenous cyclic AMP phosphodiesterase. The reaction is initiated by the addition of enzyme and terminated by inactivating the enzyme with a solution of sodium dodecyl sulphate (2%) containing carrier ATP and cyclic AMP. The reaction product [$^{32}$P]cyclic AMP is isolated using a two-step chromatographic procedure. The first step uses Dowex 50 cation exchanger and the second used neutral aluminium oxide according to Salomon *et al.*[12, 21]

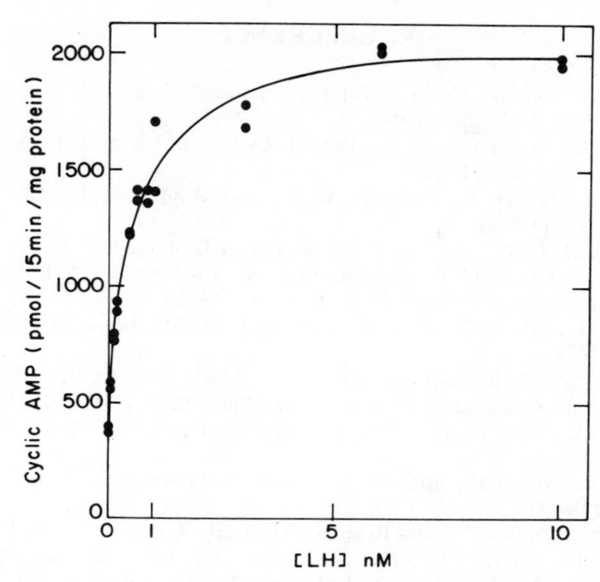


Figure 5.1. Rat ovarian adenylate cyclase activity in the presence of increasing concentrations of LH

The separation procedure permits nearly complete elimination of $^{32}$P-containing contaminants to the limit of $1 \times 10^{-4}$ % and therefore provides extremely high sensitivity. Figure 5.1 describes a typical assay performed according to this procedure. It can be seen that in addition to determining the activity of the enzyme, this method is a convenient tool for the study of hormone receptor interactions and the consequent coupling of the receptor–hormone complex to the enzyme. The reaction can also be applied for the estimation of the biological activity of hormones to which a very good correlation exists as shown by Birnbaumer *et al.*[22]

The assay procedure is simple and conveniently accomplished in a few hours. Other methods utilizing similar approaches and from which the present method emerged have been described by Krishna *et al.*;[11] White and Zenser;[23] and Ramachandran.[24]

## ACKNOWLEDGEMENT

The author thanks Professor H.R. Lindner for support and helpful discussions. This work was supported in part by the Ford Foundation and the Population Council Inc., New York, and in part by the U.S. Israel Binational Science Foundation (BSF), Jerusalem, Israel. Yoram Salomon is an incumbent of the Charles W. and Tillie K. Lubin Career Development Chair.

## V. REFERENCES

1. Steiner, A. L., Kipnis, D. M., Utinger, R., and Parker, C. W. *Proc. Natl. Acad. Sci. U.S.A.*, **64**, 367–373, (1969).
2. Steiner, A. L., Wehmann, R. E., Parker, C. W., and Kipnis, D. M. *Adv. Cyclic Nucl. Res.*, **2**, 51–61, (1972).
3. Brooker, G., Harper, F., Terasaki, W. L., and Moylan, R. D. *Adv. Cyclic Nucl. Res.*, **10**, 1–35, (1979).
4. Gilman, A. G. *Proc. Natl. Acad. Sci. U.S.A.*, **67**, 305–312, (1970).
5. Brown, B. L., Ekins, R. P., and Tampion, W. *Biochem. J.*, **120**, 8, (1970).
6. Gilman, A. *Adv. Cyclic Nucl. Res.*, **2**, 9–24, (1972).
7. Brown, B. L., Ekins, R. P., and Albano, J. D. M. *Adv. Cyclic Nucl. Res.*, **2**, 25–40, (1972).
8. Gill, G. N., and Walton, G. M. *Adv. Cyclic Nucl. Res.*, **10**, 93–106, (1979).
9. Kuo, J. F., and Greengard, P. *Adv. Cyclic Nucl. Res.*, **2**, 41–50, (1972).
10. Humes, J. L., Rounbehler, M., and Kuehl, F. A., Jr. *Anal. Biochem.*, **32**, 210–217, (1969).
11. Krishna, G., Weiss, B., and Brodie, B. B. *J. Pharmacol. Exptl. Therap.*, **163**, 379–385, (1968).
12. Salomon, Y., Londos, C., and Rodbell, M. *Anal. Biochem.*, **58**, 541–548, (1974).
13. Gill, N., and Garren, L. D. *Proc. Natl. Acad. Sci. USA*, **68**, 786–790, (1971).
14. Rubin, C. S., Erlichman, J., and Rosen, O. M. *J. Biol. Chem.*, **247**, 36–44, (1972).
15. Hofman, F., Beavo, J. A., Bechtel, P. J., and Krebs, E. G. *J. Biol. Chem.*, **250**, 7795–7801 (1975).
16. Rangel-Aldao, R., and Rosen, O. M. *J. Biol. Chem.*, **251**, 3375–3380, (1976).
17. Shaltiel, S. In *Metabolic Interconversion of Enzymes*, Vol. 13 (Fisher, E. H., Krebs, E. G., Neurath, H., and Studman, E. R. eds.), Springer Verlag, New York, pp. 379–392, (1974).
18. Corbin, J. D., Soderling, T. R., and Park, C. R. *J. Biol. Chem.*, **248**, 1813–1821, (1973).
19. Do Khac, L., Harbon, S., and Clauser, H. J. *Eur. J. Biochem.*, **40**, 177–185, (1973).
20. Dufau, M. L., Tsuruhara, T., Horner, A., Podesta, E., and Catt, K. J. *Proc. Natl. Acad. Sci. U.S.A.,* **74**, 3419–3423, (1977).
21. Salomon, Y. *Adv. Cyclic Nucl. Res.*, **10**, 35–55, (1979).
22. Birnbaumer, L., Yang, P. C., Hunzicker-Dunn, M., Bockaert, J., and Duran, J. M. *Endocrinology*, **99**, 163–184, (1976).
23. White, A. A., and Zenser, T. V. *Anal. Biochem.*, **41**, 372–396, (1971).
24. Ramachandran, J. *Anal. Biochem.*, **43**, 227–239, (1971).

SECTION II

# Intracellular hormone receptors

Cellular Receptors for Hormones and Neurotransmitters
Edited by D. Schulster and A. Levitzki


CHAPTER 6

# Steroid receptors

W. Ian P. Mainwaring
*Department of Biochemistry, University of Leeds, 9 Hyde Terrace, Leeds LS2 9LS, U.K.*

## I. INTRODUCTION

The years 1925–1940 were the beginning of a new era in endocrinology, seeing its transition from a descriptive to a quantitative discipline. During this time, the principal steroid hormones were completely characterized by a small but gifted band of organic chemists and the broad range of their biological functions was established by experimental physiologists using sophisticated methods of bioassay. As research on steroid hormones widened, many common features of their modes of action began to emerge. First, they evoke dramatic changes in intracellular metabolism at extremely low concentrations. Second, they are extremely important regulators of development from the early embryo to the sexually mature adult. Third, their effects are extremely organo- and tissue-specific. For these reasons of clinical relevance, the elucidation of the mechanism of action of steroid hormones was recognized as being of fundamental importance to biology, especially in terms of development, differentiation, reproduction, and homeostasis.

Current thinking on how steroids work owes much to the genius of Paul Ehrlich and his innovative concepts on the mode of action of drugs, first proposed about 1890. Essentially, Ehrlich suggested that the effects of drugs were expressed at the molecular level by their specific interaction with intracellular components which he termed ‘receptors’. The resultant drug–receptor complex then played a critical, indispensable part in the mode of action of the drug in question. Additional corollaries to the original receptor theory helped to explain the sensitivity and specificity of drug action. Sensitivity is imparted by the extreme specificity and high affinity of the binding of drugs to their receptors. Specificity is the result of the limited distribution of receptors, there being ‘target cells’ containing receptors for a given drug in contrast to ‘non-target cells’ lacking the appropriate receptors. The receptor concept remains the cornerstone of contemporary models for the modes of action of steroid hormones.

With the advent of radioactive or ‘tracer’ forms of steroid hormones, Jensen and Jacobsen[1] were the first to identify steroid receptors in steroid target cells, with later investigations demonstrating the presence of receptors in both the nuclear[2] and cytoplasmic[3] compartments of the cells of higher animals. Somewhat before these historic contributions on steroid receptors the classical work of Knox and Auerbach[4] and Clever and Karlson[5] had clearly indicated the importance of the syntheses of proteins and nucleic acids in hormonal responses. Based on these important foundations, the scene was now set for the exploration of the mode of action of steroid hormones with greater confidence and in more precise terms. The objective here is to present an overview of current knowledge on steroid receptors and their implication in the regulation of biochemical processes in target cells, particularly with respect to the control of macromolecular syntheses. The survey includes naturally occurring steroid hormones, synthetic steroids (diethylstilboestrol, dexamethasone), insect hormones (ecdysone), and cholecalciferol (vitamin D). Historical background may be found in more detail elsewhere.[6]

## II. STRUCTURES OF STEROIDS AND RELATED COMPOUNDS*

It is now abundantly clear that several distinctive classes of compounds work by means of similar mechanisms at the molecular level. Representative structures of these compounds are presented in Figure 6.1;

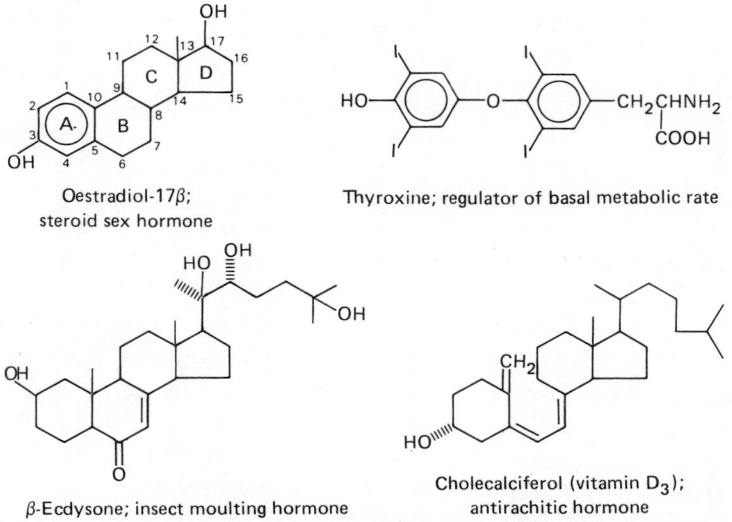

Oestradiol-17β; steroid sex hormone

Thyroxine; regulator of basal metabolic rate

β-Ecdysone; insect moulting hormone

Cholecalciferol (vitamin $D_3$); antirachitic hormone

Figure 6.1. The structural diversity of compounds acting through similar receptor mechanisms. With the exception of thyroxine, they are synthesized *in vivo* from cholesterol. The ring system and numbering of steroid molecules are shown in oestradiol (aromatic A ring)

*Trivial names for steroids and other compounds are used throughout; authentic names may be found in the appropriate section of the List of Abbreviations.

| Class of hormone | Inactive compounds | | Active analogues | Antagonist or antihormone |
|---|---|---|---|---|
| (a) Oestrogen: Oestradiol-17β | Oestradiol-17α | Oestrone | | 16-Epioestriol |
| (b) Androgen: Testosterone; 5α-Dihydro testosterone | Androstenedione | 5α-Androstane-3β, 17β-diol | R 1881 | R 2956 |

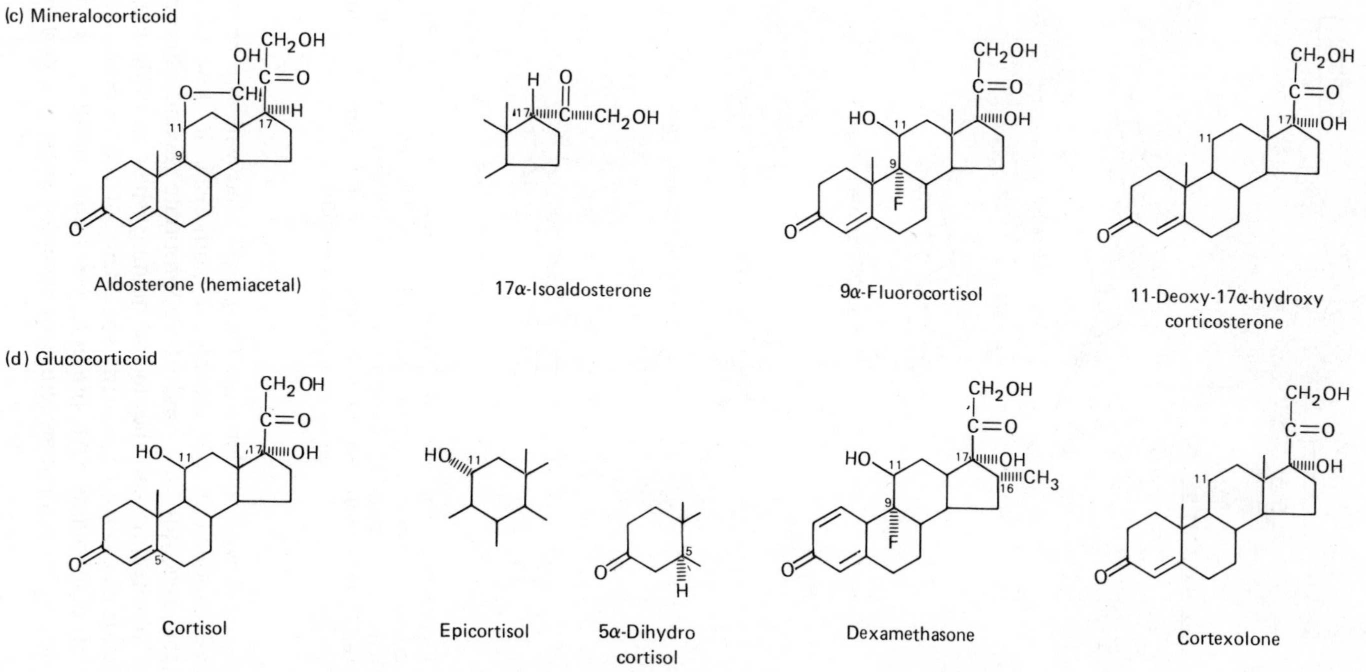

(c) Mineralocorticoid

Aldosterone (hemiacetal)

17α-Isoaldosterone

9α-Fluorocortisol

11-Deoxy-17α-hydroxy corticosterone

(d) Glucocorticoid

Cortisol

Epicortisol

5α-Dihydro cortisol

Dexamethasone

Cortexolone

Figure 6.2. The relationship between steroid structure and function. R1881 and R2956 are code numbers assigned by the Roussel Pharmaceutical Co

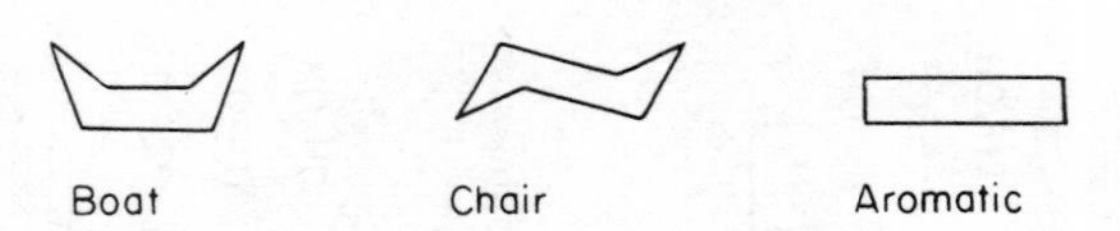


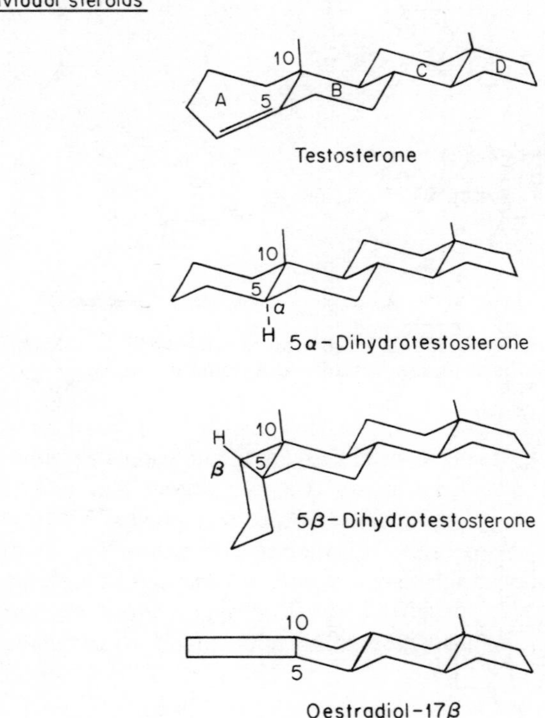


Figure 6.3. The over-all shapes of selected steroids. For clarity, substituent groups are not included, but the $C_5$–$C_{10}$ axis is indicated. Oestradiol is particularly difficult to reproduce accurately in only two dimensions; the structure shown is an approximation

thyroxine is reviewed in detail in Chapter 7 and will not be considered further here. Implicit in the receptor hypothesis is a close geometric fit between the receptor site and its appropriate ligand. Somehow, the molecular architecture of the ligand-binding sites in different receptor proteins holds the key to the acute specificity of the responses to steroids and related compounds. At present, we know nothing of the highly ordered, three-dimensional structure of receptor sites, but certainly in the case of steroids, they are clearly capably of recognizing very subtle changes in the shape of potential ligands (Figure 6.2). Steroids have rigid structures and even minor changes in either the nature or the stereochemical configuration of the substituent groups has a profound influence on biological activity. In many cases, chemical modifications lead to the loss of biological activity, but there are several synthetic analogues which are more active than the natural steroid hormones. In addition, modified steroids can act as antihormones or hormone antagonists; such compounds interact with receptors, but render them biologically inactive.

It was considered for a long time that some insight into the specificity of receptor–ligand interactions would be gained by comparing the shapes of steroids. The six-carbon rings of steroids are always in the more stable chair form, rather than the more strained boat form, and certainly steroids do display characteristic differences in their over-all shapes (Figure 6.3). Contrary to earlier expectations, however, it is difficult at present to make generalizations on the relationship between the shape of a steroid and its binding to specific receptors. On current evidence, each receptor system must be considered in the context of ligand shape and binding as an isolated, individual system. The androgen receptors preferentially bind more planar derivatives of testosterone, such as 5$\alpha$-dihydrotestosterone, rather than the more angular parent androgen, testosterone. By contrast, the planar, 5$\alpha$-reduced derivatives of all other steroids are singularly less active than the naturally secreted hormones and do not bind to receptor sites. Most of the biologically active steroids have a flat profile attributable to the *trans* configuration of the A and B rings. Extremely angular steroids resulting from the *cis* configuration of the A and B rings possess one biological activity only, the induction of haem synthesis in foetal cells (e.g. chick liver and blastoderm).

It has also proved impossible to make any generalizations on the relative importance of the lower ($\alpha$) or upper ($\beta$) faces of steroids with respect to biological function or binding. Some powerful analogues, such as dexamethasone, have characteristic substituents on the $\alpha$-face, where others, such as 7,17-dimethyl-19-nortestosterone, have bulky but essential substituents on the $\beta$-face. Taken across the board, generalizations based on steroid structures have not led to deeper insights into the specificity of receptor–steroid interactions. In terms of structure-binding relationships, each class of receptors is best considered in its own right.

## III. PHYSIOLOGICAL AND BIOCHEMICAL ASPECTS OF STEROIDS AND RELATED COMPOUNDS

The biological effects of steroids are almost bewilderingly diverse, but an attempt to summarize them, together with popular systems for their

Table 6.1. The physiological and biochemical effects of steroid hormones (synthetic analogues are in italics)

| Type | Principal steroid | Physiological response | Biochemical response | Test systems |
|---|---|---|---|---|
| (a) Steroid hormones | | | | |
| Androgen | Testosterone | Differentiation of male urogenital tract | Growth; cell division | All embryos |
| | | erythrocyte formation; haem synthesis | Induction of haem and foetal haemoglobin | Chick blastoderm: foetal liver |
| | | Differentiation of 'male' brain and liver | Induction of 'male-type' liver enzymes and non-cyclical gonadotrophin secretion | Neonatal rodents |
| | | Anabolic response in muscle | Muscle RNA and protein synthesis | All species |
| | | Growth of accessory sexual glands (male) | Cell division; induction of secretory proteins | All species, especially rat prostate and seminal vesicle |
| | | Maintenance of spermatogenesis | Induction of androgen-binding protein | All species |
| Oestrogen | Oestrodiol-17$\beta$ | Growth of accessory sexual glands (female) | Cell division; growth | All species, especially rat uterus |
| | *Diethylstilboestrol* | Production of egg proteins | Induction of ovalbumin | Chick oviduct |
| | | Maintenance of menstrual and oestrous cycles | Cell division; growth | All species; especially rat uterus |

| | | | | |
|---|---|---|---|---|
| Progestin | Progesterone | Production of egg proteins | Induction of avidin | Chick oviduct |
| | | Maintenance of pregnancy | Cell division; induction of uteroglobin | All species; especially mouse and rabbit uterus |
| Glucocorticoids | Cortisol, corticosterone | Differentiation of many developing organs | Induction retina enzymes, lung surfactant | All species |
| | *Dexamethasone* | Control of glucose metabolism | Induction of enzymes for gluconeogenesis | All liver and hepatoma cells |
| | *Triamcinolone acetonide* | Cytolysis of thymus-derived lymphocytes | Inhibition of glucose uptake and RNA synthesis | Rodent thymus |
| | | Lactation (female) | Formation of secretory epithelium, bound polyribisomes | All species |
| Mineralocorticoids | Aldosterone | Electrolyte balance; retention $Na^+$ ions | Stimulation of $Na^+$-activated ATPase | Toad bladder |
| | *9α-Fluorocortisol* | | | Rat kidney |
| (b) Steroid-related compounds | | | | |
| | Cholecalciferol | Uptake of $Ca^{2+}$ ions | Induction of $Ca^{2+}$-binding protein | Duodenum, all species |
| | Ecdysone | Metamorphosis (insect ecdysis) | Induction of dopamine decarboxylase | Most insect larvae |

experimental investigation, is presented in Table 6.1. Many responses were deliberately omitted from this table, on the grounds of either inadequate knowledge or the sheer complexity of steroid responses in certain species. For example, the intricate patterns of courtship and sexual behaviour in animals clearly involve the steroid sex hormones yet their molecular basis is currently uncertain. Even the growth of hair was omitted despite its obvious association with the androgens. In actual fact, distinctive hair, fur, plumage, or other sex-related structures (such as antlers in the deer and pheromone glands in the pig and cow) in either sex all require a complex interplay between androgens and oestrogens. Hormonal synergism is common in nature, with polypeptide hormones often playing an obligatory part. For example, evidence is accumulating to suggest that prolactin modulates the actions of all sex hormones and few, if any, cells can function properly in the absence of insulin and growth hormone. Similarly, the maintenance of spermatogenesis and oogenesis are inseparable from survival of a species, and these vital functions require the sex hormones together with the gonadotrophins, LH, and FSH. Other chapters in the book should be consulted for reviews on the receptors and modes of action of these other hormones.

The current literature contains many reports on the effects of steroids monitored *in vitro* which are difficult to reconcile with an authentic response *in vivo*. For example, oestrogens can seemingly modulate the growth of certain bacteria. Also glucocorticoids can variously act as anti-inflammatory agents, stimulators of fibroblast growth in cell culture, or promoters of certain oncogenic viruses. Responses of this kind were not included in Table 6.1.

The developmental importance of steroid hormones cannot be over emphasized. All Mammalia are inherently female and the appearance and maintenance of the male sex requires three bursts of testosterone secretion; two small bursts must occur during the embryonic and neonatal stages of development, with the third and major surge being necessary for the attainment of sexual maturity in the adult. The early embryo contains an indifferent gonad, capable of developing into either the primary sex organ of the female (ovary) or male (testis). With the female (XX) complement of sex chromosomes, the primitive gonad inevitably develops into the ovary without any other stimulus, even hormonal, being involved. By contrast, the distinctive Y chromosome of the male (XY) pair of sex chromosomes directs the synthesis of a unique plasma protein, the H-Y antigen, in the primordial structures of the embryo which in turn dictates that the indifferent gonad develops into the embryonic testis. Provided that testicular secretion of testosterone continues, then the development of the male phenotype, both in somatic and psychic terms, will occur. Embryonic production of testosterone is responsible for the differentiation of the common urogenital structures into the male reproductive tract; in its absence, the embryonic urogenital tract inexorably develops into that of the female. Immediately after birth, the neonatal secretion of testosterone is necessary for the differentiation of the brain, liver, and other organs to the 'male-type'. In the establishment of the female of mammalian species, the secretion of oestrogens is required only in the final stages of development, for the ultimate inception of the oestrous cycle (or menstrual cycle). During the final stages of differentiation of most embryonic organs, the foetal secretion of glucocorticoids is also of great importance, particularly for ensuring that the dramatic change in environment accompanying birth is met satisfactorily. Just prior to parturition, there are important changes in the structure and biochemical machinery of the eye, lung, and liver, all promoted by glucocorticoids.

Virtually all of the hormonal responses listed in Table 6.1 involve changes in macromolecular syntheses. Consequently, it is reasonable for investigators to seek a unifying mechanism of action of steroid hormones based primarily on the hormonal regulation of genetic transcription (DNA and RNA synthesis) and genetic translation (protein synthesis). It should be stressed, however, that steroid hormones can act in two ways, as switches or amplifiers. In switch mechanisms, they promote qualitative, irreversible changes, as epitomized by the embryonic differentiation of the urogenital tract. In the more usual amplification mechanism, steroids evoke quantitative changes which occur at a low or 'baseline' level even in their absence.

## IV. A GENERAL MODEL FOR THE MECHANISM OF ACTION OF STEROIDS BASED ON SPECIFIC RECEPTORS

From evidence accumulated in many laboratories, it is now possible to propose an acceptable model for the mode of action of all steroid hormones (Figure 6.4). The model is also applicable in broad principle to thyroxine and cholecalciferol. Hormone-target cell interactions are not haphazard, but rather proceed by way of an integrated sequence of ordered events. Many steps in the reaction sequence have been successfully simulated in reconstituted, cell-free systems, but some remain ill-defined and even highly controversial. The sequence of events is essentially as follows. (1) After secretion, the hormones are distributed throughout the peripheral circulation by their asssociation with selective binding proteins in the plasma. (2) The hormones enter all cells to some extent either by passive diffusion or even facilitated mechanisms of entry. (3) Depending on the hormone, extensive metabolism may occur within the target cells. (4) The hormone (or its metabolites) binds selectively and with high affinity to specific receptor proteins in the cytoplasm forming

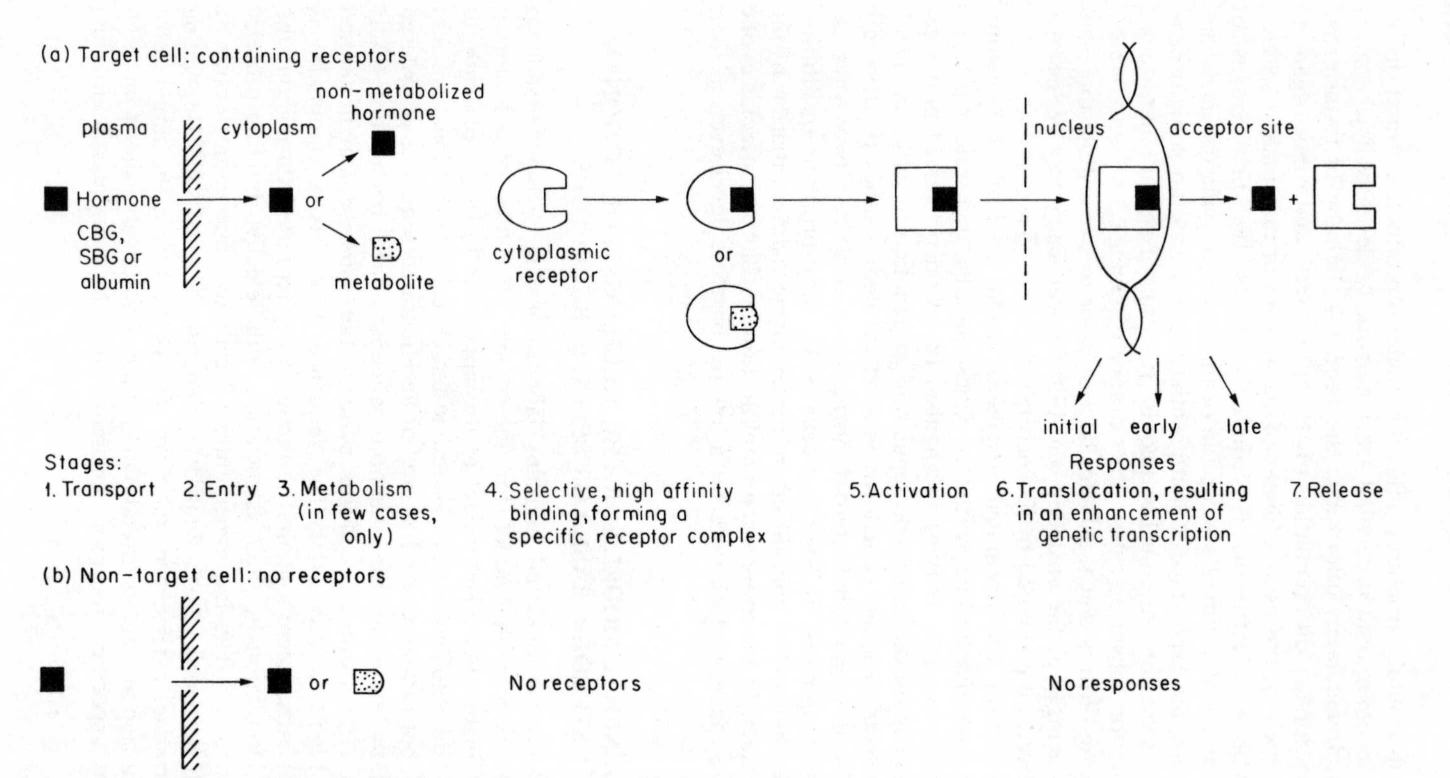


Figure 6.4. A general model for the mode of action of steroid hormones. A given target cell may have more than one type of receptor, but a similar mode of action applies to all. In certain cases, metabolism of the hormone within the target cell may be vitally important. Although shown as unidirectional many of the steps may in fact be equilibria[14]

hormone–receptor complexes. (5) The receptor complex undergoes 'activation' or some change in physiochemical configuration, thereby acquiring a particular propensity for interacting with nuclear chromatin. (6) The activated complex is translocated into the nucleus where it occupies a limited number of specific sites, usually termed as 'acceptor' sites. (7) The receptor complex remains within the nucleus for a significant but certainly not an infinite period of time, thereby triggering an enhancement of genetic transcription. (8) The receptor complex finally leaves the nucleus and in the absence of further hormone secretion, the entire process winds down.

Certain riders to the general model should be stressed. First, without exception, hormonal responses are freely reversible; it follows, therefore, that all steroid–receptor interactions must, by definition, involve non-covalent bonding only. Reports on the irreversible (covalent) attachment of steroids to intracellular components can only be viewed with suspicion in terms of normal, endocrinological responses. Only pharmacologic (or oncogenic) effects of certain steroids, such as oestrogens and diethylstilboestrol, may involve covalent interactions with DNA. Second, if the model is applicable to steroid hormones, thyroxine, and cholecalciferol, then the cells of higher animals should contain an appropriately broad spectrum of specific receptors. There is abundant evidence now to suggest that this is indeed the case. Third, if receptors are ubiquitously important, then the selective suppression of the binding of a hormone should negate virtually all expressions of its biological activity. Again, current evidence corroborates this concept.

### (i) Additional comments on the general model

#### *(a) Transport*

Discussion on the synthesis, storage, and secretion of steroid hormones is beyond the scope of the present chapter, but it should be remembered that the circulating concentrations of hormones in the blood are very low; the normal physiological range for all is $10^{-10}$–$10^{-8}$ M. As ably reviewed by Westphal,[7] all the hormones under current discussion are transported in association with well characterized proteins in the plasma (Table 6.2). Plasma albumin has an insaturable capacity to bind steroids non-specifically, but only in the case of mineralocorticoids (e.g. aldosterone) is this form of binding of biological importance. All other hormones, are transported by their association with high affinity binding proteins (dissociation constants, $K_d$, $10^{-8}$–$10^{-7}$ M. Plasma binding fulfils three functions. First, it protects hormones from premature inactivation by catabolic enzymes in the liver and elsewhere. Second, it prevents excessive

Table 6.2. The plasma proteins for the transport of steroid hormones and related compounds

| Hormone | Principal binding component in plasma |
|---|---|
| Androgens | Sex steroid-binding $\beta$-globulin (SBG) |
| Oestogens | SBG |
| | $\alpha$-Foetoprotein (foetal and neonatal stages, only) |
| Glucocorticoids | Glucocorticoid-binding $\alpha_2$-globulin (CBG) |
| Progestins | Progesterone-binding $\alpha_1$-globulin (PBG; certain species only e.g. guinea-pig) |
| | Most species, CBG |
| Mineralocorticoids | Albumin |
| Thyroxine | Thyroxine-binding $\beta$-globulin (TBG) |
| Cholecalciferol | Cholecalciferol-binding $\alpha_2$-globulin |
| Ecdysone | Not known |

and indiscriminate hormonal stimulation, as only the free (unbound) hormone in plasma is biologically active. The response of hormone-sensitive cells in culture can be impaired by high concentrations of plasma. In addition, many transport proteins (e.g. SBG; see Table 6.2) are synthesized particularly during pregnancy to provide a protective barrier for the foetus; for similar reasons, PBG (Table 6.2) and $\alpha$-foetoprotein are detectable only during and immediately after pregnancy. Finally, it even modulates hormone action, an example being the distinctive and acute sensitivity of the liver to glucocorticoids, transported by CBG (Table 6.2.). The unique sinusoid system provides a high blood flow through liver, so that free (active) glucocorticoids are more available in liver than in adjacent organs, such as the pancreas. Consequently, enzymes can be induced by glucocorticoids at a lower threshold of response in liver than in other glucocorticoid-sensitive organs.

Contamination by plasma proteins can create problems in the assay of intracellular hormone receptors, but these may be obviated by the use of radioactive analogues as model ligands; these synthetic steroids do not bind to plasma proteins. Many of the transport proteins listed in Table 6.2 can be purified to homogeneity by fractionation procedures centred on affinity chromatography. It is now possible to explore the molecular architecture of their binding sites, a particularly good example being the elegant work of Westphal[8] on PBG. One may confidently anticipate that even the determination of the amino acid sequence comprising the high affinity binding sites in some plasma proteins may soon be accomplished.

*(b) Uptake*

As steroids are lipophilic, they readily penetrate the lipid bilayers of plasma membranes and freely enter all cells. Certain human tumours, including hyperplastic prostate nodules and pituitary adenomas, demonstrate a facilitated or active entry mechanism for steroids, but the general importance of these findings to normal cells remains conjectural.

### *(c) Metabolism*

With two conspicuous exceptions, steroids and related compounds do not require further metabolism within target cells in order to elicit their biological responses. Indeed, metabolism *in vivo* created problems in receptor identification and assay until it was recognized that many synthetic analogues, while structurally appropriate for reacting with receptors, were refractory to steroid-catabolizing enzymes. In some species oestriol, a metabolite of oestradiol, may induce some uterine enzymes, such as 17$\beta$-hydroxysteroid dehydrogenase and 5$\alpha$-pregnane-3,20-dione, formed from progesterone in the chick magnum and shell gland, may prove to have a distinctive biological importance. With these minor exceptions, metabolism of nearly all hormones impairs their biological activity completely.

With testosterone and cholecalciferol, however, metabolism is a characteristic and obligatory step in their mode of action. As illustrated in Figure 6.4, the mechanism of action of androgens requires the selective conversion of testosterone to distinctive metabolites in different target cells.[9] Of these metabolites, 5$\alpha$-dihydrotestosterone formation is critically important in the maintenance of all male accessory sexual glands, such as the prostate. Classical studies by Wilson[10] indicate that the formation of 5$\alpha$-dihydrotestosterone in some primordial structures or embryonic anlagen is necessary for their differentiation. Perhaps the most unexpected but exciting aspect of the metabolism of testosterone in target cells is the necessary formation of oestradiol-17$\beta$ in the hypothalamus so that the neonatal differentiation of the brain can occur. The conversion concept of Naftolin[11] suggests that aromatization of testosterone is neccessary for the male brain to acquire ultimately a non-cyclical pattern of gonadotrophin secretion. The female pattern of secretion is clearly cyclical. The formation of androstenedione may prove to be implicated in the neonatal imprinting of steroid-metabolizing enzymes in male liver. Certain steroid hydroxylases and dehydrogenases are sexually specific, a 'male type' enzyme complement being necessary to safeguard against an excessive androgenic milieu in the mature adult. These androgen-induced changes do not appear in the female.

Many of the enzymes engaged in testosterone metabolism are concentrated in the microsomes and demonstrate a complete dependence on the cytochrome P-450 system, NADPH, $Mg^{2+}$ ions, and molecular $O_2$.

It should not be forgotten, however, that the primary role of testosterone is to maintain spermatogenesis. In addition, testosterone has

(a) Testosterone

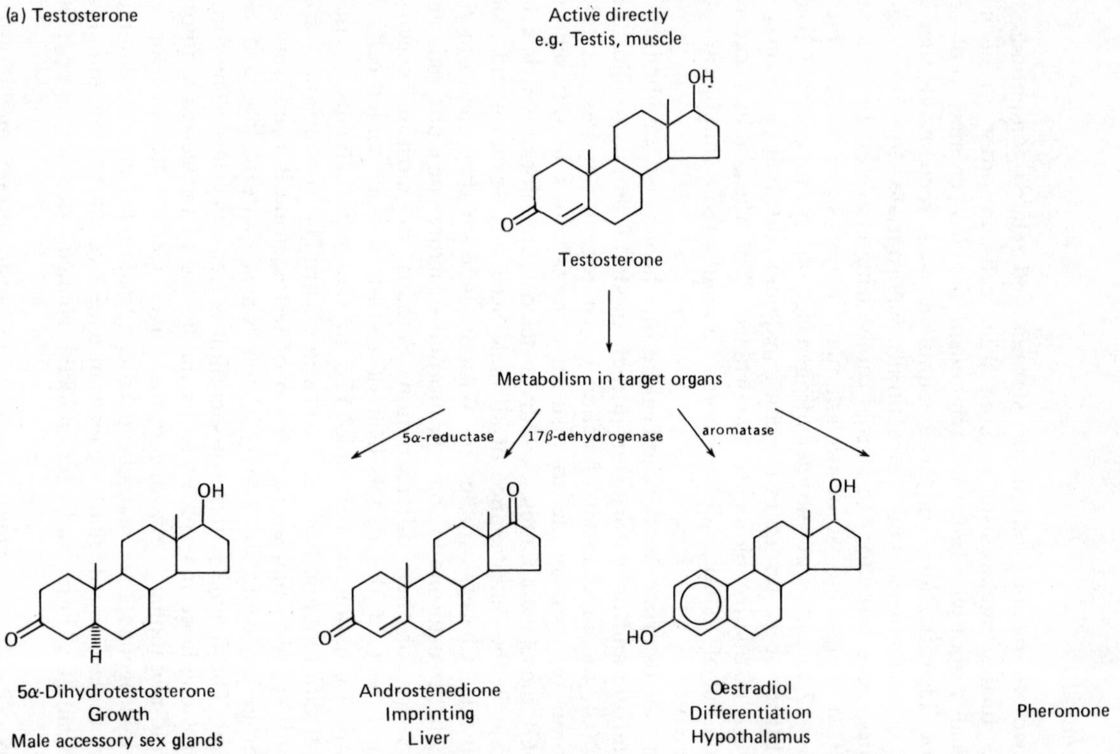

Active directly
e.g. Testis, muscle

OH

O

Testosterone

Metabolism in target organs

5α-reductase

17β-dehydrogenase

aromatase

OH

O

H

5α-Dihydrotestosterone
Growth
Male accessory sex glands

O

O

Androstenedione
Imprinting
Liver

OH

HO

Œstradiol
Differentiation
Hypothalamus

Pheromone

(b) Cholecalciferol

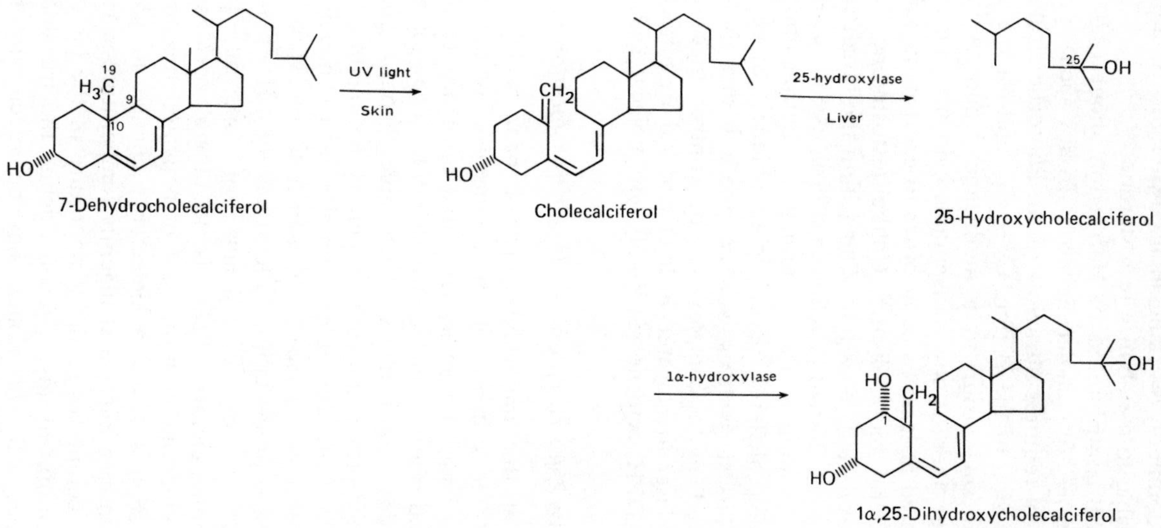

7-Dehydrocholecalciferol

UV light

Skin

Cholecalciferol

25-hydroxylase

Liver

25-Hydroxycholecalciferol

1α-hydroxylase

1α,25-Dihydroxycholecalciferol

Figure 6.5. The metabolic activation of testosterone and cholecalciferol

an anabolic function in which it promotes a slow but important growth and positive nitrogen balance in the skeletal muscles of the male. In both cases, testosterone is undoubtedly working in its own right as both adult testis and muscle are unable to metabolize testosterone.

From the classical contributions of Harrington about 1920, it is known that the formation of cholecaliciferol from such precursors as 7-dehydrocholesterol requires ultraviolet light. Photochemical reactions are involved in the scission of the B ring, $C_9 \rightarrow C_{10}$, and for the conversion of the $C_{19}$ angular methyl group to a methylene group. As illustrated in Figure 6.5, however, there is now indisputable evidence that cholecalciferol must be hydroxylated at least twice in order to exert its maximal influence on bone and mineral metabolism.[12] Unlike testosterone, however, this metabolism (hydroxylation) does not occur within the target cells. The first step, the formation of 25-hydroxycholecalciferol, occurs in the liver, and the second, forming the ultimate antirachitic hormone, $1\alpha$,25-dihydroxycholecalciferol, occurs in the kidney. Since an excess of cholecalciferol or its derivatives is toxic, these hydroxylation steps are rigorously regulated by a complex set of control mechanisms. These important hydroxylases are located in liver and kidney mitochondria, with both requiring NADPH, cytochrome P-450, and molecular $O_2$.

### *(d) Activation of receptor complexes and their translocation to the nucleus*

The stimulation of this important step has been widely attempted in reconstituted, cell-free systems without a consensus of opinion emerging. Most investigators would, however, agree that cytoplasmic receptors are in some way activated prior to nuclear translocation. The molecular basis for activation remains particularly debatable, but somehow the steroid–receptor complex acquires a propensity for binding to the nuclear acceptor sites. The confusion stems from the diverse means of activating the cytoplasmic receptor, *in vitro*, which include an excess of $Ca^{2+}$ ions, ammonium sulphate precipitation, alterations in pH or ionic strength, warming to 25 °C, or by means of specific proteases or receptor-transforming factors.[6] There is no tangible evidence that these manipulations produce an uniform type of activated receptor and except for oestrogen receptors, even a detectable change in physicochemical configuration. Translocation is a stringently temperature-sensitive step and warming receptor complexes to the temperature of the intact cell perhaps offers the most faithful means of reconstructing the translocation step, *in vitro*. In all target cells, hormones may be specifically bound in the cytoplasm even at 0 °C, yet translocation can only occur at temperatures of 20 °C and above. The present impasse stems from technical limitations. First, experimenters are attempting to reproduce a sophisticated biochemical process with crude receptor proteins, thereby opening the experiments to serious artefacts. Second, we currently have no other means than the association of the radioactive steroid to monitor receptor movement into the nucleus. Another reproducible and convincing biochemical response in the nucleus would be invaluable in resolving current difficulties.

### *(e) Acceptor sites*

The nature of the acceptor sites is another area of contentious debate. Evidence supporting both DNA and non-histone proteins in the role of acceptors has been widely documented but the issue is far from settled. There is no evidence that steroids are concentrated within the nucleolus and hence the acceptor sites must be in the nucleoplasm, nuclear membrane, or indeed both of these. New concepts on the beaded structure of chromatin[12] seem to offer little hope in the context of hormone action. The ubiquitous identification of nucleosomes ($\nu$ bodies) in eukaryotic cells has underlined the essential role of histones in packaging DNA, but not offered any advances in terms of regulation other than the observation that activated genes, including those activated by hormones, are particularly sensitive to deoxyribonuclease digestion. In sharp contrast, the demonstration of the mesh-like substructure of the nucleus, the nuclear protein matrix, may be of enormous importance in delineating the acceptor sites, The matrix is composed exclusively of non-histone proteins plus minute quantities of DNA and current evidence indicates that not only does the matrix retain steroids, but it also appears to provide some of the obligatory template (DNA) for genetic transcription.[13] Another controversial topic is whether target cells contain more acceptor sites than non-target cells. The possibility of more acceptor sites in target cells is favoured by most investigators and it would seem that differentiation may well impart target cells, perhaps by their constitution of non-histone proteins, with a penchant for the nuclear concentration of steroids. For many reasons, improved procedures for the convincing fractionation of non-histone proteins are urgently needed.

### *(f) Enhancement of biochemical processes*

In all systems studied to date, the biochemical response promoted by steroids appear in an ordered temporal sequence; consequently, responses may be classified as initial, early, and late events (Table 6.3). This classification is somewhat arbitrary and varies considerably in terms of actual response times from one system to another. Nevertheless, the over-all pattern of hormonal responses is remarkably similar and this uniformity cannot be explained by technological limitations of assay or by

Table 6.3. The temporal classification of biochemical events in cells responsive to steroids and related compounds*

| Initial events | Early events | Late events |
|---|---|---|
| Hormone binding | mRNA synthesis | Cell division |
| rRNA synthesis | Protein synthesis | Mitosis |
| Some mRNA synthesis | (Enzyme induction) | Histone synthesis |
| Some protein synthesis | | |
| Phosphorylation nuclear proteins | Membrane and phospholipid synthesis | Synthesis of enzymes associated with DNA |
| Uptake of ions | Polyribosome assembly | replication (e.g. DNA |
| ATP synthesis | Mitochondrial activation | polymerase, thymidine |
| | Polyamine synthesis | kinase) |

*The actual rate of response varies from one experimental system to another

shortcomings in experimental design. This uniformity is stressed because it has been argued by some authors that late responses, for example, proceed at barely detectable rates, even during the earliest phase of hormonal stimulation. In truth, however, this concept is untenable. The synthesis of specific species of mRNA cannot proceed without receptor occupation and cell division cannot be initiated without a prior phase of protein synthesis.

The events described in Table 6.3 strictly apply only to situations where growth is the ultimate outcome of hormone action. This is not invariably the case. For example, thel biochemical processes related to the cytolysis of lymphocytes by glucocorticoids would not be compatible with Table 6.3. Furthermore, oestrogens and androgens are generally mitogenic or growth-promoting hormones, yet they influence the brain without provoking cell division.

Inspection of Table 6.3 indicates that, unlike polypeptide hormones, steroids do not primarily express their biological function by mechanisms requiring the intracellular synthesis of cyclic nucleotides, e.g. cyclic 3′:5′-adenosine monophosphate (cyclic AMP).

### (g) *Release of receptor and steroid*

These processes remain enigmatic, for only in cultured fibroblasts has an authentic exit mechanism for steroid hormones been described; more tenuous evidence for a similar mechanism is available in the human prostate. Other possibilities include metabolism and inactivation of hormones within the nucleus or the release of receptor complexes from chromatin by their association with nascent ribonucleoprotein particles in transit to the cytoplasm. Both these schemes require wider validation. If, as seems possible, the general model of steroid action is viewed as a series of linked equilibria,[14] then thermodynamic principles related to the law of mass action would adequately explain the exit of hormones from target cells.

### (ii) Experimental support for the model

Wherever the presence of steroid receptors would be predicted on biological grounds, they have invariably been identified in steroid-responsive cells.[6] Indeed, cells such as fibroblasts contain a whole battery of different steroid receptors. The liver and kidney in most species also possess a broad spectrum of discrete receptors, but in these organs, different receptors may be specifically associated with distinctive cell types. For some time, decidualized cells in the ovary were a paradox in that they did not appear to possess progesterone receptors; this anomaly has since been rectified.[15] Priming with low doses of oestrogen was a necessary prerequisite for detecting these ovarian progesterone receptors.

#### *(a) Dose–response data*

In some systems, *in vitro*, it has proved possible to make a precise correlation between the degree of steroid binding to receptors and the extent of hormonal responses. Two such examples are presented in Figure 6.6. The precise fit of the dose–response data is impressive, providing important support for the receptor model of hormone action.

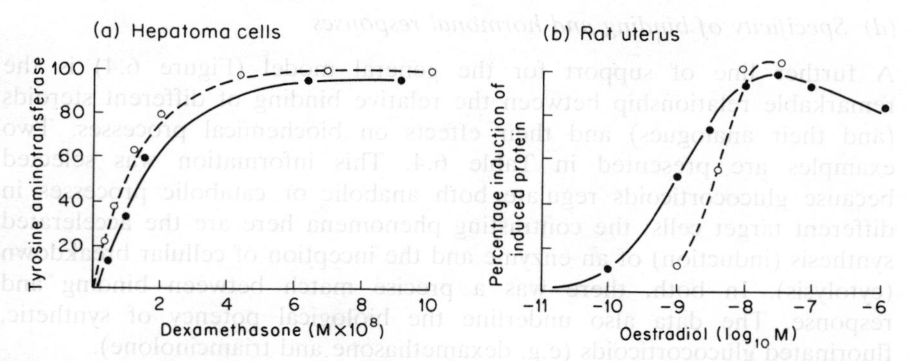


Figure 6.6. The correlation between steroid binding and hormonal responses. •, biochemical response; ○, specific binding of [$^3$H] steroids. In (a) the biochemical response is the activity of the enzyme tyrosine aminotransferase. In (b) the response is the appearance of an oestrogen-induced protein, identifiable by double isotope labelling procedures, whose function is, however, currently uncertain. Both graphs are redrawn from the originals. (a) Samuels, H. H., and Tomkins, G. M. *J. Mol. Biol.*, **52**, 57, (1970); (b) Katzenellenbogen, B. S., and Gorski, J. *J. Biol. Chem.*, **247**, 1299, (1972)

*(b) Antihormones*

For all classes of steroid hormones, antihormones (including steroid-related antagonists) are available which counter the formation of the biologically active forms of receptor complexes. Antihormone–receptor complexes may even be translocated to the nucleus, but they do not evoke an enhancement of genetic transcription. The blockade of hormonal responses by antihormones is so extensive that it is reasonable to surmise that virtually all hormonal responses are mediated by receptor mechanisms.

*(c) Experimental mutants*

The Ps (pseudohermaphrodite) rat and the Tfm (testicular feminization) mouse both have a male (XY) genotype yet a female phenotype; although genetically males, they are anatomically and physiologically infertile females. Particularly in the Tfm mouse, there are tiny, vestigial testes, high in the inguinal canal, which secrete some testosterone, yet the external genitalia and the urogenital tract are those of the female. The mutant has an absolute end-organ insensitivity to androgens, failing to respond to even massive doses of testosterone. Potential androgen target cells which do persist in these mutants such as kidney and preputial gland, do not contain androgen receptors as judged by present techniques of receptor assay. In these mutants, the absence of receptors results in a profound and permanent impairment of normal sexual development.

*(d) Specificity of binding and hormonal responses*

A further line of support for the general model (Figure 6.4) is the remarkable relationship between the relative binding of different steroids (and their analogues) and their effects on biochemical processes. Two examples are presented in Table 6.4. This information was selected because glucocorticoids regulate both anabolic or catabolic processes in different target cells; the contrasting phenomena here are the accelerated synthesis (induction) of an enzyme and the inception of cellular breakdown (cytolysis). In both, there was a precise match between binding and response. The data also underline the biological potency of synthetic, fluorinated glucocorticoids (e.g. dexamethasone and triamcinolone).

*(e) Anucleation experiments*

The mould, *Helminthosporium dematoiderum*, produces a powerful agent, cytochalasin B, which impairs microtubule formation and can even be employed for the anucleation of eukaryotic cells. Treatment of rat uterus and hepatoma cells with cytochalasin renders them virtually insensitive to

Table 6.4. Correlations between the binding of glucocorticoids and cellular responses

| Steroid | Binding | Metabolic activity |
|---|---|---|
| (a) Catabolic responses e.g. thymus-derived lymphocytes* | | |
| Cortisol (arbitrary standard) | 100 | 100‡ |
| Dexamethasone | 480 | 1000‡ |
| 9α-Fluoroprednisolone | 360 | 1800‡ |
| Prednisolone | 130 | 120‡ |
| Corticosterone | 50 | 40‡ |
| 11-Deoxycorticosterone | 50 | } |
| Progesterone | 40 | } |
| Testosterone | } | } 2‡ |
| Epicortisol | } 5 | } |
| (b) Anabolic responses e.g. cultured hepatoma (HTC) cells† | | |
| Cortisol (arbitrary standard) | 100 | 100 § |
| Dexamethasone | 570 | 600 § |
| Triamcinolone | 600 | 470 § |
| Corticosterone | 40 | 200 § |
| Progesterone | 50 | 20 § |
| 11-Deoxycorticosterone | 40 | } |
| Testosterone | } | } 2 § |
| Epicortisol | } 1 | } |

*Selected from Munck *et al*. *J. Biol. Chem.*, **243**, 5556 (1968); *Adv. Biosci.*, **7**, 301, (1971)
†Selected from Tomkins *et al*. *J. Mol. Biol.*, **52**, 57, (1970); *Proc. Natl. Acad. Sci. U.S.A.*, **68**, 932, (1971)
‡Metabolic activity = decrease in glucose uptake
§Metabolic activity = induction of tyrosine aminotransferase

their appropriate hormonal stimulators. These anucleation experiments indicate the crucial importance of nuclear events in the mode of action of steroid hormones.

### (iii) Limitations of the model

#### *(a) Anomalous receptor mechanisms*

There are certain systems which are not consistent with the general model. For instance, there is selective nuclear binding of oestrogens in foetal liver and of testosterone in bone marrow and vagina; however, in none of these can cytoplasmic receptors be identified. There is also clearly cytoplasmic (or membrane) binding of 5β-reduced steroids in chick blastoderm and of progesterone in amphibian oocytes, but in neither case does nuclear translocation occur. These findings cannot be dismissed on the grounds of experimental limitations and other mechanisms of binding may exist which nevertheless lead to a hormonal response.

There are also androgen receptors in fibroblasts and the submaxillary glands of Tfm mutant mice which seemingly have no clear function; in certain animals with obligatory delayed implantation (e.g. seals) there are also inactive yet fully occupied oestrogen receptors. It is also surprising that despite the historical importance and potential value of chromosome puffing in insects as a model system receptors for ecdysone have failed to be unequivocally identified.* Our general model is thus widely applicable, but alternative mechanisms for hormone action may exist in a limited number of target cells.

### (b) *Responses not requiring receptors or nuclear activation*

Receptor-independent responses are relatively rare. Nevertheless, they include the maturation of amphibian oocytes by progesterone, the inhibition of water into rat uterus by oestrogens, the induction of $\beta$-glucuronidase and alcohol dehydrogenase in mouse kidney by $5\alpha$-androstanediols, the induction of a limited number of enzymes of the pentose phosphate cycle by androgens and oestrogens, and the release of corticotropin-releasing factor from anucleate hypothalamic synaptosomes by glucocorticoids. These responses do not involve either receptors or transcriptional processes; they are variously insensitive to hormone antagonists (antihormones) or metabolic blockers (e.g. actinomycin D and cycloheximide). Some of them may depend on the hormonal stimulation of cyclic AMP synthesis which is not a prerequisite for the vast majority of other steroid responses.

### (c) *Hormonal synergism*

There are few, if any cells, which can function efficiently in the total absence of insulin or somatotropin and there is increasing evidence that prolactin may modulate the extent of the responses to sex hormones. The molecular basis for this hormonal synergism remains to be explained. Steroid hormones often need to work in concert. In many instances, oestrogens are needed for progesterone to exert its maximal effect; there is some evidence that the synthesis of the progesterone receptor may be under oestrogenic control, but this explanation does not apply to every progesterone response. The correct function of complex organs, such as the kidney and liver, also requires the harmonious interplay between many receptors. Again, simplistic models are not capable of explaining complex, integrated receptor function.

*Ecdysone receptors have now been identified.

### *(d) Temporal changes in hormonal responses*

Steroids are required for the initiation of developmental changes, both in the embryo and the adult. Foetal development characteristically occurs in a series of short bursts, and in terms of steroid involvement in differentiation, the process is a case of 'switch on' then 'switch off'. Regulation of receptor and acceptor synthesis provides only a partial explanation of these developmental periods of hormonal sensitivity, processes which in molecular terms, must be extremely complex over all. Similarly, the seasonal changes associated with reproduction in animals are complex. Again, it seems that fluctuations in either hormone secretion or modulation of receptor activity do not satisfactorily explain these phemomena.

### *(e) Tissue specificity of hormonal responses*

This remains the principal limitation to the simple model (Figure 6.4). Steroids promote dramatic and distinctive responses in adjacent organs which are exposed to an identical hormonal milieu. Examples include the accessory sexual glands in both sexes which each respond to the sex hormones in a unique manner, explainable now by the selective activation of a restricted number of genes in different target cells and the production of cell-specific species of mRNA. Since the receptor mechanisms are essentially similar, other components ultimately dictate the nature and extent of hormonal responses. All that really can be said at present is that differentiation imparts each target cell with the ability to respond to appropriate steroids in such a highly specific manner. Non-histone nuclear proteins may well prove to be the ultimate regulators of hormonal responses.

## V. DETAILS ON THE RECEPTORS FOR THE STEROIDS AND RELATED COMPOUNDS

### (i) Physicochemical properties

All the receptors for steroids that have been described thus far are oligomeric proteins, capable of binding their appropriate ligands with a high affinity (dissociation constants, $K_d$, are all within the range $10^{-11}$–$10^{-9}$ M). A representative survey of other physicochemical properties of many receptors is presented in Table 6.5. It cannot be overemphasized that the precise nature of the cytoplasmic receptors in the intact cell is really uncertain. All receptors have a remarkable tendency to aggregate and this problem, compounded by their notorious instability, has resulted

Table 6.5. Physicochemical properties of some cytoplasmic and nuclear receptor proteins

| Ligand | Tissue | Sedimentation coefficient ($S_{20,w}$) | Frictional ratio ($f/f_0$) | Stokes radius (nm) | Isoelectric point (pI) | Molecular weight ($\times 10^{-3}$) |
|---|---|---|---|---|---|---|
| (a) Cytoplasmic receptors* | | | | | | |
| Oestradiol | Uterus (calf, rat) | 8–8.6 | 1.65–1.69 | 6.7–7.0 | 5.8–6.2 | 220–236 |
| 5α-Dihydrotestosterone | Rat prostate | 8.0 | 1.96 | 8.4 | 5.8 | 276 |
| Progesterone | Guinea-pig uterus | 7.0 | 1.51 | 6.3 | 5.8 | |
| | Chick oviduct | 5. and 8 | 1.64 and 1.90 | | 4.0 and 4.5 | 100 and 360 |
| Glucocorticoid | Rat liver | 4 and 7 | 1.35 (4 S only) | 3.9 (4 S only) | 4.3 and 5.1 | 66 and 200 |
| Mineralocorticoid | Rat kidney | 4.5 and 8.5 | | | | |
| 1α,25-Dihydro-cholecalciferol | Rat duodenum | 3.7–5.0 | | | | |
| (b) Nuclear receptors† | | | | | | |
| Oestradiol | | 4.5–5.0 | 1.25 | | 6.4–7.0 | 70–75 |
| 5α-Dihydrotesterone | Rat Prostate | 3.5–4.0 | 1.27 | | 6.4–6.6 | 64 |
| Progesterone | Chick oviduct | 3.8 | 1.31 | | 6.2–6.5 | 66–80 |
| Mineralocorticoid | Rat kidney | 3 | | | | |
| 1α,25-Dihydroxy-cholecalciferol | Rat duodenum | 3.0–3.2 | | | | |

*Isolation and labelling at 0–4° C and low ionic strength (<0.1M KCl);
†receptor complexes extractable in 0.5M KCl

in a confusing plethora of information in the literature on their physical characteristics. In conditions of low ionic strength at temperatures near 0 °C. receptors tend to be relatively acidic, very asymmetric proteins of high molecular weight. The nuclear forms of receptors, freely extractable in 0.5 M KCl, tend to be smaller, less acidic and more spherical. The receptors for 1$\alpha$,25-dihydrocholecalciferol appear to be distinctive in that nuclear translocation is not accompanied by such profound changes in physicochemical properties. Most of the large forms of cytoplasmic receptors are dissociated in solutions of KCl exceeding 0.3 M, producing receptor–steroid complexes similar in many properties to the complexes extractable from nuclei.

Not all of the steroid associated with target cell nuclei can be extracted in 0.5 M KCl, the almost universal means for solubilizing nuclear forms of receptors. This residual binding in the nucleus is significant and may prove to be biologically important. Attempts have been made to relate the duration of acceptor occupancy by receptor complexes to the type of hormonal response. Evidence is available to suggest that only brief nuclear retention of receptor complexes is necessary for initial and early responses, whereas late responses require a protracted retention of receptor complexes. Steroid retention by the nuclear protein matrix may be essential for late responses, and nuclear binding to this subnuclear structure is more resistant to extraction into KCl.

Comparative studies on the binding of analogues and substituted derivatives of steroids to receptors suggest that multiple ligand–receptor interactions are required to attain specific, high affinity binding within the receptor site. It is envisaged that primary interactions involve the hydrocarbon skeleton and hydrophobic regions in the receptor site. These are followed by secondary interactions involving the substituent groups on the steroid with more hydrophilic regions in the receptor site, the entire process resulting in a subtle change in receptor configuration such that the steroid is enveloped or protected within the receptor protein. There are certain lines of evidence supporting the envelopment concept. In particular, receptor bound steroids become resistant to steroid-catabolizing enzymes and are also rendered unavailable to steroid-specific antibodies.

The conformational change promoted in the receptor site by the hormone is not necessarily large but it may be sufficient to evoke even further allosteric changes. The original view of Tomkins[15] that receptors can exist in either biologically active or inactive configuration has been subsequently expanded by several investigators. Steroid hormones and their agonists are predicted to stabilize the active receptor form, whereas antagonists may occupy the same receptor site yet maintain the receptor in the inactive form, only. Inactive steroids fail to bind to the receptor and again, the inactive form persists. Other sites required for the expression of

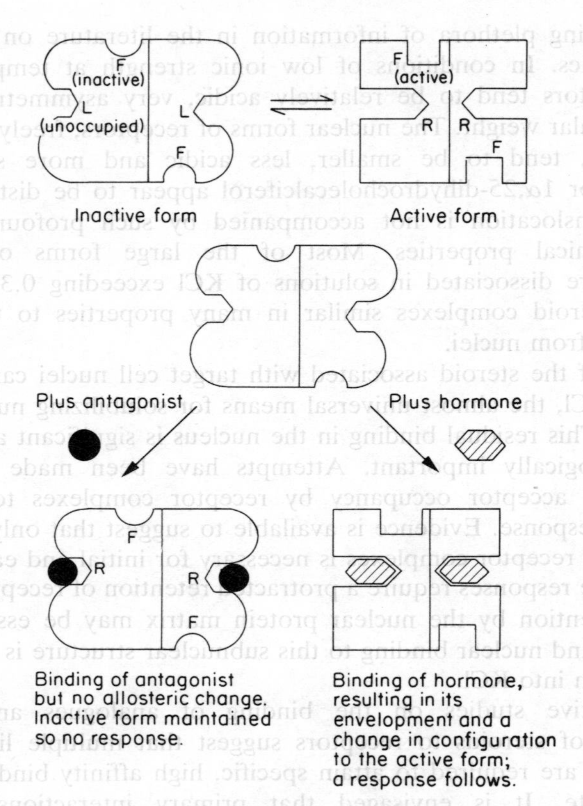


Figure 6.7. The allosteric model for ligand–receptor interactions. The original concept was formulated by Samuels, H. H., and Tomkins, G. M. *J. Mol. Biol.,* **52**, 57, (1970). The receptor, containing both a steroid-binding site and a functional site, necessary for biological activity, can exist in either inactive or active configurations. Allosteric concepts require dimer formations and cytoplasmic receptors, prior to activation, may well be dimers (see Figure 6.8). Depending on the nature of receptor–ligand interactions, the allosteric configuration is modified. L = ligand-binding site; F = functional site

hormonal responses, say, for attachment to the acceptor sites, are also influenced by the allosteric shifts set in train by the initial conformation change at the receptor site. These allosteric concepts are illustrated in Figure 6.7. This concept is supported by the following observations. First, limited digestion of receptor–steroid complexes with trypsin does not affect the receptor site as the steroids are not released but the complexes lose their ability to bind to DNA and chromatin. These experiments suggest distinct sites on the receptor, the steroid-binding site and other sites necessary for receptor function. Second, antihormones (such as the antioestrogen, nafoxidene) can bind to receptors and be translocated to the nucleus but no biological response ensues; observations such as these indicate that antihormones and steroid antagonists stabilize the receptor in the inactive configuration only.

### (ii) Receptor complex purification

A very major step forward would be made in unravelling the molecular mechanisms of steroid hormone action if receptor–steroid hormone complexes could be purified to homogeneity. Once purified, reconstitution experiments would provide insights into the activation of genetic transcription by receptor complexes and structural analysis of these important regulatory proteins could be undertaken. Receptor purification has proved to be a very daunting task, because they are labile proteins, present in only minute quantities: just a few $\mu$g per g of target cell protein. For the main part, receptor purification has been a frustrating and unrewarding venture in which all means of protein fractionation have been pushed to the limit of their resolution and specificity. After a tremendous amount of effort, in terms of time and manpower, it has been possible to purify the oestrogen receptors from calf and human uterus,[16,17] the glucocorticoid receptors from rat liver[18] and particularly the progesterone receptors from chick oviduct[19] and human uterus.[20] Some investigators report that chromatography of receptor-containing preparations on columns containing immobilized heparin was a vital contribution to their success,[21] but in the main, affinity chromatography proved to be the key procedure. In this powerful method of protein fractionation, steroid-free preparations of receptor are selectively adsorbed onto Sepharose matrices containing immobilized steroid ligands. After extensive washing to remove proteins without affinity for the matrix, the receptor complex may then be selectively eluted by applying a solution of the appropriate radioactive steroid ligand. Unfortunately, affinity chromatography does not appear to be generally applicable as a means of receptor purification, as some receptors are insufficiently stable in the absence of their favoured ligand to survive the affinity chromatography step. Somewhat surprisingly, the principle of affinity labelling, namely the covalent attachment of certain steroid derivatives to the receptor site, has not been widely exploited in receptor purification.

Through the skill and enterprise of Schrader and his collaborators,[19] the molecular architecture of the progesterone receptors of chick oviduct is

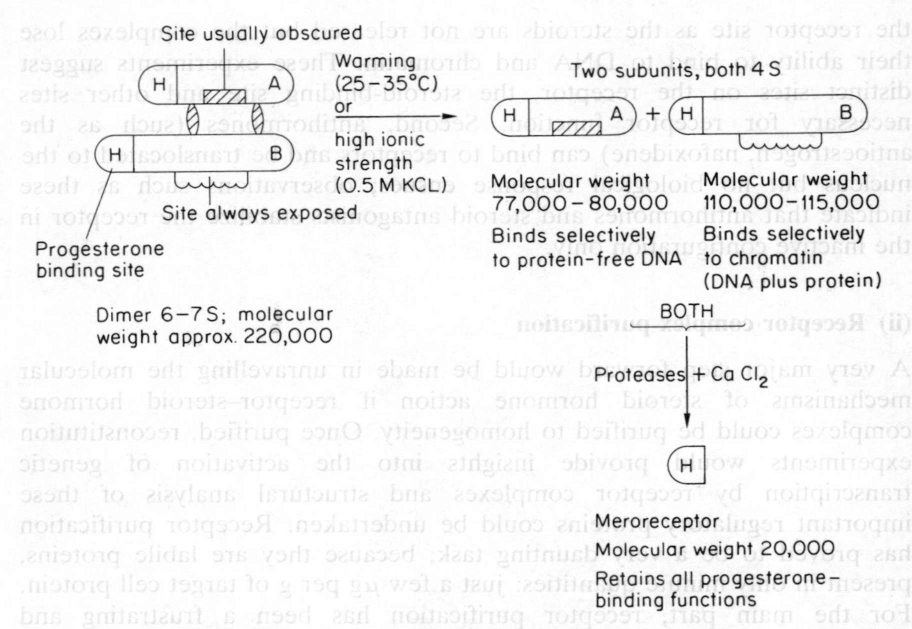


Figure 6.8. The structure of the cytoplasmic progesterone receptor from chick oviduct. The drawing is based on the findings of Schrader, W. T., Coty, W. A., Smith, R. G., and O'Malley, B. W. *Ann. N.Y. Acad. Sci.,* **286**, 64, (1977) and Sharman, M. R., Tuazon, F. B., Diaz, S. C., and Miller, L. K. *Biochemistry,* **15**, 980, (1976)

particularly well understood. Their view of the structure of the cytoplasmic form of the progesterone receptor is given in Figure 6.8. The cytoplasmic receptor is composed of two dissimilar subunits, A and B, both of which can bind progesterone selectively and which are associated together by van der Waals and ionic forces to constitute a 6–7 S dimer. Activation, either by warming to 30 °C or exposure to 0.5 M KCl, dissociates the dimer into the subunits and enables them to bind to nuclear components, the proposed acceptor sites. Both the dissociated subunits have sedimentation coefficients of 4 S, but subunit A has a molecular weight of 77,000–87,000 and binds exclusively to exposed DNA, whereas subunit B has a molecular weight of 110,000–115,000 and binds exclusively to chromatin (or DNA combined with non-histone proteins). Additional work by Sherman[22] showed that it was possible to produce a meroreceptor, molecular weight 20,000, by limited proteolysis of the receptor complex (or its component subunits) in the presence of $CaCl_2$. The meroreceptor retains all the specific features of the receptor site for progesterone but has completely lost the ability to bind to DNA or chromatin. It remains to be seen how far this exciting model can be applied to other receptor complexes, but the presence of A- and B-type subunits has not been widely confirmed to date in studies by other investigators.

### (iii) Cell-specific receptors

The remarkable range of biological responses to even a single steroid hormone raises the interesting question of whether there are in fact cell-specific receptors. The complete answer to this fundamentally important question can only be known when many more receptor complexes have been completely purified. If one adopts more indirect criteria, such as differences in ligand specificity, then it could be argued that tenuous evidence is already available to support the concept of cell-specific receptors. Androgen receptors show marked specificity in binding,[9] ranging from 5$\alpha$-dihydrotesterone in most accessory sexual glands, testosterone in muscle and mouse kidney, androstenedione in liver 5$\alpha$-androstane-3$\alpha$,17$\alpha$-diol in dog prostate. The progesterone receptors in the endometrium and myometrium of sheep uterus are strikingly different[23] and there is very persuasive evidence indeed that the glucocorticoid receptors in different areas of the brain are fundamentally different.[24]

### (iv) Regulation of receptor synthesis

Although a given cell can contain many types of receptors, there is evidence that their synthesis is controlled by unique structural genes. In the Tfm mutant mouse, the absence of androgen receptors is due to a structural modification or deletion of the Tfm locus on the X-chromosome.[25] The complement of all other hormone receptors is normal, indicating that their structural genes are distant from the Tfm locus.

It is difficult to generalize on the mechanisms controlling the receptor content of target cells. There are certain indications of autoregulation, as translocation of the oestrogen receptor to the nucleus in rat uterus provides the stimulus for oestrogen receptor synthesis.[6] The apparent loss of receptors in castrated animals has been reported[9] but this does not support the idea of autoregulation necessarily. Androgen receptors disappear from a few target cells after orchidectomy, but then reappear later. Changes such as these may reflect only the variations in proteolytic activity in atrophic cells. Certainly the restoration of androgen receptor levels in castrated animals does not require the pituitary, as it occurs in hypophysectomized-castrated animals as well.

The developmental appearance of receptors also remains a puzzle and has received little experimental investigation. The ontogeny of oestrogen receptors during the approach to sexual maturity and uterine responsiveness in female rats has been clearly established, but this process requires neither the pituitary nor the ovary.[26] Similarly, the appearance of glucocorticoid receptors in the developing lung is essential for lung development and the synthesis of the vital protein, the pulmonary surfactant, necessary for efficient respiration after birth.[27] Again, the underlying mechanisms controlling the appearance of these receptors are completely unknown.

**(v) Receptor effects on macromolecular synthesis**

This step remains the cornerstone of current ideas on the mode of steroid hormones. Despite difficulties in purifying receptors, considerable effort has been directed towards the study of the effects of receptor complexes on the biochemical activity of nuclei and chromatin in reconstituted, cell-free systems. A wide variety of receptors, albeit all far from pure, have been shown to enhance RNA synthesis in chromatin and nuclei, *in vitro*.[6] Apart from inherent difficulties imposed by the use of impure steroid–receptor complexes, these studies failed to identify clearly the species of RNA whose synthesis was accelerated under these conditions of reconstitution. Despite these limitations, the study by Jacob and coworkers[28] demonstrated unequivocally that cytoplasmic preparations from the Tfm (receptor-deficient) mouse were unable to enhance RNA synthesis in nuclei, *in vitro*, even in the presence of high concentrations of androgens. Importantly, an enhancement of RNA synthesis was detectable even in nuclei from Tfm mice if they were previously exposed to cytoplasmic fractions of normal kidney, supplemented with physiological concentrations of testosterone, *in vitro*.

O'Malley and his colleagues have produced some highly imaginative work on the effects of purified receptor complexes on the availability of the gene for ovalbumin mRNA in chick oviduct chromatin, measured under conditions, *in vitro*.[29,30] In their approach, chromatin was incubated with highly purified preparations of progesterone receptor and then transcribed with *E. coli* RNA polymerase in the presence of the antibiotic, rifamicin; nucleic acid hybridization to a radioactive, complementary DNA probe to ovalbumin mRNA was subsequently used to analyse the RNA products or transcripts. On face value, these innovative studies appear to embody all the best features of contemporary technology and the experimental findings strongly suggest that the presence of the receptor complexes accelerates selectively the initiation of RNA synthesis from the ovalbumin gene. There was a close correlation between the extent of transcription of the ovalbumin gene with input of receptor complex, asymmetry of transcription in that only one strand of the oviduct DNA served as template and even faithful transcription of totally reconstituted chromatin from DNA, its components of histones and non-histone proteins. These studies represent a major advance, but even here, certain criticisms must rightly be raised. First, however rigorously the experiments are controlled, there is now considerable mistrust in the fidelity with which bacterial RNA polymerases transcribe specific genes in mammalian chromatin.[31] Second, the direct stimulation of chromatin, *in vitro*, runs counter to the lag or latent period in all hormonal responses, of which the induction of ovalbumin is no exception.[32] Third, most investigators, including O'Malley, have used diethylstilboestrol to induce ovalbumin synthesis in chickens and yet in these critical experiments, they used purified progesterone receptors; oestrogen receptors, the ideal choice for this study, have yet to be purified from chick oviduct.

## VI. CONCLUDING REMARKS

The oustanding problem of receptor purification transcends all other considerations in terms of future progress and development. The structure of these receptor complexes must be probed in detail so that their precise function in the regulation of macromolecular synthesis may be ascertained. Despite the impressive advances which have been made, fundamental questions remain unanswered, such as the ontogeny of receptors during development to name but one. Another burning issue is why 15,000 molecules or more of receptor complex must be translocated into the target cell nucleus in order to evoke a hormonal response. This question has been impressively considered in a recent review by King.[33] It has also become rather fashionable to consider that mRNA synthesis is the *pièce de résistance* of hormone action, yet in reality, we should also be considering how hormones promote rRNA synthesis since this is the principal response in the majority of target systems. We should also perhaps not expect too much from simple reconstruction experiments in our attempts to probe hormone action. From the elegant work of Paigen,[34] for example, we know that the induction of a single enzyme, such as $\beta$-glucoronidase, is not a simple process, but involves the concerted action of several genetic loci, distributed on different chromosomes; of these, only one may be hormonally responsive and thus amenable to experimental analysis by present techniques. Finally, investigators are almost obsessed by knowing how receptors switch on biochemical events, when in biological reality, the switching off process is equally important yet widely overlooked.

## VII. REFERENCES

1. Jensen, E. V., and Jacobsen, H. I. *Recent Progr. Horm. Res.*, **18**, 387–414, (1962).

2. Edelman, I. S., Bogoroch, R., and Porter, G. A. *Proc. Natl. Acad. Sci. U.S.A.*, **50**, 1169–1175, (1963).
3. Toft, D., and Gorski, J. *Proc. Natl. Acad. Sci. U.S.A.*, **55**, 1574–1581, (1966).
4. Knox, W. E., and Auerbach, V. U. *J. Biol. Chem.*, **214**, 307–313, (1955).
5. Clever, U., and Karlson, P. *Exptl. Cell Res.*, **20**, 623–626, (1960).
6. King, R. J. B., and Mainwaring, W. I. P. *Steroid-Cell Interactions*, Butterworths, London, (1974).
7. Westphal, U. *Steroid-Protein Interactions. Monographs on Endocrinology*, Springer-Verlag, Berlin, Vol. 4., (1971).
8. Westphal, U., Stroupe, S. D., Kute, T., and Cheng, S. -I. *J. Steroid Biochem.*, **8**, 367–374, (1977).
9. Mainwaring, W. I. P. *The Mechanism of Action of Androgens. Monographs on Endocrinology*, Springer-Verlag, New York, Vol. 10, (1977).
10. Wilson, J. D. *Endocrinology,* **92**, 1192–1199, (1973).
11. Naftolin, F., Ryan, K. J., Davies, I. J., Reddy, V. V., Flores, F., Petro, Z., Kuhn, M., White, R. J., Takaoka, Y., and Molin, L. *Recent Progr. Horm. Res.*, **31**, 295–319, (1975).
12. Olins, A. L., and Olins, D. E. *Science,* **183**, 330–334, (1974).
13. Berezney, R., and Coffey, D. D. *Adv. Enzyme Regul.*, **14**, 63–100, (1978).
14. Williams, D., and Gorski, J. *Proc. Natl. Acad. Sci. U.S.A.*, **69**, 3664–3668, (1972).
15. Samuels, H. H., and Tomkins, G. M. *J. Mol. Biol.*, **52**, 57–74, (1970).
16. Sica, V., Parikh, I., Nola, E., Puca, G. A., and Cuatrecasas, P. *J. Biol. Chem.*, **248**, 6543–6558, (1973).
17. Coffer, A. I., Milton, P. J. D., Pryse-Davies, J., and King, R. J. B. *Molec. Cell Endocrinol.*, **6**, 231–246, (1977).
18. Govindan, M. V., and Sekeris, C. E. *Eur. J. Biochem.*, **89**, 95–104, (1978).
19. Schrader, W. T., Coty, W. A., Smith, R. G., and O'Malley, B. W. *Ann. N.Y. Acad. Sci.*, **286**, 64–80, (1977).
20. Smith, R. G., Iramain, C. A., Buttram, V. C., and O'Malley, B. W. *Nature New Biol.*, **253**, 271–273, (1975).
21. Molinari, A. M., Medici, N., Moncharmont, B., and Puca, G. A. *Proc. Natl. Acad. Sci. U.S.A.*, **74**, 4886–4890, (1977).
22. Sherman, M. R., Atienza, S. B. P., Shansky, J. R., and Hoffman, L. M. *J. Biol. Chem.*, **249**, 5351–5363, (1974).
23. Kontula, K. *Acta. Endocrinol. (Copenh.),* **78**, 593–603, (1975).
24. de Kloet, R., Wallach, G., and McEwen, B. S. *Endocrinology,* **96**, 598–609, (1975).
25. Bullock, L. P., and Bardin, C. W. *Endocrinology,* **97**, 1106–1111, (1975).
26. Gorski, J., Sarff, M., and Clark, J. *Adv. Biosci.*, **7**, 5–20, (1971).
27. Farell, P. M. *J. Steroid Biochem.*, **8**, 463–470, (1977).
28. Jacob, S. T., Jänne, P., Bullock, L. P., and Bardin, D. W. *Biochim. Biophys. Acta,* **418**, 330–343, (1976).
29. Kalini, M., Tsai, S. Y., Tsai, M. -J., Clark, J. H., and O'Malley, B. W. *J. Biol. Chem.*, **251**, 516–523, (1976).
30. Schwartz, R. J., Chang, C., Schrader, W. T., and O'Malley, B. W. *Ann. N.Y. Acad.Sci.*, **286**, 147–160, (1977).
31. *Cold Spring Harbor Symposium Quantitative Biology* Vol. 43: many citations and articles, (1978).
32. Palmiter, R. D., Moore, P. B., Mulvihull, E. R., and Emtage, S. *Cell,* **8**, 557–572, (1976).

33. King, R. J. B. *Essays in Biochemistry*, Academic Press, London, Vol. 12, pp. 41–76, (1976).
34. Paigen, K., Swank, R. T., Tomino, S., and Granshow, R. E. *J. Cell. Physiol.*, **85**, 379–392, (1975).

Cellular Receptors for Hormones and Neurotransmitters
Edited by D. Schulster and A. Levitzki


CHAPTER 7

# Thyroid hormone receptors

Jamshed R. Tata
*National Institute for Medical Research, Mill Hill, London NW7 1AA, U.K.*



A meaningful discussion of hormone receptors is not possible without first defining (1) the nature of the hormone, (2) the kind of biochemical and physiological actions of the hormone whose mechanism of action is sought, and (3) the context in which the term 'receptor' is used. Many workers when dealing with thyroid hormone receptors, deal with a cellular component which strongly binds the hormone with little or no relevance to the manner in which the interaction would lead to a primary biochemical change which in turn is responsible for the ultimate physiological action. It is only in the limited context of hormone binding that the term 'receptor' will be used below for thyroid hormones.

## I. NATURE OF THYROID HORMONES

As iodothyronines, thyroid hormones are unique, naturally occurring molecules with biological activity, and the relationship between their hormonal activity and the various chemical features (iodophenolic function, the diphenylether linkage, the aliphatic side chain, etc.) has attracted much attention over the last 50 years.[1] It is, however, only relatively recently that Jorgensen,[2] who established the shape of the molecules of tri-iodothyronine and thyroxine (see Figure 7.1), demonstrated their biological activities as hormones are as much a function of their individual chemical groups as their over-all shape. The importance of this spatial element of a structure–function relationship was most dramatically brought home by Jorgensen's studies which showed that the isopropyl derivative of the hormone (in which the isopropyl group replaced the iodine atom in the 3′ position) is several times more potent, in some physiological assays, than the natural hormones. Other structural analogues where the spatial arrangement was disturbed have predictably turned out to have lower or no biological activity. The availability of these synthetic analogues has been of considerable value in establishing the 'specificity' of putative receptors, as will be discussed later.

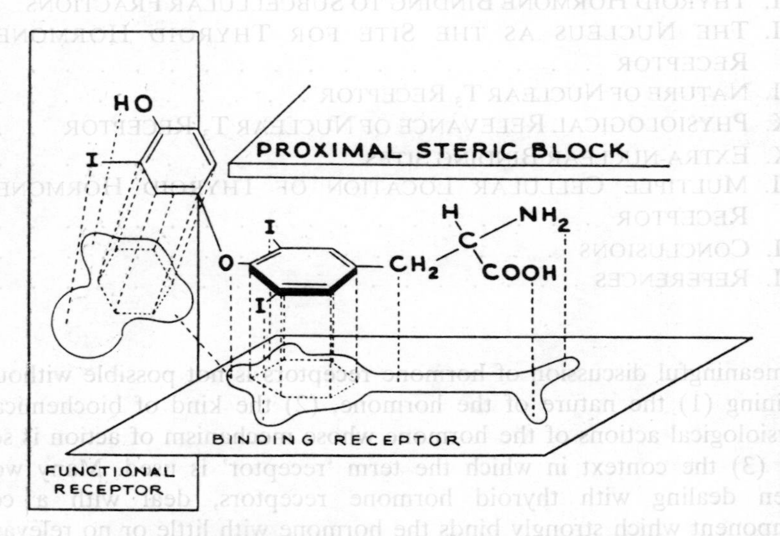


Figure 7.1. Schematic representation of the steric configuration of the molecule of tri-iodo-L-thyronine and its spatial relationship to a hypothetical receptor (Reproduced by permission from Jorgensen *et al.*[2])

## II. PHYSIOLOGICAL AND BIOCHEMICAL ACTIONS OF THYROID HORMONES

A characteristic of thyroid hormones is the wide range of physiological processes they influence.[1,3] For example, every single tissue in an amphibian tadpole responds to the metamorphic stimulus of thyroid hormone but the physiological or biochemical changes are not identical in each tissue. Thus, thyroid hormones induce urea cycle enzymes in tadpole liver, activation of lysosomal enzymes in regressing tissues such as the tail, gills, and the gut, and deposition of collagen by epidermal cells.[4,5] On the other hand, thyroid hormones have a particularly marked effect on mitochondrial respiratory enzymes in mammalian liver but do not significantly alter the activity of urea cycle enzymes. Table 7.1 lists some of the well-known features of the physiological and biochemical actions of thyroid hormones.

At the biochemical level, for a number of years the calorigenic action of thyroid hormones, as elicited by the regulation of basal metabolic rate, was considered to be of prime importance.[1,6] For this reason, it was assumed that the site of hormone action lay at the level of the regulation of cellular respiration or energy metabolism, i.e. in mitochondria.[1,6,7] More recently, the emphasis has shifted to the growth and developmental actions of thyroid hormones so that much of the current effort on thyroid hormone receptors is focused at the level of the cell nucleus.[3,8,9] This emphasis largely stems from the discovery that stimulation of RNA synthesis is one of the early responses to thyroid hormones and that actinomycin D blocks most of the physiological responses to thyroid hormones.[3,10,11]

Table 7.1. Multiple physiological actions of thyroid hormones

| Growth-promoting and developmental actions | Metabolic effects |
|---|---|
| Rate of growth of many mammalian and avian tissues | Regulation of basal metabolic rate |
| Maturation of central nervous system and bones | Regulation of water and ion transport |
| Requirement for all processes of amphibian metamorphosis | Calcium and phosphorus metabolism |
| Regulation of synthesis of some mitochondrial respiratory enzymes and structural elements | Regulation of cholesterol and fat metabolism |
| | Nitrogen (urea, creatine) metabolism |

For details, see References 1, 3, 4, 5, 6

## III. BINDING OF THYROID HORMONES TO BLOOD PROTEINS

There are three major thyroxine-binding proteins in the blood of most vertebrates: thyroxine-binding globulin (TBG), thyroxine-binding pre-albumin (TBPA), and albumin. Of these, the first two are thought to be of particular physiological importance in the dynamics of cellular availability and therefore in the eventual action or metabolism of the hormones.[12] In view of the importance of the spatial configuration of the hormone molecule in explaining its primary action,[2] it is obvious that information on three-dimensional topology of the interacting protein, or at least the binding site region, would be equally valuable. Whereas the elucidation of the primary structure of TBG is under way in Robbins' laboratory, not only has the primary sequence been worked out for TBPA[13] but Blake and his colleagues have made some striking progress with its three-dimensional structure by X-ray crystallographic analysis.[14,15] Indeed, Blake and Oatley[15] have recently suggested that the nature of the $T_4$-binding site and the configuration of TBPA-hormone complex may have considerable bearing on the hormone–receptor interaction.

Earlier studies with low resolution X-ray analysis had shown a preponderance of $\beta$-structure in TBPA, whose two subunits would each compromise four-stranded $\beta$-sheets.[14] The $T_4$-binding site or 'channel' would be generated only if the two eight-stranded sheets of the dimer were arranged face-to-face. More recently, data obtained with high resolution X-ray analysis have revealed a very interesting structural complementarity between TBPA and the double helical arrangement of DNA.[15] This has led Blake and Oatley[15] to suggest that the TBPA-$T_4$ interaction can be considered as a model for the hormone receptor in the nucleus.

As shown in Figure 7.2, which depicts the Blake and Oatley model for the DNA–protein–hormone complex, a possible DNA–TBPA binding site would be composed of two 'symmetry-related' $\beta$-sheets of TBPA comprising two helical arms which would be available for interaction with the large groove of the DNA molecule. The diagram also shows the part of the TBPA molecule which comprises the half binding site for $T_4$ and two such thyroid hormone binding sites would be included in a groove running through the entire length of the molecule.

To what extent can such a pre-albumin–DNA–$T_4$ interaction be considered a valid model for the nuclear receptor for thyroid hormone? Although the Blake and Oatley model is perhaps the most original model for the nuclear receptor at a detailed molecular level, there is no direct evidence as yet to support it. However, the model makes predictions which can be tested and therefore merits careful consideration. An interesting prediction is that the symmetry of the protein suggests a palindromic sequence in the part of DNA interacting with TBPA, inspired by the interaction between the *lac* repressor and DNA in bacteria. Perhaps the

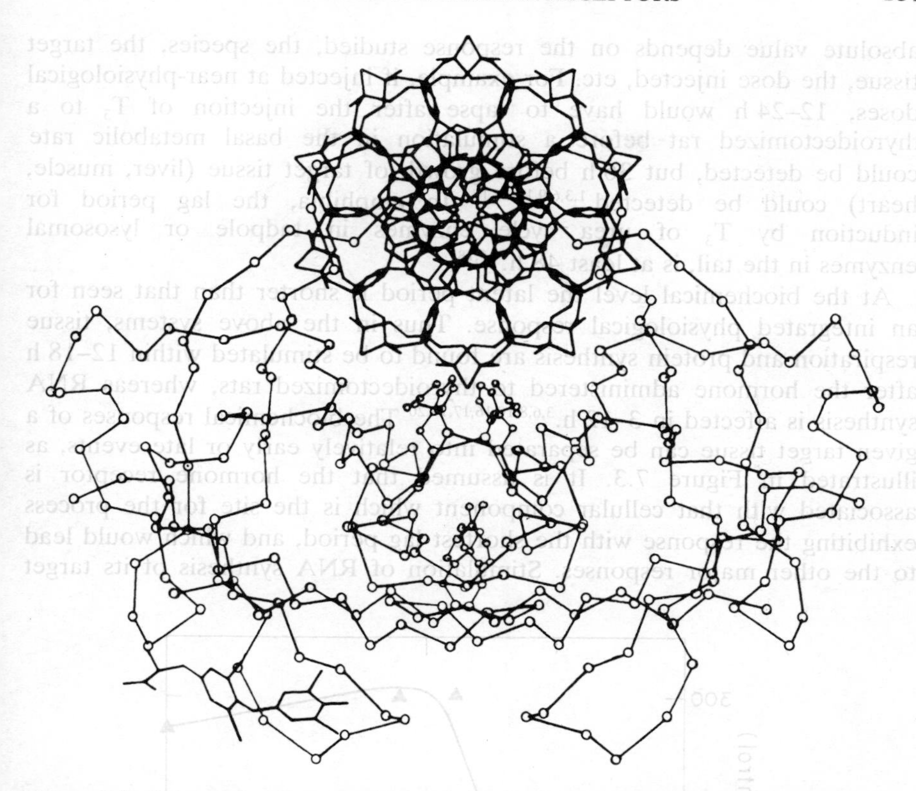

Figure 7.2. Schematic representation of the interaction between thyroid hormone, pre-albumin and DNA as proposed by Blake and Oatley[15] from their high resolution X-ray analysis. The major chain or pre-albumin dimer is shown in light line, the circles representing the $\alpha$-carbon positions. The symmetry-related pair of tyrosines 116 of the pre-albumin molecule are shown at the lower centre, and two pairs of side chains that point into the DNA binding site are shown in the centre. The DNA molecule itself is depicted as a double helix shown end-on in heavy line at the top and one thyroxine molecule bound to its half site is drawn in heavy line at the lower left

most obvious direct test of the hypothesis would be first to verify if pre-albumin, with or without the hormone, would bind to DNA. As will be mentioned below, it is not absolutely certain as to whether the nuclear receptor for $T_3$ would be expected to interact directly with DNA or with chromatin proteins (cf. Chapter 6).

## IV. LAG PERIOD OF RESPONSE TO THYROID HORMONES

There is a different and substantial latent period preceding each of the physiological responses to thyroid hormones listed in Table 7.1. The absolute value depends on the response studied, the species, the target tissue, the dose injected, etc. For example, if injected at near-physiological doses, 12–24 h would have to lapse after the injection of $T_3$ to a thyroidectomized rat before a stimulation in the basal metabolic rate could be detected, but 36 h before growth of target tissue (liver, muscle, heart) could be detected.[1,3,6,9,10,16,17] In amphibia, the lag period for induction by $T_3$ of urea cycle enzymes in tadpole or lysosomal enzymes in the tail, is at least 48 h.[4,5,18]

At the biochemical level the latent period is shorter than that seen for an integrated physiological response. Thus in the above systems, tissue respiration and protein synthesis are found to be stimulated within 12–18 h after the hormone administered to thyroidectomized rats, whereas RNA synthesis is affected in 3–12 h.[3,6,8,10,16,17,19,20] The biochemical responses of a given target tissue can be separated into relatively early or late events, as illustrated in Figure 7.3. It is assumed that the hormone receptor is associated with that cellular component which is the site for the process exhibiting the response with the shortest lag period, and which would lead to the other major responses. Stimulation of RNA synthesis of its target

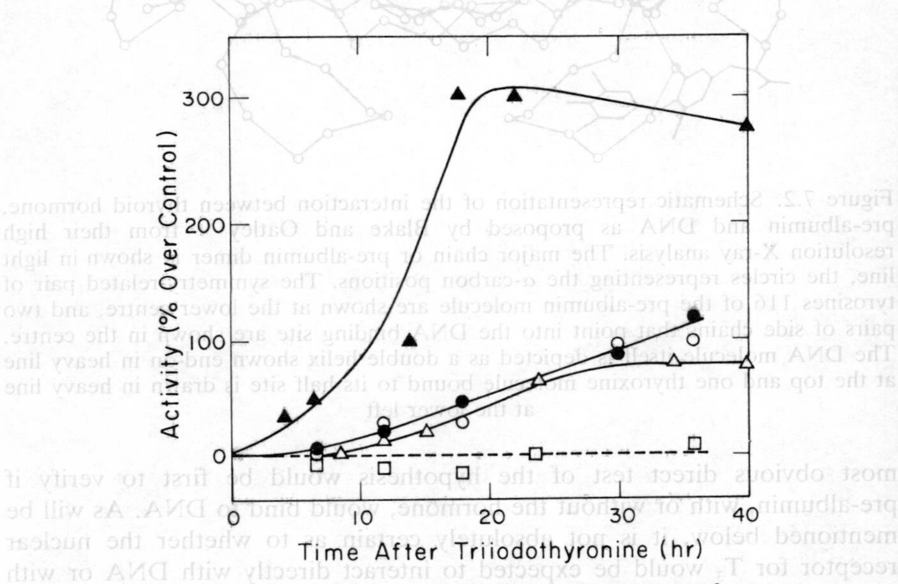


Figure 7.3. Time-course of biochemical response of rat liver nuclei to the administration *in vivo* of tri-iodothyronine. The various biochemical changes depicted include: rate of synthesis of RNA (▲); activity of RNA polymerase II (△); accumulation of newly synthesized total (○) and ribosomal protein (●); ratio of histone: DNA (□)

cells is therefore now considered to be a major early response to $T_3$ and $T_4$, and consequently much attention has been focused on the nucleus as the site of receptor for thyroid hormones.

## V. TISSUE DISTRIBUTION OF THYROID HORMONES

There is no quantitative relationship between the distribution of thyroid hormones in different tissues and their relative importance as target tissues. The major part of thyroid hormone leaving the circulation (as free hormone) is recovered in those tissues which play a major role in the metabolism, degradation or excretion of the hormone and not necessarily reflect the major hormonal target sites.[1,21] It is therefore not surprising to find a more extensive uptake of thyroid hormone in tissues such as liver, kidney, and intestine than in muscle or brain, all of which are responsive to the hormone. For this reason, it may be a disadvantage to work with tissues such as the liver, which are active in metabolizing the hormone, as models for receptors. For example, a strong binding of the hormone to the microsomal fraction of the liver may be more a reflection of hormone metabolism or degradation than a true receptor site for its action.

In the context of metabolism of thyroid hormones, it has been suggested that $T_3$ is the intracellular active form of the hormone and that $T_4$ has first to be converted to $T_3$ for its physiological activity.[9,17] There is still no general agreement on this issue and since most receptor studies are carried out with $T_3$ (if not both $T_4$ and $T_3$), it is not considered important for the discussion below.

## VI. THYROID HORMONE BINDING TO SUBCELLULAR FRACTIONS

Since almost all receptor studies involve the binding of radioactive ligand to some cellular component, the specific radioactivity of the hormone becomes extremely crucial since the putative receptor must be present in very limited number of molecules per cell, i.e. be easily saturable. Early investigations with thyroid hormones of low specific radioactivity therefore only revealed the presence of abundant or non-saturable sites of relatively low affinity.

The first studies on binding of $T_4$ and $T_3$ to subcellular fractions revealed a component in the cytosol of rat-liver and skeletal muscle which bound the hormone with moderate affinity.[21-24] The relatively low level of blood in muscle tissue made it possible to show that the cytosol protein was not TBG, TBPA or albumin.[21,22] However, the non-saturability or high binding capacity of such sites made it unlikely that one was dealing with a true receptor.

In a more systematic study of the intracellular distribution of thyroid hormone, all subcellular fractions of liver and muscle were found to interact with the hormones with an affinity in the range of $K_d$ of $10^{-7}$–$10^{-6}$ M.[24,25] The particulate fractions, i.e. nuclei, mitochondria, and microsomes, were found to account for more binding sites than the cytosol. A disturbing feature of these studies was revealed when a comparison was made between the subcellular distribution of the labelled hormone administered *in vivo* and when it was added to isolated subcellular fractions. A vast increase in the binding capacity in the latter situation suggested the danger of artifactually creating additional binding sites of relatively low affinity during the fractionation procedure.[21,24,25]

## VII. THE NUCLEUS AS THE SITE FOR THYROID HORMONE RECEPTOR

During the decade that followed the above studies (1963–1973), radioactive thyroid hormones of increasing specific activity were becoming available for binding studies. Meanwhile, work on the action of thyroid hormones clearly established the importance of the cell nucleus, in its capacity as the major site of transcription, as underlying not only the growth and developmental actions of these hormones but also their regulation of basal metabolic rate.[3,6,10,11,16,18,19,20] With the availability of hormone of higher specific radioactivity, it soon became clear from the studies from many laboratories, particularly those of Oppenheimer,[9,17,27,28,29] Samuels,[30,31,32] Baxter,[33,34,35] and DeGroot,[36,37,38] that the high affinity–low capacity binding sites of possible physiological relevance reside in the nucleus.

Oppenheimer, *et al.*[28] have studied the intracellular distribution *in vivo* of tracer $^{125}I$–$T_3$ injected into rats. As shown in Figure 7.4, the concentration of labelled $T_3$ fell rapidly in the plasma and cytosol, whereas the nuclear concentration rose during this period, before it too declined in parallel with the cytosol and plasma radioactivity. These workers calculated the fractional rates of entrance and exit from the nuclear compartment and also determined the nuclear binding capacity by injecting increasing doses of unlabelled $T_3$ together with tracer amounts of labelled hormone. It is worth mentioning that Oppenheimer's group have for some time regarded $T_3$ as the physiologically active form of thyroid hormone and that $T_4$ has to be first converted to $T_3$ to exert its action. Although this issue is still controversial, many observations, such as the relationship between nuclear binding capacity and the physiological dose–response curve, and the temporal relationship between nuclear retention of labelled hormone and changes in rates of RNA synthesis,[9,17] suggest that the nuclear binding of $T_3$ is closely linked to its physiological action.

When liver nuclei obtained from rats injected with $^{131}I$–$T_3$ were then

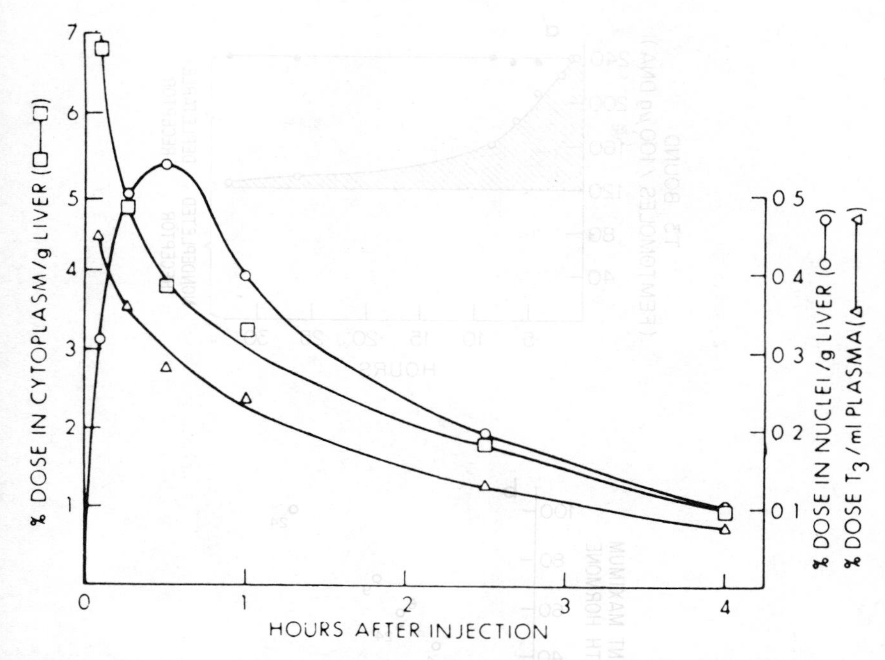


Figure 7.4. Time-course of distribution of $^{125}$I-labelled tri-iodothyronine in rat plasma (△) and liver cytoplasm (□), and nuclei (○) following the injection of a tracer dose of the labelled hormone. (Reproduced by permission from Oppenheimer *et al.*[28])

incubated *in vitro* with $^{125}$I–$T_3$ and subsequently extracted with 0.4 M KCl, the hormone with the two labels was equally distributed in the chromatin proteins.[9,17,40] Although the chromatin proteins are very heterogeneous, several studies have now suggested that the same binding sites are involved when the hormone is administered *in vivo* or directly added to isolated nuclei or sub-nuclear preparations, which will be discussed below in further detail.

At this point, it is worth turning to a cell culture system which has turned out to be very informative about the functional aspects of nuclear receptor. Samuels' group has recently studied the nuclear $T_3$ receptor in a cell line derived from rat pituitary tumours ($GH_1$) which responds to $T_3$ by synthesizing large quantities of growth hormone.[31,32,41] The enhanced rate of synthesis of growth hormone has been shown to result from a rapid and preferential increase in the transcription of its messenger RNA by $T_3$.[42,43] It is generally accepted that the functional significance of binding of $T_3$ (or $T_4$) to nuclear sites is the regulation of transcription. Whereas in rat liver no such highly selective transcription of genes takes place upon $T_3$

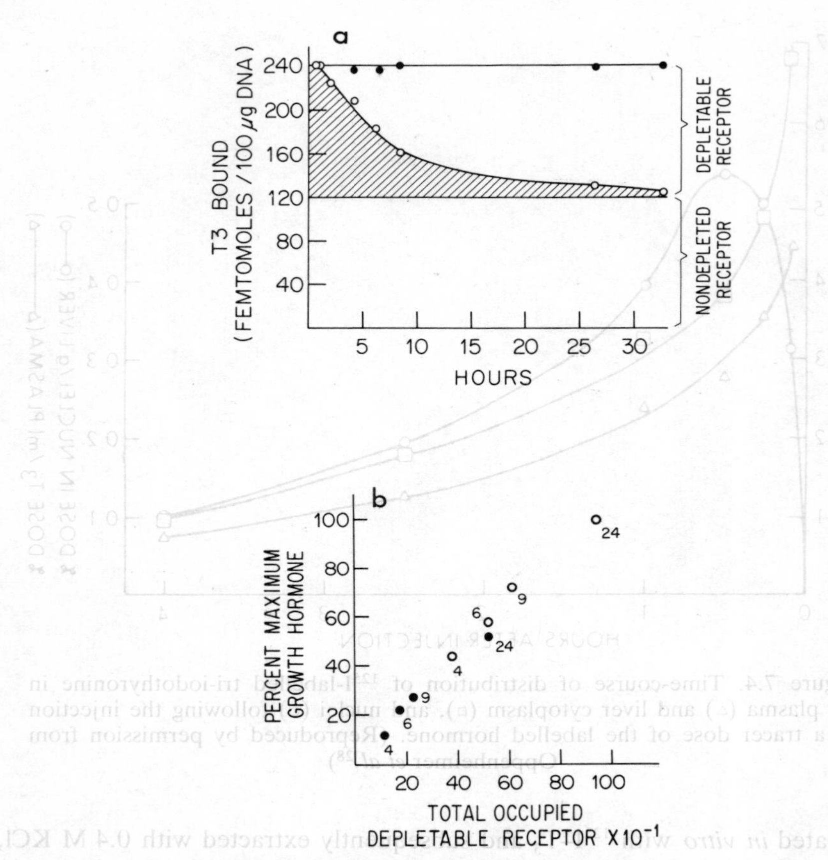


Figure 7.5. (a) Schematic representation of the influence of 5 nM tri-iodothyronine ($T_3$), as a function of time, on total 'depletable' nuclear $T_3$ receptor in $GH_1$ rat pituitary tumour cell line. Shaded area represents the residual depletable receptor at different times. Data of Samuels, *et al*.[31]

(b) Relationship of the rate of growth hormone synthesis and total depletable nuclear $T_3$ receptor in rat pituitary tumour $GH_1$ cells at different times after exposure to 0.2 nM (•) and 5 nM (○) tri-iodothyronine. The numbers adjacent to the points denote the number of hours after $T_3$ was added to the culture medium. Experimental details and calculations can be found in Reference 32. (Reproduced by permission of the American Society of Biological Chemists, Inc.)

administration, the $GH_1$ pituitary tumour cells make it possible to quantitatively establish the relationship between $T_3$ binding to nuclear sites and the transcription of a well-defined gene.

Based on their work on the correlation between nuclear binding of $T_3$ and growth hormone production, Samuels' group have suggested that there are two types of nuclear $T_3$ receptor sites in $GH_1$ cells.[30,31] This suggestion stems from an observation that $T_3$ causes a time- and dose-dependent depletion of a part of the nuclear bound receptor, the other part being unaffected. The depletion of the receptor was dependent on hormone binding and the depletable and non-depletable forms of the receptor are thought to be present in equal amounts. That the process of receptor depletion was physiologically meaningful was inferred from a better fit obtained between the rate of growth hormone synthesis and level of occupied depletable receptor than with total receptor occupancy (see Figure 7.5). Such an heterogeneity of nuclear receptors has not been found in rat liver where the action of the hormone is thought to be a function of total nuclear receptor occupancy.[9,17] On the other hand, DeGroot *et al.*[38] find that $T_3$ receptor sites in rat liver chromatin are non-saturable and suggest that it is the concentration of receptor in the nucleus which is the most important factor in thyroid hormone action, whereas Spindler *et al.*[34] do not find that $T_3$ itself influences receptor content of the nuclei.

## VIII. NATURE OF NUCLEAR $T_3$ RECEPTOR

Most investigators now report similar $T_3$-binding characteristics for the nuclear component in a variety of target cells, such as rat liver, brain, and pituitary tumour cell lines, with a $K_d$ in the range $0.5–5.0 \times 10^{-10}$ M. There seems to be less agreement about the number of receptor sites per nucleus, although a figure of 10,000–20,000 molecules per cell is most commonly reported. This value would be within the range also found for nuclear receptors for steroid hormones. The nuclear $T_3$ receptor is firmly bound to chromatin but can be extracted or solubilized with 0.4 M KCl.[9,17,35,38,39,40,44,45,46,47] By gel filtration or sucrose density gradient centrifugation it is thought to be a protein of molecular weight of about 60,000 or 3.5 S. Thus, the receptor would qualify as a typical non-histone chromatin protein. In view of the enormous heterogeneity of non-histone proteins of the nucleus, further characterization merely on the basis of $T_3$-binding activity will be a most difficult problem, especially if one accepts only a limited number of physiologically relevant binding sites per nucleus. A few attempts have been made at localization of the receptor at the sub-nuclear but the results of such nuclear fractionation do not show any striking localization or even mutual agreement. Thus, depending on the procedure used for fractionating nuclei or chromatin, an enrichment of $T_3$-binding component has been observed in what is thought to correspond to the transcriptionally active fraction of chromatin, in the transcriptionally inactive fraction, in nucleoli and in ribonucleoprotein particles.[32,35,49,50,51]

An extension of the pre-albumin model to the cellular thyroid hormone-binding protein would predict a strong interaction between DNA and $T_3$ receptor (see Section III). Indeed some laboratories, particularly Baxter's, have devoted much effort to this problem and conclude that the nuclear $T_3$ receptor is a DNA binding protein.[33,48,52,53] In the absence of a purified receptor, it is not possible to draw conclusions about the nature of the DNA–protein interaction, sequence specificity of DNA, etc. However, by studying the properties of a solubilized nuclear $T_3$-binding protein, Torresani *et al.*[48] have concluded that the hormone itself may not be required for the protein to bind DNA. Finally, one has also to take account of the fact that multiple forms of nuclear $T_3$ receptors have been reported.[31,32,44,48,54]

## IX. PHYSIOLOGICAL RELEVANCE OF NUCLEAR $T_3$ RECEPTOR

There are several lines of indirect evidence which support the notion that the binding of thyroid hormones to nuclear components is related to the physiological action of the hormones.

First, the transcriptional activity of the nucleus is an important locus of action of thyroid hormones. Although a few relatively minor effects of thyroid hormones may depend on non-nuclear site of action, the majority of their growth, developmental and even metabolic activities can be explained by hormonal regulation of nuclear RNA synthesis.[3,6,9,11,16,18] The stimulation of the synthesis of all classes of RNA is among the earliest responses of target cells to the hormone.[10,20,26,43]

Second, the nuclear sites are limited in number and their high affinity and limited capacity is compatible with the low doses required to bring about their physiological responses.[1,3,4,6,9,17,19,31] Since there is no absolute yardstick of the affinity and binding capacity required for a binding component to be termed a receptor in its functional sense, the values observed do not necessarily confer physiological relevance to the binding of thyroid hormones to the nucleus. However, a limited number of high affinity sites is generally accepted as a major criterion for satisfying the requirements of a receptor, so that the $T_3$-binding characteristics of cytosol, for example, rule out the possibility of a cytoplasmic receptor, analogous to that found for steroid hormones[55] (see Chapter 6). Indeed, it is now well established that $T_3$ need not interact with any cytoplasmic components in order to be bound to the nucleus.[8,9,17,54]

Table 7.2. Nuclear binding and biological activity of 3′-substituted 3,5-di-iodo-thyronines relative to those of 3,5,3′-tri-iodo-L-thyronine. (Data of Koerner, et al.[29])

| Substitution in 3′ position | Relative nuclear binding | Relative biological activity (antigoitre activity) |
|---|---|---|
| I | 100 | 100 |
| H | 0.3 | 1.2 |
| Cl | 6.2 | 3.4 |
| $CH_3$ | 13.5 | 10.6 |
| $CH_2CH_3$ | 21.0 | 69 |
| $CH(CH_3)_2$ | 104.0 | 125 |
| $C(CH_3)_3$ | 38.5 | 15 |
| $CH_2CH(CH_3)_2$ | 20.0 | 7.5 |
| $C_6H_5$ | 2.0 | 1.4 |

Third, competitive binding studies reveal a reasonably good agreement between the relative nuclear binding affinities and physiological potencies of a large number of structural analogues of thyroxine and tri-iodothyronine.[9,17,29,53] As shown in Table 7.2, the low physiological activity of tetra-iodothyroacetic acid matches the low nuclear binding affinity and, conversely, the high biological potency of tri-iodothyroacetic acid and the isopropyl derivative of $T_3$ are reflected in the ease with which they displace bound $T_3$.

Fourth, there is good correlation between $T_3$ binding to nuclei isolated from target and non-target tissues. Thus nuclei from liver, heart, and kidney which are well known to respond to thyroid hormones in adult mammals contain high affinity–low capacity sites whereas those from tissues like brain, testis, and spleen do not.[9,17]

Finally, there is good temporal relationship between nuclear retention of $T_3$ (or the binding to depletable nuclear sites) and the stimulation of transcription by thyroid hormones. This argument has been particularly effectively made by Oppenheimer's group for the kinetics of lag period, the decay of stimulation of RNA synthesis following $T_3$ administration and relative rates of analogue retention[9,17,53,56] and by Samuels[30,31] for the onset and decline of growth hormone messenger RNA synthesis.

None of the above considerations individually establishes the nucleus, or some nuclear component, as the physiologically relevant site of action of thyroid hormones. However, when considered together they argue very strongly in its favour but without necessarily excluding the role of possible extra-nuclear sites of action.

## X. EXTRA-NUCLEAR BINDING SITES

Earlier studies with thyroid hormones of relatively low specific radioactivity had shown that, not just the nucleus, but all major cytoplasmic subcellular fractions had $T_3$- and $T_4$-binding sites with affinities of $K_d$ about $10^{-8}$ M but which were difficult to saturate.[7,8,21,24,25,57,58] More recently, however, there have been reports of extra-nuclear high affinity, saturable sites in mitochondria,[59] plasma membranes,[60] and cytosol.[61] Sterling[59] (also see discussion in Reference 9), in particular, has argued that there may be several possible mechanisms by which thyroid hormones may act and that mitochondria represent one of these loci of action. They find that liver and kidney mitochondria extracts contain a protein which has an affinity for $T_3$ that is '500 times' that of nuclei ($\sim 2.5 \times 10^{11}$) when measured under identical conditions. There is also a good correlation between physiological potency and relative mitochondrial binding of $T_3$ analogues.

Historically, since the recognition of the regulation of basal metabolic rate as a major action of thyroid hormones, mitochondria have been considered as a major site of their action.[1,6] However, it is now recognized that the regulation of basal metabolic rate is indirectly achieved by a regulation of protein synthesis via nuclear transcription[3,5,10,11] and that most mitochondrial proteins involved in respiratory activity are synthesized on cytoplasmic polysomes coded for by nuclear mRNA.[62] In this context, it is worth considering the work from Edelman's laboratory[63] on the regulation of $Na^+/K^+$–ATPase by thyroid hormones. These authors have suggested that $T_3$ regulates the level of this plasma membrane enzyme via a nuclear transcriptional mechanism, and not by a direct interaction with plasma membranes, and which in turn would explain the high respiratory rates achieved by a target cell under the influence of thyroid hormones. It is therefore difficult to envisage a hormonal control of the limited protein synthesis that occurs within mitochondria as the basis for explaining physiological effects of thyroid hormones that involve changes in the levels of mitochondrial enzymes.[1,6,16,17] Furthermore, a comparison of absolute values of binding affinities of different subcellular components can be misleading since these are a function of the conditions (method of isolation, ionic strength, pH of medium, etc.) used for their determination. It is also worth recalling the earlier work on low affinity binding of thyroid hormones showing that the process of isolation of different subcellular particles itself can cause an artifactual generation of a large number of binding sites.[21,34,25] Admittedly this objection is equally valid for nuclear sites as for mitochondria, but the demonstration of the close relationship between occupancy of $T_3$ on nuclear sites[9,17,53] or depletable nuclear receptors,[31,32] studied *in vivo*, and the regulation of RNA synthesis or *de novo* induction of proteins are strong arguments in favour of physiological relevance of nuclear receptors. Similar demonstration of a close association between the occupancy of mitochondrial binding sites *in vivo* and a major physiological function is not yet available.

Recently, Pliam and Goldfine[60] have described a high affinity saturable $T_3$ receptor in plasma membrane preparations of rat liver ($K_d \sim 3.2 \times 10^{-9}$ M). Analogues with low physiological potencies were found to bind to the membranes less firmly. These authors do not assign a central role to such receptors for explaining specific physiological actions of thyroid hormones but propose that the interaction with plasma membranes may have a role in the regulation of transport of small molecules, such as amino acids and sugars, into cells,[64,65] or in the transport of thyroid hormone itself into the cell. In another study, on the uptake of thyroid hormone by isolated rat liver cells,[66] two sites for the transport of $T_3$ into cells ($K_d$ values of 52 and 1,446 nM) were demonstrated.

## XI. MULTIPLE CELLULAR LOCATION OF THYROID HORMONE RECEPTOR

A characteristic of thyroid hormone is the variety of physiological and biochemical responses it provokes in its target cells.[1,3,4,5] Whereas the physiological responses are hormone-, species- or tissue-specific, most biochemical responses are non-specific and common to all hormones which regulate protein synthesis. These have been referred to as 'pleiotypic' responses by Tomkins,[67] some of which are described as 'early' and 'late' responses in Table 7.3.

Two major questions arise from a consideration of the responses in Table 7.3 which are relevant to receptors in general. First, are all the late responses a consequence of early events provoked by the hormone?

Table 7.3. Early and late responses of target cells to thyroid hormones

| Early events | Late events |
|---|---|
| Enhanced uptake of ions, amino acids, nucleosides | Accelerated rate of synthesis of all RNAs |
| Redistribution or modification of acidic chromatin proteins | Stimulation of RNA polymerases A and B |
| Small increase in RNA polymerase B activity | Higher content of polysomes and enhanced over-all amino acid incorporation |
| Small increase in amino acid incorporation into protein | Redistribution of free and membrane-bound ribosomes |
| Increase in synthesis of ornithine decarboxylase and nuclear polyamines | Increase in $Na^+/K^+$-ATPase levels |

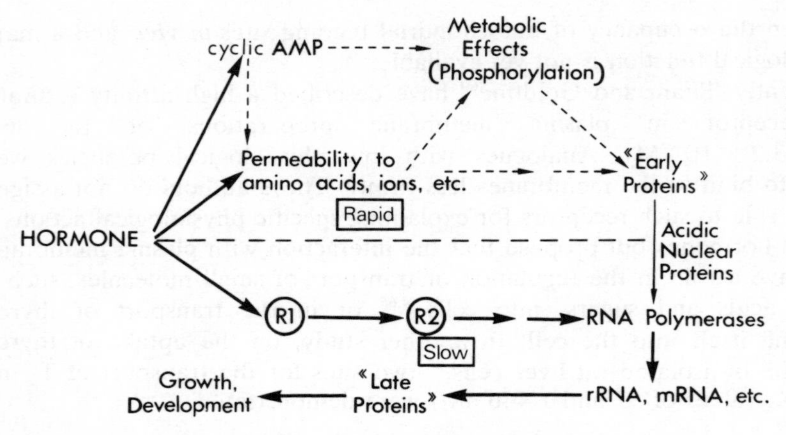


Figure 7.6. General scheme showing the integration of 'early' and 'late' responses to a hormone having a growth and developmental activity, including thyroid hormones

Second, are the different early events the result of an interaction between the hormone and a single type of receptor or do they reflect an interaction with multiple receptors? Whereas the answer to the first question is in the affirmative in most instances, that for the second remains unknown yet. Figure 7.6 attempts to 'rationalize' the multiple biochemical actions of thyroid hormones, irrespective of a single or multiple species of receptor molecules.

The major target cell responses are divided into relatively 'early', or independent of protein synthesis (upper part of Figure 7.6), and relatively 'late' actions which would be mediated via a control of protein synthesis (lower part of Figure 7.6). Thus, the rapid effects on amino acid or sugar transport could result from an interaction of the type recently demonstrated in plasma membrane[60] and which would be independent of protein synthesis.[68] Phosphorylation and dephosphorylation of proteins is now considered to be a major cellular regulatory mechanism also not dependent on protein synthetic activity.[69] Although thyroid hormone action is not considered to act via cyclic AMP or GMP, they may generate some other 'second messengers' which would catalyse phosphorylation of regulatory proteins. What is particularly interesting is that these are acidic non-histone nuclear proteins which are thought to facilitate selectively nuclear transcription.[70] The latter would be the major biochemical action of thyroid hormones underlying most physiological actions which are known to be dependent on protein synthesis. We have considered much evidence in this chapter that supports the notion that nuclear receptors play a major role in bringing about many of these actions.

## XII. CONCLUSIONS

The availability of radioactive thyroid hormones of high specific activity has enabled rapid progress to be made in characterizing cellular components that could be classified as receptors. Recent studies have helped clarify some of the earlier anomalies of thyroid hormone binding to various subcellular fractions of their target cells and strongly suggest that binding to nuclei may have sufficient physiological relevance to explain the action of these hormones. In this context, it is worth recalling the role of the rat pituitary tumour cell system since it has made it possible to quantitatively relate nuclear binding of thyroid hormone to the transcription of a specific gene, that of growth hormone. Unlike steroid hormones which also interact with the cell nucleus and regulate transcription, thyroid hormones appear to interact directly with the nucleus without prior binding to a cytoplasmic receptor. The nature of nuclear $T_3$ binding sites, whether or not these are heterogeneous, whether the receptor protein interacts with a chromosomal protein or directly with DNA, are questions which remain to be answered unequivocally.

Regulation of transcription in target cells is now considered to be a key step underlying not only the growth and developmental actions of thyroid hormones but also many of their metabolic regulatory functions, such as control over basal metabolic rate and the movement of sodium ions. However, the wide variety of biochemical and physiological actions of thyroid hormones have prompted some investigators to search for possible extra-nuclear sites for thyroid hormone receptor. In particular, $T_3$ binding sites, with affinities comparable to those of nuclear receptor, have been reported for mitochondria and plasma membranes. It has been suggested that the interaction with these sites would underlie some of the special hormonal effects, such as mitochondrial respiration or the transport of amino acids across plasma membranes. If the same kind of evidence can be provided in favour of the physiological relevance of these extra-nuclear $T_3$-binding sites as for nuclei the multiplicity of cellular sites poses some fundamental questions for future work on receptor function. For example, it suggests that multiple hormonal actions are not a consequence of interaction with a unique site of action but indicate responses to separate classes of receptors. A further question to be solved would be whether the hormone-binding sites present in different cellular locations represent qualitatively the same component distributed in several subcellular fractions or whether they are distinct entities. It can only be answered ultimately by purifying receptors from the different fractions.

Finally, it is important to reconsider the functional definition of 'receptor'. In this chapter, the term receptor has been used interchangeably with a $T_3$-binding protein. Whereas the techniques for studying binding are now well established, we do not yet have a functional assay or a 'bio-assay' for receptors, i.e. it is not possible to confirm that the immediate consequence of the interaction between the hormone and the binding component would necessarily directly lead to a biochemical event which would be the primary link in the chain of responses underlying a given physiological action of the hormone. Until such a functional test is devised, all the work described above can strictly be considered as a study of hormonal binding to various cellular fractions or components. This fundamental reservation does not apply only to thyroid hormones but to other hormones as well and, indeed, to all biologically active substances.

## XIII. REFERENCES

1. Pitt-Rivers, R., and Tata, J. R. *The Thyroid Hormones*, Pergamon Press. London, (1959).
2. Jorgensen, E. C., Lehman, P. A., Greenberg, C., and Zenker, N. *J. Biol. Chem.,* **237**, 3832–3838, (1962).
3. Tata, J. R. *Handbook of Physiology*, Section 7, Vol. 3, pp. 469–478. (1974).
4. Frieden, E., and Just, J. J. in *Biochemical Actions of Hormones* (Litwack, G., ed.) Academic Press, New York, Vol. 1, pp. 1–52, (1970).
5. Beckingham Smith K., and Tata, J. R. in *Developmental Biology of Plants and Animals*, (Graham, C. F., and Wareing, P. F., eds.) Blackwell, Oxford, pp. 232–245, (1976).
6. Tata, J. R. *Adv. Metabol. Dis.,* **1**, 153–189, (1964).
7. Sterling, K., Saldhana, V. F., Brenner, M. A., and Milch, P. O. *Nature (London),* **250**, 661–663, (1974).
8. Tata, J. R. *Nature (London),* **257**, 18–23, (1975).
9. Oppenheimer, J. H., Schwartz, H. L., Surks, M. I., Koerner, D., and Dillmann, W. H. *Recent Progr. Hormone Res.,* **32**, 529–553, (1976).
10. Tata, J. R., and Widnell, C. C. *Biochem. J.,* **98**, 604–620, (1966).
11. Tata, J. R. *Nature (London),* **197**, 1167–1168, (1963).
12. Robbins, J., Cheng, S., Gershengorn, M., Glinoer, D., Cahnmann, H. J., and Edelhoch, H. *Recent Progr. Hormone Res.*, **34**, 477–516, (1978).
13. Kanda, Y., Goodman, D. S., Canfield, R. E. and Morgan, F. J. *J. Biol. Chem.,* **249**, 6796–6805, (1974).
14. Blake, C. C. F., Swan, I. D. A., Rerat, C., Berthou, J., Laurent, A., and Rerat, B. *J. Mol. Biol.,* **61**, 217–224, (1971).
15. Blake, C. C. F., and Oatley, S. J. *Nature (London),* **268**, 115–120, (1977).
16. Tata, J. R. in *Actions of Hormones on Molecular Processes* (Litwack, G., and Kritchevsky, D., eds.), John Wiley, New York, pp. 58–131, (1964).
17. Oppenheimer, J. H., and Surks, M. I. in *Biochemical Actions of Hormones* (Litwack, G. ed.) Academic Press, New York, Vol. 3, pp. 119–157, (1975).
18. Tata, J. R. *Current Topics Develop. Biol.,* **6**, 79–110, (1971).
19. Tata, J. R., Ernster, L., Lindberg, O., Arrhenius, E., Pedersen, S., and Hedman, R. *Biochem. J.,* **86**, 408–428, (1963).
20. De Groot, L. J., Rue, P., Robertson, M. Bernal, J., and Scherberg, N. *Endocrinology,* **101**, 1690–1700, (1977).
21. Tata, J. R. *Recent Progr. Hormone Res.,* **18**, 221–259, (1962).
22. Tata, J. R. *Biochim. Biophys. Acta,* **28**, 91–94, (1958).
23. Lissitzky, S., Roques, M., and Benevent, M. T. *Biochim. Biophys. Acta,* **41**, 252–263, (1960).

24. Tata, J. R., Ernster, L., and Suranyi, E. M. *Biochim. Biophys. Acta,* **60**, 461–479, (1962).
25. Tata, J. R., Ernster, L., and Suranyi, E. M. *Biochim. Biophys. Acta,* **60**, 480–491, (1962).
26. Kurtz, D. T., Sippel, A. E., and Feigelson, P. *Biochemistry,* **15**, 1031–1036, (1976).
27. Oppenheimer, J. H., Koerner, D., Schwartz, H. L., and Surks, M. I. *J. Clin. Endocrin. Metab.,* **35**, 330–333, (1972).
28. Oppenheimer, J. H., Schwartz, H. L., Koerner, D., and Surks, M. I. *J. Clin. Invest.,* **53**, 768–777, (1974).
29. Koerner, D., Schwartz, H. L., Surks, M. I., Oppenheimer, J.H., and Jorgensen, E. C. *J. Biol. Chem.,* **250**, 6417–6423, (1975).
30. Samuels, H. H., and Tsai, J. S. *Proc. Natl. Acad. Sci. U.S.A.,* **70**, 3488–3492, (1973).
31. Samuels, H. H., Stanley, F., and Shapiro, L. E., *Proc. Natl. Acad. Sci. U.S.A.,* **73**, 3877–3881, (1976).
32. Samuels, H. H., Stanley, F., and Shapiro, L. E. *J. Biol. Chem.,* **252**, 6052–6060, (1977).
33. MacLeod, K. M., and Baxter, J. D. *Biochem. Biophys. Res. Commun.,* **62**, 577–583 (1975).
34. Spindler, B. J., MacLeod, K.M., Ring, J., and Baxter, J. D. *J. Biol. Chem.,* **250**, 4113–4119, (1975).
35. Charles, M. A., Ryffel, G. U., Obinata, M., McCarthy, B. J., and Baxter, J. D. *Proc. Natl. Acad. Sci. U.S.A.,* **72**, 1787–1791, (1975).
36. De Groot, L. J., and Strausser, J. L. *Endocrinology,* **95**, 74–83, (1974).
37. De Groot, L. J., Refetoff, S., Strausser, J., and Barsano, C. *Proc. Natl. Acad. Sci. U.S.A.,* **71**, 4042–4046, (1974).
38. De Groot, L. J., Hill, L., and Rue, P. *Endocrinology,* **99**, 1605–1611, (1976).
39. De Groot, L. J., and Torresani, J. *Endocrinology,* **75**, 357–369, (1975).
40. Surks, M. I., Koerner, D., Dillman, W., and Oppenheimer, J. H. *J. Biol. Chem.,* **248**, 7066–7072, (1973).
41. Tsai, J. S., and Samuels, H. H. *Biochem. Biophys. Res. Commun.,* **59**, 420–428, (1974).
42. Bancroft, F. C., Wu, G. -J., and Zubay, G. *Proc. Natl. Acad. Sci. U.S.A.,* **70**, 3646–3649, (1973).
43. Martial, J. A., Baxter, J. D., Goodman, H. M., and Seeburg, P. H. *Proc. Natl. Acad. Sci. U.S.A.,* **74**, 1816–1820, (1977).
44. Latham, K. R., Ring, J. C., and Baxter, J. D. *J. Biol. Chem.,* **251**, 7388–7397, (1976).
45. Samuels, H. H., Tsai, J. S., and Casanova, J. *Science,* **184**, 1188–1191, (1974).
46. Thomopoulos, P., Dastugue, B., and Defer, N. *Biochem. Biophys, Res. Commun.,* **58**, 499–506, (1974).
47. Surks, M. I., Koerner, D. H., and Oppenheimer, J. H. *J. Clin. Invest.,* **55**, 50–60, (1975).
48. Torresani, J., Anselmet, A., and Wahl, R. *Molec. Cell. Endocrinol.,* **9**, 321–333, (1978).
49. Gardner, R. S. *Biochem. Biophys, Res. Commun.,* **67**, 625–633, (1975).
50. Levy, W. B., and Baxter, J. D. *Biochem. Biophys. Res. Commun.,* **68**, 1045–1051, (1976).
51. Defer, N., Sabatier, M. M., Kruh, J., Dabauvalle, M. C., Creuset, C., and Loeb, J. *FEBS Letters,* **76**, 320–324, (1977).
52. MacLeod, K. M., and Baxter, J. D. *J. Biol. Chem.,* **251**, 7388–7397, (1976).

53. Surks, M. I., and Oppenheimer, J. H. in *Hormone–Receptor Interaction: Molecular Aspects* (Levy, G. S., ed.). Marcel Dekker, New York, pp. 373–384, (1976).
54. Docter, R., Visser, T. J., Stinis, J. T., van den Hout-Gaemaat, N. L., and Hennemann, G. *Acta Endocrinol.,* **81**, 82–95, (1976).
55. King, R. J. B., and Mainwaring, W. I. P. *Steroid-Cell Interactions*, Butterworths, London, (1974).
56. Goslings, B., Schwartz, H. L., Dillman, W., Surks, M .I., and Oppenheimer, J. H. *Endocrinology,* **98**, 666–675, (1976).
57. Sufi, S. B., Tocafondi, R. S., Malan, P. G., and Ekins, R. P. *J. Endocrin.,* **58**, 41–52, (1973).
58. Davis, P. J., Handwerger, B. S., and Glaser, F. *J. Biol. Chem.,* **249**, 6208–6219, (1974).
59. Sterling, K., and Milch, P. O. *Proc. Natl. Acad. Sci. U.S.A.,* **72**, 3225–3229, (1975).
60. Pliam, N. B., and Goldfine, I. D. *Biochem. Biophys. Res. Commun.,* **79**, 166–172, (1977).
61. Kistler, A., Yoshizato, K., and Frieden, E. *Endocrinology,* **100**, 134–137, (1977).
62. Shore, G. C., and Tata, J. R. *Biochim. Biophys. Acta,* **472**, 197–236, (1977).
63. Edelman, I. S., and Ismail-Beigi, F. *Recent Progr. Hormone Res.,* **30**, 235–257, (1974).
64. Goldfine, I. D., Simons, C. G., Smith, G. J., and Ingbar, S. H. *Endocrinology,* **96**, 1030–1037, (1975).
65. Segal, J., Schwartz, H., and Gordon, A. *Endocrinology,* **101**, 143–149, (1977).
66. Rao. G. S., Eckel, J., Rao. M. L., and Breuer, H. *Biochem. Biophys. Res. Commun.,* **73**, 98–104, (1976).
67. Herschko, A., Mamont, P., Shields, R., and Tomkins, G. M. *Nature (London),* **232**, 206–211, (1971).
68. Segal, J., and Gordon, A. *Endocrinology,* **101**, 150–156, (1977).
69. Rubin, C. S., and Rosen, O. M. *Ann. Rev. Biochem.,* **44**, 831–887, (1975).
70. O'Malley, B. W., Towle, H. C., and Schwartz, R. J. *Ann. Rev. Genet.,* **11**, 239–275, (1977).

SECTION III

# Cell membrane–surface receptors for hormones

Cellular Receptors for Hormones and Neurotransmitters
Edited by D. Schulster and A. Levitzki


CHAPTER 8

# Receptors for the glycoprotein hormones: LH, FSH, hCG, and TSH

Yoram Salomon
*Department of Hormone Research, Weizmann Institute of Science, Rehovot, Israel*



## I. INTRODUCTION

The glycoprotein hormones belong to the class of hormones which control target cell function by acutely modulating adenylate cyclase activity in their respective target cells.[1-3] They comprise a distinct family of hormones characterized by their high molecular weight, 28,000–43,000, and by common features in their molecular structure. Four hormones are included in this group: follitropin (FSH)*, lutropin (LH)†, thyrotropin (TSH)‡, all secreted by the anterior pituitary, and chorionic gonadotropin which is of

*also termed follicle stimulating hormone
†also termed luteinizing hormone
‡also termed thyroid stimulating hormone

placental origin and found only in pregnant females of higher primates. FSH, LH, and human chorionic gonadotropin (hCG) regulate gonadal functions and are therefore termed gonadotropins. TSH maintains the function of the thyroid gland.

Complex feedback mechanisms which involve the participation of the target organs control the secretion of the pituitary glycoprotein hormones. The secreted hormones migrate to their target tissues via the bloodstream.

The endocrine control exerted by glycoprotein hormones upon the respective target cells is made possible by specific cellular receptors located on the cell surface. These receptors are the building blocks of the cellular control switchboard which senses and translates the externally applied endocrine input into intracellular signals. One such signal, but not necessarily the only one, is represented by cyclic AMP. Its role as a 'second messenger' was originally proposed by Sutherland and Rall[4] in their studies on the hormonal control of liver glycogen metabolism (see Chapter 4).

The increase in cyclic AMP levels following hormonal stimulation is perceived by the cell as a metabolically meaningful signal which initiates a series of pre-programmed biochemical reactions that finally lead to the physiological response characteristic for each cell type. There is still some debate on the question of whether additional intracellular signals may be generated as a result of glycoprotein hormone interaction with their cell specific receptors. It has recently been shown by Grollman *et al.*[5] that the interaction of TSH with thyroid cells induces hyperpolarization of the cell membrane. These authors postulate that the electrical gradient formed is a primary event resulting from the receptor hormone interaction and may thus represent the transmembrane signal. These highly interesting but preliminary observations still need further experimental support which should clearly establish the exact sequence of molecular events which follow the primary interaction of glycoprotein hormones with their respective membrane receptors.

In this chapter, discussion is confined to the hormone receptor interaction and the consequent regulation of the adenylate cyclase for which a large body of evidence is available. Also included is a short description of the major physiological events occurring in the respective target cells/organs in response to hormonal stimulation.

## II. STRUCTURE AND CHEMISTRY OF GLYCOPROTEIN HORMONES

The complexity of the structure of glycoprotein hormones stands in contrast to the simplicity of other regulatory ligands such as biogenic amines, prostaglandins, and even low molecular weight peptide hormones which also stimulate adenylate cyclase activity in their respective target cells. The relatively complex structure of the glycoprotein hormones must somehow be

Table 8.1. Some molecular properties of human glycoprotein hormones

| Hormone | Subunit | Molecular weight | No. of amino acid residues (res mol$^{-1}$) | Carbohydrate content (res mol$^{-1}$) |
|---|---|---|---|---|
| FSH | | 36,300[6] | 204[7,8] | 26[6]* |
| | $\alpha$ | | 89 | |
| | $\beta$ | | 115 | |
| LH | | 28,850[9] | 218[10] | 23[10] |
| | $\alpha$ | 15,750 | 89 | 16 |
| | $\beta$ | 15,350 | 129 | 7 |
| hCG | | 36,700[11] | 231[10] | 45[10] |
| | $\alpha$ | 14,500 | 92 | 16 |
| | $\beta$ | 22,200 | 139 | 29 |
| TSH | | 25,000[6] | 215[10] | 20[10] |
| | $\alpha$ | | 90 | 14 |
| | $\beta$ | | 125 | 6 |

Superior numbers give the reference from which data were taken
*Percentage of total dry weight

reflected in the complementary structure of the specific cell receptor. It is therefore necessary to highlight some aspects of their basic structure.

The glycoprotein hormones are intimately related and share many structural and functional properties. Each molecule is composed of two non-identical polypeptide chains designated as $\alpha$ or $\beta$ subunits. As glycoproteins they also contain a carbohydrate moiety representing 7–30% of the total molecular mass (Table 8.1). The carbohydrate residues are linked to the polypeptide chains through *N*-glycosidic linkages to either the amide group of asparagine residues or the hydroxyl groups of serine or threonine residues. Heterogeneity in total carbohydrate content, as well as in the composition of its monosaccharide constituents, has been found in hormone preparations of the same hormone derived from different species. Microheterogeneity has also been described in preparations of the same hormone obtained from the same species. The most frequently represented monosaccharides in glycoprotein hormone molecules are: L-fucose, D-glucose, D-mannose, *N*-acetylglucosamine, *N*-acetylgalactosamine, and sialic acid.

The comparison of the physicochemical properties of the two subunits derived from different glycoprotein hormones revealed that a striking homology exist in the primary sequence of the $\alpha$ subunits, whereas the $\beta$ subunits differ markedly in this respect. Dissociation of the hormone into subunits can be achieved by various agents, including acidification to below pH 2.0, with instantaneous loss of biological activity. By using adequate chromatographic procedures, separation of the individual subunits may be achieved. Reassociation of the subunits may be accomplished by incubation of the separated subunits under proper conditions after removal of the dissociating agent. This slow reassociation process, which might take as long as five days at 4 °C, has been reported to restore much of the original biological activity. For example, reassociation of $\alpha$ and $\beta$ LH subunits may result in the restoration of 85% of the original biological activity. Because of the homology between the $\alpha$ subunits of LH, FSH, TSH, and hCG, it was possible to form hybrid molecules between the $\alpha$ subunit of one hormone and a $\beta$ subunit originating from another hormone. Experiments of this kind have shown that the biological activity of the hybrid molecule was only partially restored and was qualitatively similar to that of the native hormone which contributed the $\beta$ subunit. It was suggested that the $\beta$ subunit of the glycoprotein hormone determines the biological specificity of the molecule but, this can only be expressed when associated with the $\alpha$ subunit. The relative contribution of each subunit to the binding of the hormone molecule and how the direct and intimate contact with the membrane receptor is established has not yet been determined. However, increasing amounts of evidence suggest that specific gangliosides present in the plasma membrane may play an important role in glycoprotein hormone–receptor interaction. This has been suggested for the TSH receptor present in thyroid cells,[12,13,14] but may also be true for the other glycoprotein hormones. hCG, LH, FSH, and TSH have been shown to interact with specific gangliosides. Free ganglioside has been shown to interfere with binding activity of these hormones. These findings suggest that binding to gangliosides on the cell membrane may be important for the normal action of the glycoprotein hormone molecule.[12]

Antibodies raised against glycoprotein hormones show some cross-reactivity which is related to the similarity of their $\alpha$ subunit structure. Cross-reactivity among antibodies raised against the same hormone, but derived from different species, is explained by the existence of common sequences in their $\beta$ subunit. The $\beta$ subunits are believed to contain most of the antigenic determinants of the native hormone. Antibodies with greater specificity have therefore been produced by using only $\beta$ subunits for immunization.

The participation of the carbohydrate moiety in the activity of the glycoprotein hormones has been extensively studied. It has been shown that removal of the sialic acid residues from hCG affected primarily the half-life of the hormone in the circulation, thus reducing its biological potency *in vivo*. Such treatment slightly enhanced the ability of the hormone to bind to testicular membranes containing specific hCG receptors, but had only a small effect on the immunological properties of the hormone. (For further details see References 6, 9, 15, 16, 17, 18, and 19.)

## III. THE TSH RECEPTOR

The TSH receptor has been extensively studied in purified plasma membranes from bovine thyroid. The specificity of the receptor for TSH was assessed by determining its ability to bind iodinated bovine TSH. Increasing concentrations of non-labelled TSH added to the assay mixture were found to displace effectively the labelled hormone on the membrane receptors,[20] thus proving that the bound label has properties identical to authentic TSH.

The relative specificity of the membrane receptor was determined by comparing the concentrations of other non-labelled hormones required for the displacement of 50% of the bound $^{125}$I-TSH. As seen in Figure 8.1, it required ten times more LH than TSH to displace bound $^{125}$I-TSH. Other peptide hormones such as glucagon, growth hormone, insulin, ACTH, prolactin, and even albumin were totally ineffective in this regard, indicating that these are not recognized and therefore do not interact with the thyroid membrane receptor. These results show that the thyroid membrane receptor has a ten-fold higher affinity for TSH than for LH, and infinitely higher affinity for TSH as compared to all the other protein and peptide hormones tested. That the highest specificity for the membrane receptor is exerted by TSH would be a qualitative term to express the relative relationship of the receptor with this series of proteins.

The contribution of the $\alpha$ and $\beta$ subunits of TSH to its binding and biological activity was studied by comparing the ability of the individual peptide chains to displace labelled hormone from the thyrotropin receptor. As calculated from Figure 8.1, a 50- and 200-fold excess of $\beta$ and $\alpha$ subunits,

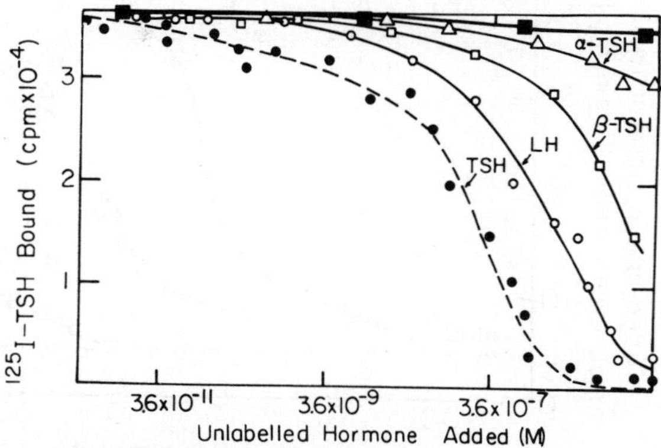


Figure 8.1. Inhibition of $^{125}$I-TSH binding to beef thyroid plasma membranes by unlabelled TSH (•), LH (○), $\beta$TSH (□), $\alpha$TSH (△), and albumin, prolactin, adrenocorticotropic hormone, insulin, glucagon, and growth hormone (■). Modified from Tate *et al.*[20]

respectively were needed to displace 50% of the bound hormone. A similar relationship between the parent hormone and its subunits was established when the activation of TSH-sensitive adenylate cyclase was studied (Figure 8.2). Here the hormone was found to be 1,000-times and twelve-times more effective in its ability to stimulate adenylate cyclase activity than were the individual $\alpha$ and $\beta$ subunits respectively. These results were shown to correspond very well with the biological activity of the hormone.[21] However, in interpreting these results one should not exclude the possibility that they may be explained in part by extremely small levels of TSH present as contaminants in preparations of LH, $\alpha$ and $\beta$ TSH.

Recent studies in which thyroid membranes were solubilized with lithium di-iodosalicilate have indicated that the molecular weight of the thyrotropin receptor of bovine thyroid is in the range of 15,000–30,000. In the soluble form the receptor retains its specificity for thyrotropin. Interestingly treatment of thyroid plasma membrane with trypsin also yielded a soluble protein with similar properties, presumably an active fragment of the thyrotropin receptor.[20,22]

The biochemical processes coupled to the activity of the thyrotropin receptor are extremely complex and extend beyond the scope of this chapter. In brief, the thyroid gland secretes two biologically active hormones: 3,5,3′-L-tri-iodothyronine ($T_3$) and L-thyroxine ($T_4$), both iodinated derivatives of the amino acid tyrosine. The biosynthesis of these

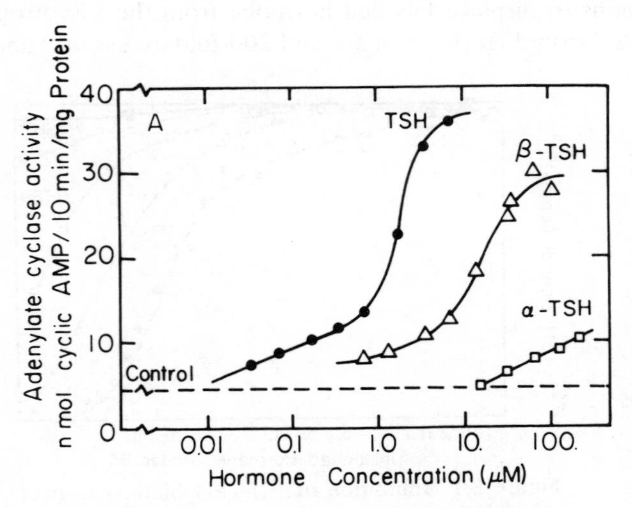


Figure 8.2. The effect of TSH, and its subunits on adenylate cyclase activity of beef thyroid plasma membranes. Control level indicates enzyme activity in the absence of added stimulant. Modified from Wolff *et al.*[21]

hormones is accomplished as follows: a large and unique protein (M.W. 670,000), thyroglobulin, is synthesized in the thyroid cell and secreted into the follicular lumen at the apical side of the cell, where it is stored in a colloidal form. At that stage, iodination of covalently linked amino acids (mainly tyrosine) within the protein takes place. The secretion of the thyroid hormones into the bloodstream involves uptake of the iodinated thyroglobulin colloid, proteolysis of the protein (presumably by lysosomal proteases), and release of $T_3$ and $T_4$ into the bloodstream at the basal cell surface. All the steps involved in the biosynthesis and release of thyroid hormones are believed to be controlled by TSH in a distinct sequence and relation to each other. (For a more detailed review the reader is referred to references 23 and 24.)

## IV. GONADOTROPIN RECEPTORS

The development and normal function of testes and ovaries is controlled and maintained by the hormones FSH and LH. It is well known that hCG can substitute for LH. It effectively competes with LH for the membrane receptor site in testes and ovaries and also resembles LH in its biological activity.[25,26,27] Due to its greater availability in highly purified form, hCG is widely used as an LH substitute for experimental as well as therapeutic purposes.

The gonadotropin receptors in gonadal tissue are distributed among a variety of cell types, as follows: FSH interacts with cell membrane receptors in granulosa cells of the Graafian follicle and in epithelial cells of the seminiferous tubules (probably Sertoli cells). The receptors that specifically bind LH and hCG are found in the cell membrane of Leydig (or interstitial cells of the testes and in the cell membrane of several types of cells in the ovary: thecal cells and granulosa cells of follicles approaching maturity, cells of the interstitial tissue, and cells of the corpus luteum.

### (i) FSH receptors

FSH receptors present in seminiferous epithelial cells of immature male rats have been reported to bind FSH with high affinity $K_D = 10^{-10}$ M. LH at 100-fold excess did not compete for bound labelled FSH. The binding sites appear to be associated with the plasma membrane and are situated on the outer surface of these membranes. Binding activity of the receptor is lost upon treatment of the membranes with trypsin or pronase indicating that the receptor contains an essential protein component. Treatment with phospholipase A abolishes receptor activity, suggesting phospholipid participation in receptor function.[28] Specific FSH receptors (about 1,000 per cell) have also been reported in rat granulosa cells, with an apparent

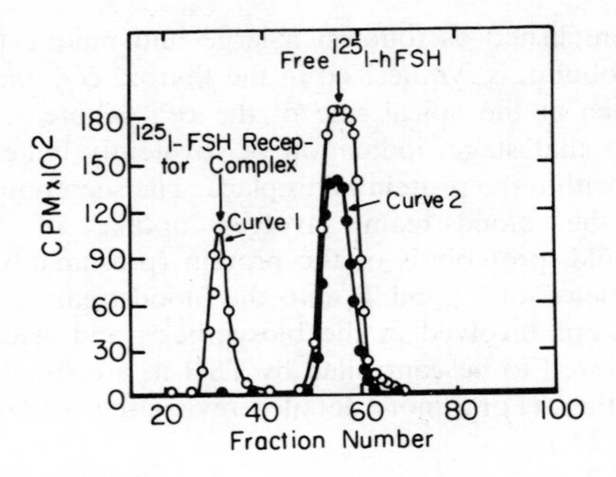


Figure 8.3. Gel filtration of soluble calf testes FSH receptor. Curve 1 (open symbols) shows that the receptor hormone complex migrates well ahead (fraction 30) of the unbound $^{125}$I-FSH (fraction 55). Curve 2 shows an identical experiment performed in the presence of an excess (1,000-fold) of non-labelled FSH which effectively displaces the bound hormone on the receptor and eliminates the first peak. Modified from Abou-Issa and Reichart[31]

dissociation constant of $K_D = 10^{-10}$ M.[29] Treatment of membrane particles from calf testes containing FSH receptors with the non-ionic detergent Triton X-100 has yielded a soluble protein (M.W. 146,000) which was capable of binding $^{125}$I-hFSH. The receptor $^{125}$I-FSH complex (M.W. 183,000) could be separated from the non-bound hormone (M.W. 30,000) by gel filtration chromatography (Figure 8.3). Analysis according to Scatchard[30] indicated one class of binding sites ($K_D = 5 \times 10^{-10}$ M) essentially identical to that obtained for the particulate receptor. The binding capacity of the solubilized receptor was $19 \times 10^{-14}$ mol per mg protein. Binding was found to be temperature dependent. Optimal binding was observed at 24 °C and at this temperature equilibrium binding was reached within 2–4 h.[31]

As a consequence of FSH binding, cyclic AMP levels in seminiferous cells, as well as in isolated granulosa cells increase instantaneously. Adenylate cyclase in purified plasma membranes from testicular tubule cells has been shown to be stimulated by FSH, but to be only slightly responsive to LH. The magnitude of stimulation was similar to that obtained with NaF.[28]

**(ii) LH receptors**

Among the membrane receptors that specifically bind gonadotropins, the LH receptor is probably the best described.[25,32,33] The LH-receptor has been obtained in soluble form by extracting testicular or ovarian homogenates with non-ionic detergents, Lubrol PX or Triton X-100. The extraction procedure using Triton X-100 yielded uniform and reproducible preparations which were subjected to further phsyicochemical analysis. Table 8.2 summarizes some of the basic properties of LH-receptors in preparations derived from rat testes and ovaries.[25]

Bound $^{125}$I-hCG was effectively displaced only by non-labelled hCG or LH, but not by FSH, TSH, prolactin, or ACTH. This finding shows that the high specificity of the LH receptor was not altered by the solubilization procedure. Maximal binding was obtained at 24 °C (pH 7.4), and under these conditions equilibrium was reached within 4–6 h. The physicochemical properties of the receptor were determined following gel filtration and sucrose density gradient centrifugation. The receptor molecule was found to be a highly asymmetrical structure as suggested by a prolate axial ratio of 12.0. Purification of the testicular LH receptor by affinity chromatography on hCG Sepharose yielded a 15,000-fold purification of a protein that had a binding capacity of 2,500 pmol hCG per mg protein. Electrophoresis in SDS polyacrylamide gel revealed a single component with an estimated molecular weight of 90,000, suggesting that the receptor may exist naturally in a dimeric form with a molecular weight of about 200,000 (Table 8.2). The soluble hCG receptor complex dissociated upon treatment with 2 M urea or guanidine HCl or upon acidification to pH 3.5.[34]

The protein nature of the receptor is indicated by the finding that its activity is rapidly lost following proteolytic treatment with trypsin. Treatment of the solubilized receptor with phospholipase A, but not with phospholipase C, resulted in a substantial loss of receptor activity suggesting that certain phospholipids are essential for normal receptor activity. The

Table 8.2. Comparison of physicochemical properties of Triton X-100 solubilized hCG receptors from rat testes and ovaries

| Parameter | Ovarian receptors | Testicular receptors |
|---|---|---|
| Dissociation constant ($K_D$) | $1.5 \times 10^{-10}$ M | $1.7 \times 10^{-10}$ M |
| Stokes radius | 60 Å | 64 Å |
| Sedimentation constant | 7.5 S | 7.5 S |
| Apparent molecular weight | 225,000 | 194,300 |

Data from Dufau and Catt[25]

strong adsorption of the receptor to concanavalin A indicates that the receptor itself is a glycoprotein.[25,33,35]

The interaction of LH with the membrane receptor leads to stimulation of adenylate cyclase. The LH-sensitive adenylate cyclase of the Graafian follicle provides a special case in that it was shown to acquire sensitivity to LH during the process of follicular maturation. The response of the enzyme to LH therefore changes in relation to the oestrous cycle in the rat. Responsiveness of the enzyme increases gradually and reaches its highest levels shortly before the midcycle LH surge.[36] It was shown that acquisition of responsiveness to LH in follicular tissue is a process that results from an increase in the number of tissue LH receptors.[37,38,39] This rise is accompanied by a parallel rise in the response of adenylate cyclase to LH.[40,41] In contrast, this process is not accompanied by a rise in the number of catalytic units of adenylate cyclase. This has been suggested by the finding that stimulation of the enzyme by non-hormonal activators (NaF, p(NH)ppG) remained unchanged despite a four or five-fold increase in the response of the enzyme to LH.[42] The sensitivity of the enzyme of LH remains unchanged throughout this process (half maximal stimulation by LH at $4 \times 10^{-10}$ M). This value is essentially identical to the dissociation constant described for the LH receptor (Table 8.2). Stimulation of the enzyme by LH in purified ovarian plasma membranes is largely dependent on the presence of GTP.[41,42]

### (iii) Biological activities of gonadotropins

The absolute dependence of the gonads on the continuous presence of LH and FSH can be demonstrated by removal of the pituitary gland. Elimination of circulating gonadotropins in the whole animal leads to cessation of gonadal steroidogenesis. The gonads degenerate followed by termination of gametogenesis.

Testes and ovaries are multicellular organs, the function of which in the respective sexes, is controlled in concert and interdependently by LH and FSH. It is therefore more difficult to establish the exact molecular events which follow cyclic AMP-dependent protein kinase activation by the individual hormones in the respective target cells. The major physiological consequences connected with gametogenesis are described below.

#### *(a) Testes*

LH stimulates testosterone production in Leydig cells. This androgen is secreted and acts as the principal stimulus of spermatogenesis and sperm maturation. The Sertoli cell produces androgen binding protein and secretes it into the seminiferous tubule as a result of FSH stimulation. By combining with androgen binding protein within the seminiferous tubules, testosterone is retained in the tubular compartment in close proximity to the androgen-dependent germ cells. The germ cells undergo an extremely complicated process of differentiation. During this developmental process the maturing sperm cells are embedded in and supported by the Sertoli cells. The fully mature spermatozoa are finally released into the seminiferous tubule. (For review: see References 28, 43, 44.)

*(b) Ovary*

FSH stimulates follicular growth and development by stimulating the differentiation and proliferation of granulosa cells in the immature Graafian follicle. During follicular maturation, the antral cavity is formed and the granulosa cells acquire sensitivity to LH. (For review see Richards and Midgley[39]). Following the midcycle LH surge, steroid production by the mature follicle is stimulated. This is followed by rupture of the follicle, completion of the meiotic division in the oocyte, release of the ovum, and formation of the corpus luteum. (For review see Lindner *et al.*[45]). This ovarian organelle is stimulated to produce and secrete progestins by LH or hCG. The complex process which leads to ovulation and luteinization has recently been reviewed by Zor and Lamprecht.[46]

## V. CONCLUSIONS AND FUTURE PROSPECT

There is no doubt today that gonadotropins activate target cell adenylate cyclase by combining with specific cell surface receptors. These receptors are by far the best characterized (in biochemical terms) among those receptors which potentially couple to adenylate cyclase. However, it is not entirely clear whether activation of this enzyme system is the one and only transmembrane signal mediated by these receptors. Reconstitution of solubilized components (receptor, enzyme, and possibly other essential factors) and restoration of the regulatory properties of the adenylate cyclase system have not yet been achieved and require further studies. It remains to be established whether the long-term, trophic effects of glycoprotein hormones as compared to the short-term acute effects are all mediated by cyclic AMP. Future studies must elucidate the detailed molecular events which are involved in proper function and response of the respective target cells.

Finally, elucidation of the properties and mechanism of action of glycoprotein hormone–receptors may provide new methods for treatment of certain forms of human endocrine disfunction. In the case of gonadotropin receptors such knowledge may also contribute to the development of safer methods of family planning and provide new means of increasing livestock productivity.

## ACKNOWLEDGEMENT

The author thanks Professor H. R. Lindner for support and helpful discussions. This work was supported in part by the Ford Foundation and the Population Council Inc., New York, and in part by the U.S. Israel Binational Science Foundation (BSF), Jerusalem, Israel. Yoram Salomon is the incumbent of the Charles W. and Tillie K. Lubin Career Development Chair. I wish to thank Mrs M. Kopolowitz for excellent secretarial assistance.

## VI. REFERENCES

1. Rommerts, F. F. G., Cooke, B. A., and van der Molen, H. J. *J. Steroid Biochem.*, **5**, 279–285, (1974).
2. Marsh, J. M. *Adv. Cyclic Nucl. Res.*, **6**, 137–199, (1975).
3. Cooke, B. A., Lindh, M. L., and Janszem, F. H. A. *Biochem. J.*, **160**, 439–446, (1976).
4. Sutherland, E. W., and Rall, T. W. *Pharmacol. Rev.*, **12**, 265–299, (1960).
5. Grollman, E. F., Lee, G., Ambesi-Impiombato, F. S., Meldolesi, M. F., Aloj, S. M., Coon, H. G., Kaback, H. R., and Kohn, L. D. *Proc. Natl. Acad. Sci. U.S.A.*, **74**, 2352–2356, (1977).
6. Liu, W. K., and Ward, D. N. *Pharmac. Therap. B. B.*, **1**, 545–570, (1975).
7. Shome, B., and Parlow A. F. *J. Clin. Endocrinol. Metab.*, **39**, 199–202, (1974).
8. Shome, B., and Parlow, A. F. *J. Clin. Endocrinol. Metab.*, **39**, 203–205, (1974).
9. Jutisz, M., and Tertrin-Clary, C. *Curr. Topics Exp. Endocrinol.*, **2**, 195–246, (1975).
10. Cornell, J. S., and Pierce, J. G. *J. Biol. Chem.*, **248**, 4327–4333, (1973).
11. Canfield, R. E., Birken, S., Morse, J. H., and Morgan, F. J. in *Peptide Hormones* (Parsons, J. A., ed.), Macmillan Press Ltd., London, pp. 299–315, (1974).
12. Kohn, D. L. in *Receptors and Recognition Series A*, (Cuatrecasas, P. and Greaves, M. F., eds.), Chapman and Hall, London, Vol. 5, pp. 134–212, (1978).
13. Mullin, B. R., Pacuszka, T., Lee, G., Kohn, L. D., Brady, R. O., and Fichman, P. H. *Science*, **199**, 77–79, (1978).
14. Fishman, P. H., and Brady, R. O. *Science*, **194**, 906–915, (1976).
15. Vaitukaitis, J. L., Ross, G. T., Braunstein, G. D., and Rayford, P. L. *Recent Prog. Horm. Res.*, **32**, 289–331, (1976).
16. Papkoff, H., Sairam, M. R., Farmer, S. W., and Li, C. H. *Recent Prog. Horm. Res.*, **29**, 563–590, (1973).
17. Pierce, J. G. *Endocrinology*, **89**, 1331–1344, (1972).
18. Mondgal, N. R. (ed.) *Gonadotropins and Gonadal Function* Academic Press, New York, San Francisco and London, (1974).
19. Bahl, O. P. *Fed. Proc.*, **36**, 2119–2127, (1977).
20. Tate, R. L., Schwartz, H. I., Holmes, J. M., Kohn, L. D., and Winand, R. J. *J. Biol. Chem.*, **250**, 6509–6515, (1975).
21. Wolff, J., Winand, R. J., and Kohn, L. D. *Proc. Natl. Acad. Sci. U.S.A.*, **71**, 3460–3464, (1974).
22. Tate, R. L., Holmes, J. M., Kohn, L. D., and Winand, R. J. *J. Biol. Chem.*, **250**, 6527–6533, (1975).
23. Ingbar, S. H., and Woeber, K. A. in *Textbook of Endocrinology* 5th Edn.

(Williams, R. H., ed.), W. B. Saunders Co., Philadelphia, London, and Toronto, pp. 95–232, (1975).
24. Greep, R. O., and Astwood, G. B. (eds.) *Handbook of Physiology, Section 7, Endocrinology* American Physiological Society, Washington D.C., Vol. 3, (1974).
25. Dufau, M. L., and Catt, K. J. in *Cell Membrane Receptors for Viruses Antigens and Antibodies, Polypeptide Hormones and Small Molecules* (Beers, R. F., Jr., and Bassett, E. G., eds.), Raven Press, New York, pp. 135—163, (1976).
26. Koch, Y., Zor, U., Chobsieng, P., Lamprecht, S. A., Pomerantz, S., and Lindner, H. R. *J. Endocr.,* **61**, 179–191, (1974).
27. Knobil, E. *Biol. Reprod.*, **8**, 246–258, (1973).
28. Means, A. R. in *Handbook of Physiology Section* 7 (Greep, R. O, Astwood, G. B., eds.), American Physiological Society, Washington D.C., Vol. 5, pp. 203–218, (1975).
29. Nimrod, A., Erickson, G. F., and Ryan, J. K. *Endocrinology*, **98**, 56–64, (1976).
30. Scatchard, G. *Ann. N.Y. Acad. Sci.*, **51**, 660–672, (1949).
31. Abou-Issa, H., and Reichart, L. E., Jr. *J. Biol. Chem.*, **252**, 4166–4174, (1977).
32. Lee, C. Y., and Rayan, R. J. in *Gonadotropins and Gonadal Function* (Moudgal, N. R., ed.), Academic Press, New York, San Francisco, and London, pp. 444–459, (1974).
33. Dufau, M. L., Charreau, E. H., and Catt, K. J. *J. Biol. Chem.*, **248**, 6973–6982, (1973).
34. Dufau, M. L., Ryan, D. W., Baukal, A. J., and Catt, K. J. *J. Biol. Chem.*, **250**, 4822–4824, (1975).
35. Ryan, R. J., and Lee, C. Y. *Biol. Reprod.*, **14**, 16–29, (1976).
36. Hunzicker-Dunn, M., and Birnbaumer, L. *Endocrinology*, **99**, 198–210, (1976).
37. Lee, C. Y., and Ryan, R. J. *Endocrinology*, **95**, 1691–1693, (1974).
38. Channing, C. P., and Kammerman, S. *Endocrinology*, **92**, 531–540, (1973).
39. Richards, J. S., and Midgley, A. R. *Biol. Reprod.*, **14**, 82–94, (1976).
40. Lee, C. Y. *Endocrinology*, **99**, 42–48, (1976).
41. Salomon, Y., Yanovsky, A., Mintz, Y., Amir, Y., and Lindner, H. R. *J. Cyclic Nuc. Res.,* **3**, 163–176, (1977).
42. Salomon, Y. in *Membrane proteins* (Nicholls, P., Møller, J. V., Jørgensen, P. L., and Moody, A. J., eds.), Fed. of European Biochemical Soc., Pergamon Press, Oxford and New York, Vol. 45, pp. 299–308, (1978).
43. Hansson, V., Ritzen, E. M., French, F. S., and Nayfeh, S. N. in *Handbook of Physiology Section 7*, (Greep, R. O., and Astwood, G. B., eds.), American Physiological Society, Washington, D.C., Vol. 5, pp. 173–201, (1975).
44. Means, A. R., Fakunding, J. L., Huckins, C., Tindall, D. J., and Vitale, R. *Recent Prog. Horm. Res.*, **32**, 477–527, (1974).
45. Lindner, H. R., Tsafriri, A., Lieberman, M. E., Zor, U., Koch, Y., Bauminger, S., and Barnea, A. *Recent. Prog. Horm. Res.*, **30**, 79–138, (194).
46. Zor, U., and Lamprecht, S. A. in *Biochemical Actions of Hormones* (Litwack, G., ed.), Academic Press, New York, San Francisco, and London, Vol. 4, pp. 85–133, (1977).

Cellular Receptors for Hormones and Neurotransmitters
Edited by D. Schulster and A. Levitzki


CHAPTER 9

# Receptors for growth hormone, prolactin, and the somatomedins

Michael Wallis
*School of Biological Sciences, University of Sussex, Falmer, Brighton BN1 9QG, U.K.*

## I. INTRODUCTION

Growth hormone (somatotropin) and prolactin are structurally related protein hormones, produced in different cells in the anterior lobe of the pituitary gland, which show some overlap in biological actions.[1] It is therefore convenient to consider their interactions with cellular receptors under a single heading, together with those of the structurally related lactogens produced by the placenta. Some of the actions of growth hormone (and perhaps prolactin) are mediated by a family of insulin-like proteins, the somatomedins, and receptors for these factors will also be considered in this chapter. In many respects, however, somatomedins should be considered as related to the insulin family rather than the growth hormone–prolactin group.

In some ways our knowledge about the cellular receptors discussed in this chapter is less advanced than that of many other hormone receptors. Little can be said about the relationship between hormone structure and receptor binding because the three-dimensional structure of growth hormone and prolactin is not known. Detailed discussion of the biochemical events linking hormone–receptor binding to the actions of the hormones is restricted by lack of knowledge about such actions at the biochemical level. For these reasons, this chapter will concentrate mainly on what is known about the distribution and characteristics of specific cellular receptors for growth hormone and prolactin; consideration of the role of these receptors will necessarily be limited.

## II. GROWTH HORMONE, PROLACTIN, AND PLACENTAL LACTOGEN

### (i) Structure

These hormones are all proteins containing a single polypeptide chain of molecular weight about 20,000–22,000. In some cases dimers may be formed, but it is unlikely that these predominate under physiological conditions. Growth hormone contains two disulphide bridges and prolactin contains three such cross-links. Various slightly modified, naturally occurring, forms of the hormones have been identified, and it is possible that some of these have enhanced biological activity and play an important role in the physiological actions of the hormones. Both prolactin and growth hormone have a stable three-dimensional tertiary structure, and it is clear that this is all-important in determining the hormones' biological actions, including their interaction with receptors. Unfortunately the three-dimensional structures are unknown, largely due to lack of suitable crystals for X-ray crystallographic studies.

Growth hormone and prolactin show considerable species variation with respect to both primary structure and biological actions.[1] In particular, growth hormone from man and other primates is very different from that of other mammalian groups (differing at about 35% of all amino acid residues). This probably relates to the fact (see later) that growth hormones from non-primate mammals are not active in man. Such species differences must be borne in mind when receptor-binding activity is considered.

Growth hormone and prolactin are structurally homologous proteins; their primary structures are identical at about 25% of all residues. The three-dimensional structures may also be very similar, and this probably underlies various biological similarities. Human placental lactogen is structurally very similar to human growth hormone (more similar than human growth hormone is to non-primate growth hormones), though its biological properties differ markedly.

### (ii) Biological actions

The main biological action of growth hormone is to promote and regulate somatic growth in the young animal. Some of its actions (e.g. some on liver and skeletal muscle) appear to be direct and involve stimulation of anabolic processes including protein synthesis, while others (e.g. those on cartilage) are mediated by somatomedins. Somatomedins themselves are produced by the liver in response to growth hormone, but little is known about the biochemistry of this process. The relative importance of direct and somatomedin-mediated actions in the over-all physiological control of growth is not yet clear (see also Section III). Growth is of course a complex process and for this reason the physiological and biochemical actions of growth hormones remain poorly understood, despite a good deal of research effort. Growth hormone also has a wide range of actions on carbohydrate and lipid metabolism, but the physiological importance of these effects has not been fully defined.

The main actions of prolactin in the female mammal are concerned with promotion of mammary development and lactation. It induces the production of specific milk proteins in mammary epithelial cells, and also promotes protein and lipid synthesis generally in such tissue. The hormone also has luteotrophic effects in some mammals and there have been suggestions of other effects on mammalian metabolism. In lower vertebrates prolactin has a range of actions, mainly on water and electrolyte balance and on various secondary sexual aspects of reproduction.

Placental lactogen has been most fully characterized in humans, but there is now good evidence for a placental lactogen in many other mammalian groups. The biological function of this hormone, which is produced in very large amounts during pregnancy, is not fully understood. It may play an important role in mammary development, in foetal growth and nutrition and/or in several aspects of maternal metabolism.

There is considerable overlap between the biological properties of the hormones of this family, which is not surprising in view of the structural homologies. Of particular note is the possession of considerable lactogenic activity by human growth hormone (unlike non-primate growth hormones). These overlapping actions reflect overlapping hormone–receptor interactions, and must be borne in mind whenever such interactions are being considered. Similarly, the considerable species specificity shown by the hormones in this family (well exemplified by the lack of growth-promoting activity in man of non-primate growth hormones) can be an important factor when hormone–receptor interactions are considered.

### (iii) Experimental approaches to the study of receptors of growth hormone and prolactin

As in the case of other hormones, studies on receptors for growth hormone and prolactin have been dependent on development of two main experimental techniques—preparation of labelled hormone and production of suitable receptor preparations. Both have presented problems. Hormone labelling is normally achieved using modification with radioactive iodine, but considerable doubts have been expressed about the biological activity of preparations of growth hormone and prolactin labelled in this way. Some authors have successfully used conventional iodination procedures, using the oxidizing agent chloramine T to convert $I^-$ to $I^+$ (first devised for iodination of human growth hormone for radioimmunoassay), but most workers have preferred to use milder methods. Use of enzymic methods (particularly with lactoperoxidase) to achieve labelling has been most widely adopted but an alternative method using very low concentrations of chloramine T has also been successful.[2]

As with other polypeptide hormones, receptors for growth hormone and prolactin are associated mainly with the plasma membrane (see later) and most work on the receptors for these hormones has involved use of relatively crude membrane preparations. Some work has been done with whole cells, however, and also with solubilized and partially purified receptor preparations.

### (iv) Distribution of receptors for growth hormone and prolactin within the body

#### *(a) Distribution of labelled hormone* in vivo

Some information about the distribution of hormone receptors within the body can be gained by injecting labelled hormone and detecting its presence in the various organs and tissues, using autoradiography where appropriate. Such studies are inevitably limited because of degradation of the labelled hormone and the very real possibility that much of the radioactivity detected subsequently is no longer associated with intact hormone. Nevertheless, this type of approach may give some useful information about the main locations of hormone receptors, and in fact the results obtained by this method (for growth hormone and prolactin) are in fairly good agreement with others.

Injection of $^{125}$I-labelled prolactin into rats,[3] mice,[3,4] and rabbits[5] followed by location of the hormone by autoradiography showed that the label was concentrated mainly in the liver (over both parenchymal and Kupffer cells) and kidney (proximal tubular cells), with some in the ventral prostate, testis (mainly around the Leydig cells), mammary glands (especially during lactation), and corpus luteum. In accordance with the high binding to the kidney, label (injected as $^{125}$I-labelled prolactin) was also concentrated in the urine, but little of this was associated with intact prolactin.[5] $^{125}$I-labelled prolactin associated with mammary tissue in rabbits (after intravenous or intraductal administration) was localized mainly on or near the alveolar secretory membrane, on the side adjacent to the vascular supply.[5]

When $^{125}$I-labelled human growth hormone was injected into male rats, radioactivity became associated primarily with the kidney (proximal convoluted tubule) and to a lesser extent the liver, adipose tissue and adrenal cortex.[6] $^{3}$H-labelled human growth hormone was also concentrated in the kidney, liver, and adrenal glands after injection into rats or guinea-pigs[7] though other tissues also bound the hormone (and in terms of 'bulk' localization, more label was bound by skeletal muscle than any other tissue).

Distribution of $^{14}$C-labelled (guanidinated) or $^{125}$I-labelled bovine growth hormone after injection into rats has been studied recently.[8] Concentration in liver, kidney, and spleen was rapid, but subsequently declined; concentration in heart and skeletal muscle, pancreas and intestine occurred more slowly, but was less transitory. After 2 h the bulk of the label was associated with skeletal muscle (Figure 9.1).

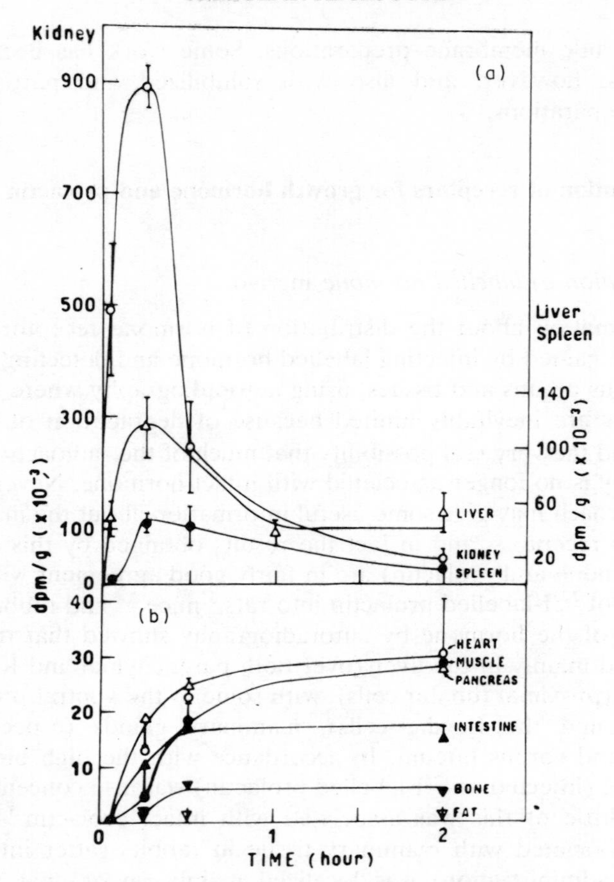


Figure 9.1. Localization of radioactivity at different times after the administration of $^{14}$C-guanidinated bovine growth hormone to the rat (50 $\mu$g per 100 g body weight, intravenously). Means of 3–4 observations ± S.E.M. Notice in (a) the different values of the ordinate for kidney and for liver and spleen. Reproduced with permission of Dr W. Junk b.v., publishers from Retegui-Sardou, L. A., Scaramel, L. O., Dellacha, J. M., and Paladini, A. C. *Mol. Cell. Biochem.*, **16**, 91 (1977)

These studies suggest that cellular receptors for growth hormone and prolactin are rather widely distributed. Concentration of the labelled hormones in liver, kidney, and spleen may reflect a role of these organs in metabolizing the hormones and removing them from the circulation; the specific receptors that have been detected in liver and kidney may be concerned with such processes rather than with mediating the biological actions of the hormones.

### *(b) Hormone binding to tissues and cells* in vitro

A good deal of work has been carried out on the ability of tissue and cell preparations to bind growth hormone and prolactin *in vitro*. Binding of $^{125}$I-labelled prolactin to rabbit mammary tissue slices[5] and dispersed mouse mammary cells[4] has been described, and binding to a range of other rat tissues has also been investigated, using autoradiography.[9] Specific binding (i.e. binding displaceable by excess unlabelled prolactin) was found in the liver, adrenal gland (cortex), kidney (cortex), mammary gland, ovary, testis (Leydig cells), and prostate gland (secretory epithelium), but not in fat, muscle, heart, lung, spleen, or uterus.[9] The fairly wide distribution of prolactin receptors is thus illustrated.

Binding of $^{125}$I-labelled human growth hormone to cultured human lymphocytes has been studied in detail.[10] Binding was rapid and bound hormone was readily displaceable by excess unlabelled human growth hormone, but not by non-primate growth hormone or unrelated hormones. Approximately 4,000 binding sites per cell could be detected, with an affinity constant of $1.3 \times 10^9$ $M^{-1}$. Binding of $^{125}$I-labelled human growth hormone to dispersed rat hepatocytes has also been demonstrated[11] but this probably reflected binding to lactogenic rather than growth-promoting receptors, since the labelled hormone could be displaced by prolactin but not by non-primate growth hormone. Other authors have found specific growth hormone receptors in dispersed rat hepatocytes in addition (in cells from female rats) to lactogenic receptors.[12]

### *(c) Membrane-bound receptors*

Cellular receptors for growth hormone and prolactin are associated mainly with cell membranes (see below). Ability of membrane preparations from a variety of tissues to bind the labelled hormones has been studied[13] and provides a direct indication of the distribution of receptors. Binding varied greatly according to the tissue and species from which the membrane preparation was obtained. Low specific binding was found in a wide range of tissues. The greatest binding of $^{125}$I-labelled prolactin and human growth hormone was in rabbit and rat liver, frog kidney, rabbit adrenal gland, and several sheep tissues. In most of these tissues (except rabbit liver) it appeared that the binding sites detected were primarily lactogenic rather than growth-promoting.

### *(d) Ontogeny—factors influencing levels of growth hormone and prolactin receptors in the target tissues*

The levels of receptors for growth hormone and prolactin have been studied in several tissue of developing rats and rabbits.[14] Binding of labelled human

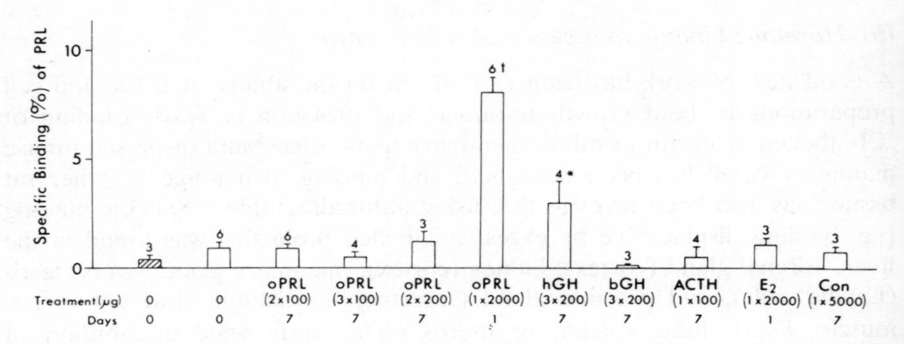


Figure 9.2. Specific binding (%) for prolactin in particulate fractions obtained from livers of hypophysectomized female rats (or hypophysectomized, adrenalectomized female rats—shaded bars) treated with various polypeptide or steroid hormones. oPRL = ovine prolactin, hGH = human growth hormone, bGH = bovine growth hormone, $E_2$ = oestradiol, Con = cortisone, *significantly different from controls, $p < 0.05$; †significantly different from control, $p < 0.01$. Reproduced with permission of Marcel Dekker, Inc., from Bohnet, H. G., Aragona, C., and Friesen, H. G. *Endorc. Res. Commun.*, **3**, 193 (1976)

growth hormone and ovine prolactin was very low in liver membranes from female foetal or young rats. A nine-fold (ovine prolactin) and 3.5-fold (human growth hormone) increase in binding was seen between 20 and 40 days of age, with further increases during pregnancy. A similar pattern was seen using bovine growth hormone as the labelled ligand. Lactogenic receptors in the rat mammary gland increase rapidly during the first two days of lactation.[15]

A major factor involved in regulating the level of lactogenic receptors in the liver and mammary gland appears to be prolactin itself.[16–20] Thus, when hypophysectomized rats (in the livers of which the levels of prolactin receptors are drastically reduced) were given a range of hormones, the only effective inducers of lactogenic receptors in the liver were prolactin and human growth hormone. ACTH and oestradiol appeared to potentiate the effects of these lactogenic hormones[16–20] (Figure 9.2). A very low level of prolactin receptors in hepatocytes from hypophysectomized rats has also been observed.[12] Prolactin appears to induce its own receptors in rat mammary gland also[18] and here too the effect is potentiated by oestradiol. These studies suggest that, even in hypophysectomized animals, a very small number of prolactin receptors must be present, a major function of which is to mediate induction of increased levels of the receptors in the target tissue.

In the case of growth hormone receptors, on the other hand, there is evidence for 'down-regulation'. If cultured human lymphocytes are incubated with human growth hormone (or very high levels of bovine growth hormone) the levels of receptors are decreased, the decrease being proportional to the concentration of growth hormone in the medium.[21] Hepatocytes from hypophysectomized rats do not appear to have decreased levels of growth hormone receptors, compared with those from normal animals.[12]

Binding of human growth hormone to liver receptors was markedly reduced in rats with streptozotocin induced diabetes,[22] due to loss of binding sites. Insulin injection restored the number of binding sites, and it thus seems likely that insulin is involved in the regulation of such receptors. It was considered likely that these were in fact lactogenic receptors rather than 'growth-promoting' ones.

### (v) Characterization of receptors for growth hormone and prolactin

#### *(a) Membrane-bound receptors*

When target tissues for growth hormone (liver) or prolactin (mammary gland or liver) are subjected to homogenization and subcellular fractionation the receptors for these hormones are found associated primarily with particulate fractions, especially microsomes. This is generally thought to be because the receptors are associated with plasma membrane fragments which fractionate mainly with the microsomes. Much work on the characterization of the receptors for growth hormone and prolactin has been carried out using crude microsomal fractions of this kind.

Characterization of membrane-bound receptors for prolactin has been studied using a preparation of plasma-membrane-containing subcellular particles isolated from rabbit mammary glands.[23] $^{125}$I-labelled ovine prolactin was bound specifically to such particles and could be displaced by unlabelled prolactin obtained from several species. Binding was pH dependent (optimum at pH 7.5–8.5), required calcium and magnesium ions, but was unaffected by a wide range of low molecular weight compounds tested. Receptor activity was susceptible to digestion with trypsin and phospholipase C, suggesting that protein and phospholipid are required for binding. Scatchard analysis indicated an association constant of $3 \times 10^9\ M^{-1}$.

Binding of $^{125}$I-labelled human growth hormone to microsomal membranes from rat liver has been studied in detail.[24] Specific binding of the labelled hormone was demonstrated, and was dependent on time, temperature, and pH. The labelled hormone could be displaced by unlabelled human growth hormone, and by ovine prolactin, but not by non-primate growth hormones (or insulin, thyrotropin, LH, or FSH). Scatchard analysis showed two types of binding site, with association constants of $0.64 \times 10^{10}\ M^{-1}$ and $0.03 \times 10^{10}\ M^{-1}$ and capacities of 98 fmol mg$^{-1}$ and 315 fmol mg$^{-1}$ of protein respectively. In view of the specificity of binding of human growth hormone to this liver–membrane preparation, it seems likely that the hormone is binding primarily to lactogenic receptors rather than growth-promoting receptors in the liver. Other workers have obtained similar results, but whether rat liver really lacks true growth hormone receptors or whether their apparent absence is due to technical difficulties is not clear.

Growth hormone receptors appear to exist in much greater quantities in rabbit liver[25] and membrane-containing preparations have been obtained from this source which bind labelled bovine growth hormone specifically. Bound $^{125}$I-labelled growth hormone (bovine or human) was displaced by unlabelled growth hormone from several species, but not by prolactin.

### (b) *Solubilization and purification of receptors for growth hormone and prolactin*

Growth hormone specific receptors from rabbit liver membranes have been solubilized using the detergent Triton X-100.[26] The solubilized receptor preparation retained ability to bind iodinated bovine or human growth hormone (bound and unbound hormone could be separated by gel filtration or precipitation of the hormone–receptor complex with polyethylene glycol). $^{125}$I-labelled human growth hormone could be displaced readily from receptors by unlabelled human, bovine, or ovine growth hormone. Prolactins and human placental lactogen showed only a slight ability to displace the labelled hormone. The binding sites are thus predominantly growth hormone rather than lactogenic receptors. $^{125}$I-labelled bovine growth hormone was also bound by the solubilized receptor, but to a lesser extent than the human hormone.

Scatchard analysis of the binding of $^{125}$I-labelled human growth hormone to the solubilized receptors showed that there may be two types of receptor present, one of which binds human growth hormone and bovine growth hormone (though affinity for the former would be about five times greater than that for the latter) and one which binds only the human hormone. These results were taken to indicate that solubilization of the liver receptors had led to a preparation in which only receptors for growth hormone were retained, while those for prolactin were lost for some reason. An alternative explanation might be, however, that the relatively small quantities of Triton X-100 retained in the assay (to keep the receptors soluble) may cause inactivation of prolactins and some growth hormones and slight inactivation of bovine growth hormone, while leaving human growth hormone unaffected.

Considerable purification of Triton-solubilized growth hormone receptors from rabbit liver has been achieved, using affinity chromatography on immobilized human and bovine growth hormones.[27] Separation of receptors for growth hormone and prolactin was achieved, the former being eluted preferentially with 4.5 M urea, the latter with 5 M $MgCl_2$. The growth hormone receptor was further purified by preparative isoelectric focusing and gel filtration to give a final purification of over 1,000-fold.

Antibodies to the purified growth hormone receptor were generated in guinea-pigs. These inhibited the binding of $^{125}$I-labelled ovine growth hormone to rabbit liver membranes, but had no effect on binding of $^{125}$I-labelled ovine prolactin. Binding of growth hormone to liver membranes from rat, sheep, mouse, and human was also inhibited, though to a lesser extent. These studies represent a major step towards complete purification and characterization of the growth hormone receptor.

Solubilization of prolactin receptors from rabbit mammary glands has also been achieved, using 1% Triton X-100.[28] Presence of Triton X-100 affected the physical properties of $^{125}$I-labelled ovine prolactin, and $^{125}$I-labelled human growth hormone was therefore used for most of the binding studies with these lactogenic receptors (confusion with growth hormone binding receptors was not encountered since there are few such receptors in the rabbit mammary gland). Binding of labelled human growth hormone to soluble receptors was detected by gel filtration or precipitation of the hormone–receptor complex with polyethylene glycol. The binding specificity of the solubilized receptor was similar to that of the membrane bound one, but the affinity of the soluble receptor ($K_a$ $16 \times 10^9 M^{-1}$) was five-fold greater than that of the particulate receptor ($K_a$ $3 \times 10^9 m^{-1}$). Solubilization of the receptor thus gives a preparation usable for a very sensitive radio-receptor assay.

The soluble prolactin receptor was purified about 1,500-fold by affinity chromatography using human growth hormone coupled to 'Affi-Gel 10' (the *N*-hydroxysuccinimide ester of 3, 3′-diaminodipropylaminosuccinyl agarose). The receptor could be eluted by 5 M $MgCl_2$, and 0.5 mg of partially purified receptor protein was obtained from 100 g of mammary tissue. The purified receptor appeared to have a molecular weight of about 220,000 (by gel filtration on Sepharose 6B) and polyacrylamide gel electrophoresis revealed several protein-containing bands, one or two of which appeared to have receptor activity.

The partially purified prolactin receptor was used to raise antibodies in guinea-pigs.[29] These gave a precipitin reaction with the purified receptor and specifically blocked the binding of labelled ovine prolactin (but not labelled insulin) to a membrane fraction from rabbit mammary glands and to solubilized, purified prolactin receptors. The antiserum blocked binding of labelled prolactin to its receptors in various tissues, from various species. It also blocked the actions of prolactin on cultured explants of rabbit mammary gland[30] (see below).

### *(c) Radioreceptor assays for growth hormone and prolactin*

Radioreceptor assays have been established for growth hormone and prolactin using several types of receptor preparation. Human growth hormone can be assayed specifically using intact cultured lymphocytes.[31] The characteristics of binding to such cells have been described above. The usable region of the dose–response curve covered a range of 1–100 ng $ml^{-1}$ (0.05–5 nM) of human growth hormone, and displacement of the labelled hormone was very specific, being unaffected by non-primate growth hormones or by prolactin.

A radioreceptor assay for growth hormone using a membrane preparation from rabbit liver has also been described.[25] Human growth hormone or bovine growth hormone were used as labelled ligands, and displacement could be effected by primate or non-primate growth hormones, but not by prolactin. Again, a dose–response curve with a useful range of about 1–100 ng $ml^{-1}$ (0.05–5 nM) of growth hormone was established. Some workers have found it difficult to set up a satisfactory assay of this type, partly because many rabbit liver membrane preparations contain prolactin receptors in addition to those for growth hormone.

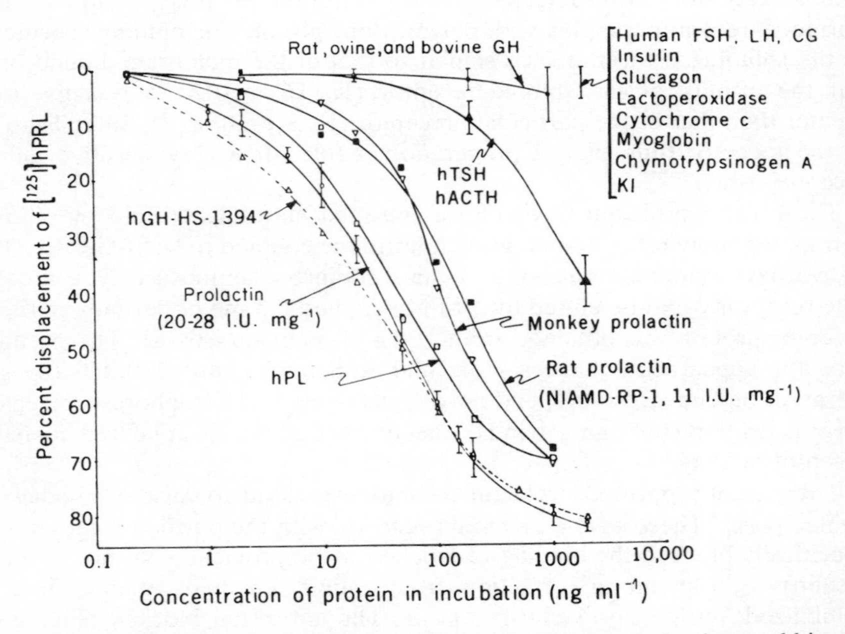


Figure 9.3. Specificity of binding of $^{125}$I-labelled human prolactin to rabbit mammary receptors. hGH = human growth hormone and hPL = human placental lactogen. Reproduced with permission of the American Association for the Advancement of Science, ©1973, from Shiu, R. P. C., Kelly, P. A., and Friesen, H. G. *Science,* **180**, 969 (1973)

A radioreceptor assay for prolactin has been described, based on a membrane preparation made from rabbit mammary gland.[32] The assay is specific for prolactin and other lactogenic hormones (placental lactogen and human growth hormone) but shows little species specificity. Dose–response curves were usable over the range 5–500 ng $ml^{-1}$ (0.25–25 nM) of prolactin (Figure 9.3).

A membrane preparation from rabbit liver has also been used as the basis for a radioreceptor assay for prolactin.[33] With [$^{125}$I]prolactin as the labelled ligand, it was specific for lactogenic receptors, and had similar characteristics to the prolactin assay based on receptors from rabbit mammary gland.

### *(d) The relationship between receptors for growth hormone and prolactin*

It will be clear from the foregoing that there is considerable overlap between the receptor interactions of growth hormone, prolactin, and the placental lactogens. It would seem worth trying to clarify this overlap at this stage.

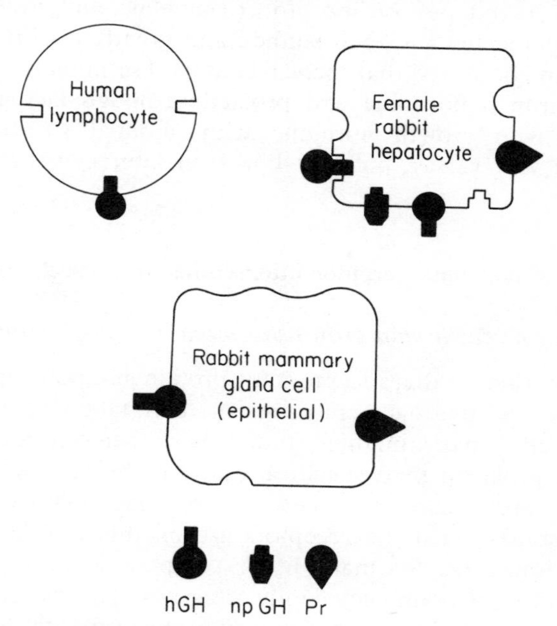


Figure 9.4. Schematic diagram of hormone binding to three types of cell. Human lymphocytes bind only human growth hormone (hGH). Hepatocytes from female rats have separate binding sites for non-primate growth hormone (npGH) and prolactin (Pr); human growth hormone can bind to both. Rabbit mammary gland cells have lactogenic binding sites, which bind prolactin and also human growth hormone

The position has been put very clearly by Roth and his colleagues.[21] The available data suggest that, with regard to receptor interactions, this hormone family can be divided into three groups and that there are three types of receptor (Figure 9.4). Receptors for growth hormone from man and other higher primates have high affinity for primate growth hormone and bovine or sheep placental lactogens, and very slight affinity for non-primate growth hormone, human placental lactogen and ovine (but probably not human) prolactin. Receptors for growth hormone from non-primate mammals have high affinity for primate and non-primate growth hormones and also for bovine placental lactogen, but none for human placental lactogen or ovine or human prolactin. Receptors for lactogenic hormones have high affinity for prolactins, placental lactogens and human growth hormone, but none for non-primate growth hormones. Why the human growth hormone receptor has evolved such a narrow specificity, while the human growth hormone itself has acquired ability to interact with non-primate lactogenic receptors is not clear.

Presence of two types of receptor (lactogenic and 'somatogenic') on female rat hepatocytes has been studied and greatly clarified by Ranke *et al.*,[12] who again confirmed that these receptors discriminate clearly between non-primate growth hormone and prolactin (though lactogenic receptors may interact with growth hormone with about 0.1% the affinity for prolactin, and vice versa), but that both can interact with human growth hormone.

### (vi) Coupling of hormone–receptor interactions to intracellular events

#### *(a) The sites of prolactin and growth hormone receptors within the cell*

It is usually considered that receptors for protein and polypeptide hormones are located on the plasma membranes of their target cells. As has been indicated already, many binding studies have demonstrated that growth hormone and prolactin bind specifically to membrane fractions, but these fractions have rarely been established as comprising mainly (let alone solely) plasma membranes. That the receptors are on the outside of the cell has been demonstrated (as for many other polypeptide hormones) using the appropriate hormone bound covalently to agarose particles. Such complexes have been shown to be active on adipocytes (growth hormone[34]) and mammary gland epithelial cells (prolactin[35]). This suggests strongly that the hormones can bring about their actions without entering their target cells; they must presumably act, therefore, at receptors on the outside of the cell.

Studies of this kind have been subjected to some criticism, particularly in view of the possibility that the hormone may be released from the agarose beads by proteolysis. There is at least one observation which suggests that prolactin may have to enter its target cell in order to bring about some of its actions—the hormone has been reported to increase RNA synthesis in isolated mammary epithelial nuclei.[36]

*(b) Intracellular consequences of hormone–receptor binding*

Cellular events following binding of prolactin and growth hormone to receptors in their target cells include alterations of transport of amino acids and other metabolites, induction of specific proteins (such as casein and $\alpha$-lactalbumin for prolactin or ornithine decarboxylase and macroglobulin$_{2u}$ for growth hormone) and increase in the general machinery for protein and nucleic acid synthesis.

Little can be said about the biochemical events which occur in the cell immediately following binding of growth hormone or prolactin to its receptor. A few of the actions of growth hormone may be mediated by cyclic AMP[37] but for most actions there is no clearcut evidence that this or other cyclic nucleotides are directly involved as second messengers for these hormones. Other molecules have been proposed as second messengers (prostaglandin $F_{2\alpha}$, polyamines etc.) but in no case is the evidence entirely convincing.

In view of the paucity of information about the linking of the formation of hormone–receptor complexes to later biochemical events in target cells, some doubts must remain as to whether the binding sites identified for growth hormone and prolactin are really involved in mediating the biological actions. Such doubts apply particularly to receptors in the liver and kidney, which are probably major locations for the removal of the hormones from the circulation. Binding sites in these tissues could be involved primarily in mediating such removal.

In the case of mammary-gland receptors for prolactin, however, these doubts have been elegantly dispelled by Shiu and Friesen.[30] These authors prepared antibodies against partially purified prolactin receptors and showed that these would block the binding of $^{125}$I-labelled prolactin to receptors in rabbit mammary gland membranes. The antibodies would also inhibit biological actions of prolactin in rabbit mammary gland explants incubated *in vitro* (stimulation of [$^{3}$H]leucine incorporation into casein; stimulation of [$^{14}$C]aminoisobutyric acid transport). Binding of $^{125}$I-labelled insulin was not affected, and nor were the actions of insulin blocked. Thus, it appears that the binding of prolactin to its receptors, in this tissue at least, can be clearly and specifically related to the actions of the hormone.

**(vii) Prolactin receptors in mammary tumours**

There has been considerable interest in the possibility that prolactin may be involved in regulating the growth of some mammary tumours. Thus, dimethylbenzanthracene-induced rat mammary tumours may be either prolactin responsive or prolactin independent (on the basis of growth response to prolactin administration). The prolactin responsive tumours have more receptors for this hormone (after treatment of the animals with prolactin) than the prolactin independent tumours.[38]

Such observations have led to the hope that it might be possible to detect prolactin-responsive human mammary tumours (using biopsy samples) and thus to determine those patients most likely to respond to treatment designed to lower prolactin levels. Approximately 20% of human malignant mammary tumours possess significant levels of prolactin receptors,[39] but it is not yet clear whether identification of patients with such tumours can lead to improvements in treatment of the disease.

## III. SOMATOMEDINS

The somatomedins are a family of insulin-like peptides which are produced in the liver, kidney, and possibly other organs, partly under the influence of growth hormone. They are closely related to (and possibly identical with) other growth-promoting factors in plasma, such as insulin-like growth factor (non-suppressible insulin-like activity). At least three different somatomedins have been described (A, B, and C). Somatomedins A and C are rather similar, but somatomedin B differs considerably from these and there remains some doubt as to whether it is correctly termed a somatomedin. Characterization of these factors is as yet rather incomplete, and information about their receptors is correspondingly scarce.

The somatomedins mediate the actions of growth hormone on cartilage and possibly other tissues. Growth hormone itself has little direct effect on cartilage but somatomedins A and C have marked anabolic effects on this tissue, increasing synthesis of polysaccharides, DNA, RNA, and protein. Thus, by promoting somatomedin production growth hormone has a marked *indirect* effect on cartilage growth.

### (i) Receptors for somatomedin B

No success was achieved in studies designed to investigate binding of somatomedin B to membrane preparations[40]—possibly because iodination of the hormone destroys its biological activity.

### (ii) Receptors for somatomedins A and C

Specific receptors for somatomedin A are found in the placenta, from which a convenient receptor-containing membrane preparation can be obtained (though little is known about action of somatomedins on the placenta). This has been used as the basis of a radioreceptor assay.[41] $^{125}$I-labelled preparations of somatomedin A could be bound by such receptors, and displaced by either of two somatomedin A fractions ($A_1$ and $A_2$). Somatomedin C was equally effective in displacing the labelled somatomedin A, and insulin displaced at high doses (and was approximately 100 times less potent than somatomedin A). Somatomedin B, ACTH, human growth hormone, calcitonin, and somatostatin did not cause any displacement in the radioreceptor assay.

$^{125}$I-labelled somatomedin A bound specifically to chick cartilage membranes and also to membranes prepared from a variety of rat and monkey tissues, including lung, kidney, liver, brain, thymus, spleen, pancreas, heart, and fat.[42] Whether this binding represents a biologically significant effect is not yet known; so far somatomedin A has been shown to have effects primarily on adipose tissue and cartilage (and there are doubts as to whether those on adipose tissue are really of physiological significance).

Interactions between $^{125}$I-labelled somatomedin C and membrane receptors made from placenta have also been studied.[43,44] Again, a radioreceptor assay was set up, and insulin could displace the labelled somatomedin slightly (potency about 0.15% that of somatomedin C itself). It was demonstrated that the placenta possesses distinct plasma–membrane receptors for insulin and somatomedin C, and that cross reaction between the two is slight. Similar somatomedin C receptors were also shown to occur in liver, kidney, lung, brain, muscle, and thymus.

Binding of $^{125}$I-labelled somatomedin C to placental membranes was shown to be maximal at about pH 8 and to be relatively independent of magnesium or calcium ions.[44] Binding decreased with increasing ionic strength. Scatchard analysis showed that two types of somatomedin receptor may be present on placental membranes, high affinity ($K_a$ $6.3 \times 10^8$ $M^{-1}$) and low affinity ($K_a$ $5.6 \times 10^7$ $M^{-1}$) in a ratio of about 1:5.

The somatomedin C receptor on placental membranes was partially destroyed by digestion with proteolytic enzymes, especially pronase, but was more resistant to such digestion than the corresponding insulin receptor. Hyaluronidase, neuraminidase, and phospholipase C has no effect on the receptor.[44] Binding of $^{125}$I-labelled somatomedin C to placental membranes was not inhibited by a wide range of other hormonal factors, including somatomedin B, brain fibroblast growth factor, and epidermal growth factor. Large doses of insulin and proinsulin did inhibit binding. A somatomedin A preparation was about 6% as effective as somatomedin C. Multiplication stimulating activity (MSA) also competed with labelled somatomedin C but produced a non-parallel dose–response curve.[44] Serum from several mammals competed with somatomedin C in the receptor binding assay, the order of effectiveness being rat > pig > human = monkey = rabbit = guinea-pig > dog > sheep. Serum from a fish was about 2% as effective as human serum.

Binding of $^{125}$I-labelled somatomedin C by membranes from a variety of tissues has been studied.[43,44] The highest specific binding was found in the liver and placenta, though in each of these cases binding of $^{125}$I-labelled insulin was greater than that of $^{125}$I-labelled somatomedin C. Significant binding was also found to membranes made from kidney, chondrocytes, brain, lung, muscle, and thymus and, in most of these cases, binding of somatomedin C was greater than that of insulin. Low binding of somatomedin C to adipose tissue was detected, but it seems likely that this represents binding to insulin receptors, and that the effects of somatomedin C on adipose tissue that have been described reflect an interaction with insulin receptors and not a physiological effect.

Binding of labelled somatomedin C to foetal tissue is comparable to that in adult tissue, except for the lung; the foetal lung binds more somatomedin C, with a greater affinity than the corresponding adult tissue.

### (iii) The actions of somatomedins on cartilage—involvement of adenylate cyclase

At least some of the actions of somatomedins A and C on cartilage may be mediated by cyclic AMP.[45] 5% serum from normal (but not hypophysectomized) rats increased amino acid transport, chondroitin sulphate synthesis, protein synthesis, and RNA synthesis in chick cartilage and also elevated cyclic AMP levels in such tissue.[45] Dibutyryl cyclic AMP

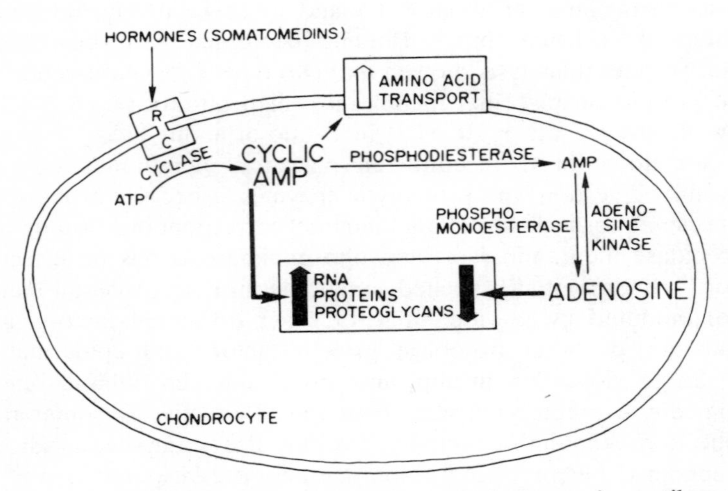


Figure 9.5. Proposed model for regulation of cartilage metabolism by cyclic AMP and adenosine. Reproduced by permission of Excerpta Medica, from Lebovitz, H. E., Drezner, M. K., and Neelon, F. A. in *Growth Hormone and Related Peptides* (Pecile, A., and Müller, E. E., eds.) p. 214 (1976)

and theophylline produced the same effects. The effects are inhibited, however, by adenosine, and it is possible that cartilage metabolism is subject to a dual control by cyclic AMP and adenosine, as illustrated in Figure 9.5. It is thus possible that in the case of these actions of somatomedin, at least, the formation of a hormone–receptor complex is linked to activation of adenylate cyclase, as in the case of many other polypeptide hormones. The details of such linkage are as yet poorly understood in the case of cartilage, and not all authors agree with the model shown in Figure 9.5.

## IV. CONCLUSIONS

It is evident from the foregoing that considerable progress has been made in characterizing the receptors for the protein hormones considered in this chapter. Radioreceptor assays have been developed, distribution and ontogeny of receptors has been determined, and the properties of membrane-bound, and in some cases solubilized, receptors have been investigated. Inevitably in a chapter of this kind many of the problems are passed over, and difficulties remain with regard to the stability and activity of both some labelled ligands and some receptor preparations. Nevertheless, the further purification and fuller characterization of individual receptors now seem within reach. Perhaps the largest remaining gap is in our knowledge of the coupling of hormone–receptor binding to the observable effects of these hormones within their target cells and tissues. This reflects our lack of knowledge about the mechanism of action of these hormones at the biochemical level (especially in the case of growth hormone) and it is to be hoped that it will be here that major advances are made in the next few years.

## V. REFERENCES


1. Wallis, M. *Biol. Rev.*, **50**, 35–98, (1975).
2. Roth, J. *Methods Enzymol.*, **37**, 223–233, (1975).
3. Rajaniemi, H., Oksanen, A., and Vanha-Perttula, T. *Hormone Res.*, **5**, 6–20, (1974).
4. Bullough, W. A., and Wallis, M. *Hormone Res.*, **8**, 37–50, (1977).
5. Birkinshaw, M., and Falconer, I. R. *J. Endocr.*, **55**, 323–334, (1972).
6. De Kretser, D. M., Catt, K. J., Burger, H. G., and Smith, G. C. *J. Endocr.*, **43**, 105–111, (1969).
7. Collipp, P. J., Patrick, J. R., Goodheart, C., and Kaplan, S. A. *Proc. Soc. Exp. Med.*, **121**, 173–177, (1966).
8. Retegui-Sardou, L. A., Scaramel, L. O., Dellacha, J. M., and Paladini, A. C. *Mol. Cell. Biochem.*, **16**, 87–96, (1977).
9. Costlow, M. E., and McGuire, W. L. *J. Endocr.*, **75**, 221–226, (1977).
10. Lesniak, M. A., Gorden, P., Roth, J., and Gavin, J. R. *J. Biol. Chem.*, **249**, 1661–1667, (1974).

11. Herington, A. C., and Veith, N. M. *J. Endocr.*, **74**, 323–334, (1977).
12. Ranke, M. B., Stanley, C. A., Tenore, A., Rodbard, D., Bongiovanni, A. M., and Parks, J. S. *Endocrinology*, **99**, 1033–1045, (1976).
13. Posner, B. I., Kelly, P. A., Shiu, R. P. C., and Friesen, H. G. *Endocrinology*, **95**, 521–531, (1974).
14. Kelly, P. A., Posner, B. I., Tsushima, T., and Friesen, H. G. *Endocrinology*, **95**, 532–539, (1974).
15. Bohnet, H. G., Gómez, F., and Friesen, H. G. *Endocrinology*, **101**, 1111–1121, (1977).
16. Bohnet, H. G., Aragona, C., and Friesen, H. G. *Endocr. Res. Commun.*, **3**, 187–198, (1976).
17. Aragona, C., Bohnet, H. G., Fang, V. S., and Friesen, H. G. *Endocr. Res. Commun.*, **3**, 199–208, (1976).
18. Posner, B. I., Kelly, P. A., and Friesen, H. G. *Science*, **188**, 57–59, (1975).
19. Posner, B. I. *Endocrinology*, **99**, 1168–1177, (1976).
20. Djiane, J., Durand, P., and Kelly, P. A. *Endocrinology*, **100** 1348–1356, (1977).
21. Lesniak, M. A., Gorden, P., and Roth, J. *J. Clin. Endocr. Metab.*, **44**, 838–849, (1977).
22. Baxter, R. C., and Turtle, J. R. *Biochem. Biophys. Res. Commun.*, **84**, 350–357, (1978).
23. Shiu, R. P. C., and Friesen, H. G. *Biochem. J.*, **140**, 301–311, (1974).
24. Herington, A. C., Veith, N., and Burger, H. G. *Biochem. J.*, **158**, 61–69, (1976).
25. Tsushima, T., and Friesen, H. G. *J. Clin. Endocr. Metab.*, **37**, 334–337, (1973).
26. Herington, A. C., and Veith, N. M. *Endocrinology*, **101**, 984–987, (1977).
27. Waters, M. J., and Friesen, H. G. *Proc. Austr. Biochem. Soc.*, **11**, 101, (1978).
28. Shiu, R. P. C., and Friesen, H. G. *J. Biol. Chem.*, **249**, 7902–7911, (1974).
29. Shiu, R. P. C., and Friesen, H. G. *Biochem. J.*, **157**, 619–626, (1976).
30. Shiu, R. P. C., and Friesen, H. G. *Science*, **192**, 259–261, (1976),
31. Lesniak, M. A., Roth, J., Gorden, P., and Gavin, J. R. *Nature, New Biol.*, **241**, 20–22, (1973).
32. Shiu, R. P. C., Kelly, P. A., and Friesen, H. G. *Science*, **180**, 968–971, (1973).
33. Parke, L., and Forsyth, I. A. *Endocr. Res. Commun.*, **2**, 137–149, (1975).
34. Hecht, J. P., Dellacha, J. M., Santomé, J. A., Paladini, A. C., Hurwitz, E., and Sela, M. *FEBS Letters*, **20**, 83–86, (1972).
35. Turkington, R. W. *Biochem. Biophys. Res. Commun.*, **41**, 1362–1367, (1970).
36. Chomczynski, P., and Topper, Y. J. *Biochem. Biophys. Res. Commun.*, **60**, 56–63, (1974).
37. Ahrén, K., Albertsson-Wikland, K., Isaksson, O., and Kostyo, J. L. in *Growth Hormone and Related Peptides* (Pecile, A., and Müller, E. E., eds.), Excerpta Medica, Amsterdam, pp. 94–103, (1976).
38. Holdaway, I. M., and Friesen, H. G. *Cancer Res.*, **36**, 1562–1567, (1976).
39. Friesen, H. G. *Rec. Results Cancer Res.*, **57**, 143–149, (1976).
40. Hall, K., Takano, K., Enberg, G., and Fryklund, L. in *Growth Hormone and Related Peptides* (Pecile, A., and Müller, E. E., eds.), Excerpta Medica, Amsterdam, pp. 178–189, (1976).
41. Hall, K., Takano, K., Fryklund, L., and Sievertsson, H. *Adv. Metabolic Disorders*, **8**, 61–71, (1975).
42. Takano, K., Hall, K., Fryklund, L., and Sievertsson, H. *Hormone Metabolic Res.*, **8**, 16–24, (1976).
43. Van Wyk, J. J., Underwood, L. E., Baseman, J. B., Hintz, R. L., Clemmons, D. R., and Marshall, R. N. *Adv. Metabolic Disorders*, **8**, 127–150, (1975).

44. D'Ercole, A. J., Underwood, L. E., Van Wyk, J. J., Decedue, C. J., and Foushee, D. B. in *Growth Hormone and Related Peptides* (Pecile, A., and Müller, E. E., eds.), Excerpta Medica, Amsterdam, pp. 190–201, (1976).
45. Lebovitz, H. E., Drezner, M. K., and Neelon, F. A. in *Growth Hormone and Related Peptides* (Pecile, A., and Müller, E. E., eds.), Excerpta Medica, Amsterdam, pp. 202–215, (1976).

Cellular Receptors for Hormones and Neurotransmitters
Edited by D. Schulster and A. Levitzki


CHAPTER 10

# Insulin and glucagon receptors

Lothar Kuehn
*Biochemische Abteilung, Diabetes-Forschungsinstitut an der Universität Düsseldorf, Hennekamp 65, Düsseldorf, W. Germany*
Tom Blundell
*Laboratory of Molecular Biology, Department of Crystallography, Birkbeck College, University of London, London WC1E 7HX, U.K.*



## I. INTRODUCTION

There is now convincing evidence that the first step in the action of insulin and glucagon is their interaction with specific receptor sites on the surface of the cell plasma membrane.

Indirect support for this notion initially came from *in vitro* studies with rat hemidiaphragms.[1] After brief incubation of the tissue with insulin and subsequent removal, a persisting effect of the hormone on glycogen synthesis could be observed. Pastan[2] later showed that the effect was due to the presence of the hormone on the cell surface, since it could be reversed by treating the tissue with insulin-specific antiserum.

Adenylate cyclase has been demonstrated to function as the effector system for glucagon.[3] The hormone stimulates the activity of the enzyme and the increased intracellular level of cyclic AMP, the product of this enzyme, leads to most of the known physiological actions of glucagon in liver. Both the glucagon recognition system and the effector system are co-purified with other plasma membrane markers during subcellular fractionation.[4]

In the past few years, direct evidence for the localization of insulin and glucagon receptors on the plasma membrane of the cell has been obtained from binding studies utilizing isolated cells or plasma membranes and carefully validated $^{125}$I-labelled hormone of high specific activity. Since the receptors have not yet been isolated as pure chemical entities, it is necessary to define them by their binding characteristics. These include specificity, high affinity, rapid and reversible binding, saturability, and direct or indirect correlation of binding to a biological response.

This chapter will focus on the physicochemical properties of the hormone-receptor interaction and how this interaction is thought to be related to the known biological effects. (For comprehensive reviews, see References 5 and 6). We first review the structures of the hormones, then describe the characteristics of their receptors and finally discuss models for hormone receptor interactions.

## II. STRUCTURE

### (i) Glucagon

Glucagon is a single polypeptide of 29 amino acids, which is probably synthesized as a larger precursor.[7] The conformation of the glucagon molecule as determined by X-ray analysis of crystals[8] is shown in Figure 10.1. The molecule is approximately $\alpha$-helical between residues 6 and 27 resulting in Phe$^{6}$,Tyr$^{10}$, Tyr$^{13}$, and Leu$^{14}$ forming one mainly hydrophobic region and Ala$^{19}$, Phe$^{22}$, Val$^{23}$, Trp$^{25}$, Leu$^{26}$, and Met$^{27}$ another. These hydrophobic regions interact to form trimers and the trimers are further associated as oligomers of cubic symmetry in crystals. Residues 1–5 are not constrained by intermolecular interactions and appear to be flexible. Circular dichroism[9] and nuclear magnetic resonance (K. Wütrich, personal communication) indicate that glucagon has little secondary structure in aqueous solution at high dilutions characteristic of those in circulation. The percentage of the helical conformers must be very low, but it is increased at high concentrations by self association to trimers[10] and in the presence of certain aliphatic alcohols.[11]

### (ii) Insulin

Insulin is synthesized as a preprohormone, containing a hydrophobic N-terminal extension responsible for directing the nascent polypeptide to

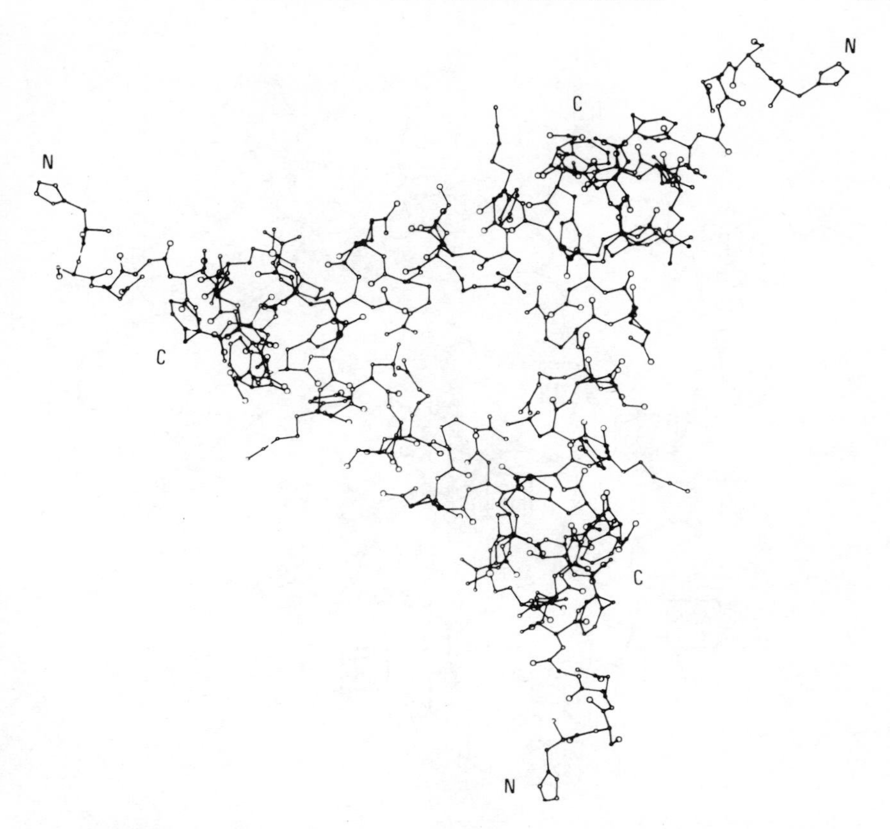


Figure 10.1. The structure of the glucagon trimers found in cubic crystals.[8] One glucagon molecule forms each side of the triangular arrangement and intermolecular interactions are through hydrophobic residues also thought to be important in receptor binding. Reproduced with permission from Blundell *et al.*, *Metabolism,* **25**, 1331–1336, (1976)

the endoplasmic reticulum for export.[12] It is transported to the site of granulation in the Golgi bodies as proinsulin, a single chain polypeptide of ~80 amino acid residues.[13] Proinsulin is cleaved to give insulin, which comprises two polypeptides, the A-chain of 21 amino acids and the B-chain of 30 amino acids. The hormone is stored in zinc insulin granules, which are released into circulation by exocytosis.

X-ray studies of porcine insulin have demonstrated that it has the globular structure shown in Figure 10.2. It has two hydrophobic surfaces which are buried in 2Zn insulin hexamers. X-ray studies have demonstrated that most mammalian and fish insulins form similar hexamers with the exception of hagfish insulin which forms only dimers and guinea-pig and coypu insulins which form only monomers and probably have a rather distorted

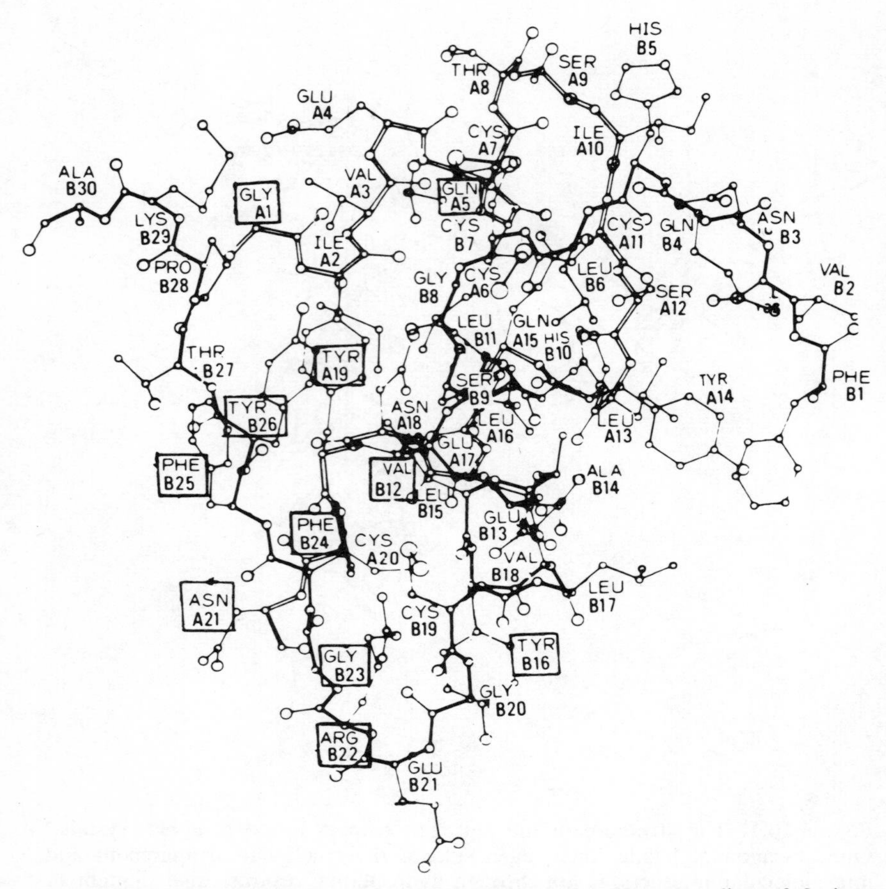


Figure 10.2. The structure of insulin based on the X-ray studies of 2-zinc insulin.[14] The residues with numbers enclosed in boxes are thought to be involved in receptor binding. Reproduced with permission from Pullen *et al.*, *Nature (London)*, **259**, 369–373 (1976)

conformation. Circular dichroism studies indicate that insulin and proinsulin have similar three-dimensional structures in solution (see Reference 14 for review).

## III. CHARACTERIZATION OF RECEPTORS

### (i) Glucagon

Rodbell and coworkers have performed detailed studies on binding of glucagon to isolated liver plasma membranes and the ensuing activation of adenylate cyclase in this tissue.[15] The binding of [$^{125}$I]glucagon has been found to be specific, in that binding can be inhibited only by unlabelled hormone and not by other polypeptide hormones including insulin or biologically inactive peptide fragments of glucagon. Glucagon interacts with its receptor in accordance with binding isotherms for a single set of homogeneous, non-interacting receptor sites, having a high affinity constant ($1 \times 10^{10}$ $M^{-1}$), approximately equal to that for the activation of adenylate cyclase. Treatment of membranes with phospholipase A, digitonin, or urea depressed both hormone binding and the biological response. The hormone–receptor dissociation is markedly accelerated by ethylenediamine tetra-acetic acid, suggesting the role of a divalent cation in binding. Liver membranes are known to inactivate glucagon very rapidly[16] which makes it impossible to measure accurately rate constants of association and dissociation. However, the fact that receptor-bound hormone, removed spontaneously or by acid[16] does bind again to fresh membranes whereas inactivated glucagon does not, suggests that receptor binding and inactivation are separate processes. It is not yet known whether the inactivation in these membrane preparations reflects a physiologically relevant process. Rodbell and coworkers have postulated an absolute requirement for either ATP or GTP in the activation of rat liver adenylate cyclase by glucagon. Phosphorylation appears not to be involved. It has also been shown that GTP stimulates the rate and degree of dissociation of the [$^{125}$I]glucagon (and of [$^{3}$H]glucagon) membrane complex and lowers the total amount of hormone bound. A chemical derivative of glucagon, lacking the N-terminal His residue, des-His-glucagon, was reported to be unable to stimulate adenylate cyclase, but to act as a competitive inhibitor of native glucagon, with an approximately ten times lower affinity. However, more recent studies indicate that it may be a partial agonist.[17]

Digestion of membranes with phospholipase C abolishes the effect of GTP on glucagon dissociation from membranes. These studies suggest that, although phospholipids do not directly participate in the binding reaction, they do play a role in the stimulation of adenylate cyclase by the hormone. Despite the specificity and affinity of glucagon–receptor interactions in liver plasma membranes and the correlation between binding and activation of adenylate cyclase, Birnbaumer and Pohl[18] have challenged their own earlier view that glucagon specific binding actually represents receptor interactions. On the basis of time studies, they showed that hormonal stimulation is maximal when only 10–20% of the binding sites are occupied, suggesting that the bulk of binding sites are not related to biological activation. Despite this discrepancy, glucagon probably does bind to specific receptors. To expect complete congruency of the kinetics of binding and enzyme activation assumes that any inactivation of receptor and adenylate cyclase that occurs during liver plasma membrane isolation occurs to the same extent for both components.[18] It is also conceivable, or even likely, that the uncoupling of receptor and enzyme is the process most susceptible to disruption, thus simulating an excess of binding sites which *in vivo* represent functional receptors. Unfortunately, it has not yet been possible to perform binding and activation studies in living cells so that this discrepancy cannot at present be resolved.

The solubilization and partial purification of glucagon receptor from rat liver plasma membranes have been reported.[19] The binding activity was extracted using Lubrol PX, a non-ionic detergent, and an approximately 3,000-fold purification of the glucagon receptor structure was achieved by affinity chromatography, although this resulted in a loss of the adenylate cyclase component.[19] The solubilized activity was assumed to be the physiological membrane receptor because of identical specificities and affinities and rapid and reversible binding. The binding activity was abolished after digestion with proteolytic enzymes,[19] indicating that a protein component is important for binding. Physicochemical studies indicate the solubilized protein to have a molecular weight of 190,000.

**(ii) Insulin**

Insulin-specific receptors have been demonstrated in a large number of tissue preparations in a variety of species. The basic properties of the insulin receptor interaction in the various tissues studied appear to be very similar. They show specificity and high affinity for insulin, rapid and reversible binding, saturability, and a nearly identical dependence on pH for optimal binding. The specificity of the receptor has been further established with the study of natural analogues and chemically modified derivatives of the hormone. The relative potency of a variety of insulin analogues to compete for binding of labelled insulin has been shown to be in direct proportion to their *in vivo* and *in vitro* activity.[20]

With high physiological hormone concentrations ($\sim 10^{-10}$ M), steady-states of binding are observed within a few minutes at 37 °C.[21] The level of binding at steady-state, however, is substantially higher at lower temperature (4–20 °C). The decreased and less stable binding above 20 °C can be explained, at least in part, by accelerated degradation of both hormone and receptor. The relationship between receptor binding and degradation has been the subject of considerable discussion.[21,22] Terris and Steiner[22] suggest that over a wide range of concentrations insulin bound to hepatocyte plasma membranes is the substrate for insulin degradation by the liver, although Freychet *et al*.[21] had previously found differences of analogue specificity and pH, ionic strength, and temperature between degradation of insulin and receptor binding.

The quantitative aspects of insulin–receptor interaction are complex and

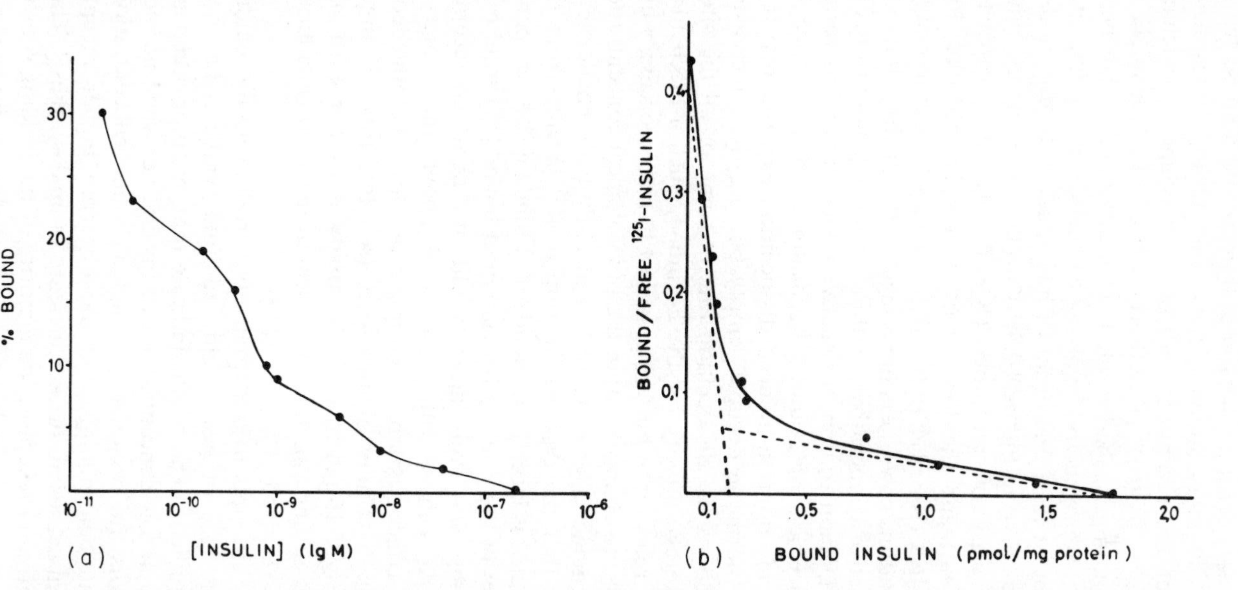


Figure 10.3. (a) Percentage of [$^{125}$I]insulin specifically bound to purified pig liver plasma membranes as a function of total insulin concentration. Membranes (0.33 mg/ml$^{-1}$) were incubated for 4 h at 15 °C with [$^{125}$I]insulin ($2 \times 10^{-11}$ M) in the presence of increasing concentrations of native insulin. Membranes were then separated by filtration on millipore filters and the membrane bound radioactivity was determined. Corrections for non-specific binding, which was less than 5% of total binding, have been made.
(b) Scatchard plot of the data in Fig 10.3(a). The bound:free ratio is plotted as a function of hormone bound to plasma membranes (•–•). The intrinsic binding plot (- - -) is obtained by resolution of the experimental curve into high and low affinity independent linear components[23]

have been the subject of some controversy.[21] Only Cuatrecasas has reported a single class of high affinity receptor sites in both liver and adipose tissue ($\sim 10^{10}$ $M^{-1}$). Other investigators studying these and other tissues have noted lower affinities for the receptor[21] and, often, curvilinear Scatchard plots are observed (Figure 10.3(b)). (References 21, 23; L. Kuehn, H. J. Bubenzer, and P. Roesen, unpublished results). This curvilinearity has been first ascribed to receptor heterogeneity, involving two classes of binding sites with affinities differing by at least one order of magnitude.[21] In this model, the upper portion of the curve in Figure 10.3(b) would represent the 'high affinity, low capacity' class of receptors ($K \sim 6 \times 10^9$ $M^{-1}$, capacity; 0.2 pmole per mg protein) and the lower part would describe the 'low affinity, high capacity' sites ($K \sim 1.4 \times 10^8$ $M^{-1}$, capacity; 1.58 pmol $mg^{-1}$ of protein). De Meyts and his coworkers have suggested that these data are also consistent with negative cooperativity within a relatively homogeneous class of binding sites.[23] Demonstration of negative cooperativity has rested on the enhanced dissociation of bound [$^{125}$I]insulin in the presence of unlabelled hormone during dilution-induced dissociation experiments. It is thought that the native hormone occupies empty binding sites, increasing through site–site interactions the dissociation rate of some or all the sites, thereby lowering the average affinity for the hormone. Such a mechanism could be physiologically relevant in buffering against hormone excess, while preserving high sensitivity to changes within the physiological concentration range of circulating hormone. This concept has recently been challenged[24] with the demonstration that the dissociation rate of [$^{125}$I]insulin from the human lymphocyte receptor is largely independent of binding site occupancy and that an enhancement of the rate of dissociation of bound [$^{125}$I]insulin by native insulin can be observed even under conditionss in which receptor occupancy decreases. At values within and well above the range of physiological hormone concentration, binding occurs to a homogeneous class of independent, high affinity binding sites (see Reference 2; upper portion of Figure 10.3(b)). If negative cooperativity does not exist, the anomalous binding of insulin at higher concentrations to membrane receptors remains to be resolved.

There have been few attempts to quantitatively correlate insulin binding to hormone action. The rate of insulin induced lipid synthesis has been determined in kinetic studies.[25] The results indicate that receptor occupancy is rate-determining only at low concentrations of insulin, i.e. when no more than 2% of the receptors are occupied. At higher insulin concentrations, other steps become rate-determining, and the observed higher receptor occupancy at equilibrium causes no further increase in lipid synthesis. Thus, as observed with the glucagon receptor, the majority of the binding sites for insulin do not appear to exert any biological effect. For a discussion of 'spare receptors' see Chapter 2; some authors have equated these 'spare' receptors to physiologically unimportant structures. An alternative possibility would be that such receptors are linked to other, as yet undetected, cellular functions. It should also be pointed out that the fat cells used for these studies have been obtained after collagenase digestion of adipose tissue, a process which may have resulted in an inactivation of membrane molecules essential for transmission of the hormonal signal or their uncoupling from the receptor structure. In all these cases, 'inactive' binding sites would be indistinguishable from true insulin receptors.

Attempts at isolating and purifying the insulin receptor structure from rat liver plasma membranes,[26] pig liver plasma membranes (L. Kuehn, H. J. Bubenzer, and P. Roesen, unpublished results), and turkey erythrocyte membranes[27] have been made. Most of the properties of the cell membrane receptors are preserved in these solubilized fractions in that insulin binding to solubilized receptors is saturable, reversible, and time and temperature dependent. The solubilized receptor appears to be a protein with carbohydrate moieties. By gel filtration and density gradient centrifugation its molecular weight has been estimated to be 300,000. It appears to be a highly eccentric molecule with a Stokes radius of 70 Å. Studies with the soluble insulin receptor from turkey erythrocyte membranes[27] suggest that the molecule is an oligomer containing four or more insulin binding sites. The purification of the insulin receptor remains a formidable task. Since a total of 1 mg of receptor protein is present in the homogenate of 1 kg of pig liver, which contains about 180 g of protein, isolation of homogeneous receptor will require 200,000-fold purification.

In contrast to the glucagon receptor in liver cell plasma membranes, adenylate cyclase appears not to be the primary effector of insulin action. The effector system(s) for insulin has (have) remained largely undefined.

## IV. THE NATURE OF HORMONE–RECEPTOR INTERACTIONS

We have shown elsewhere[14,28] that insulin and glucagon[8] are probably stored in the pancreatic cells as oligomers packed in crystalline or amorphous granules. In circulation, insulin and glucagon granules slowly dissolve and at concentrations of $\sim 10^{-10}$ M the oligomeric forms (hexamers and trimers) must dissociate to give monomers. This makes the monomer the most likely candidate for the active species, but it is still possible that reassociation occurs at or in the proximity of the receptor. However, this seems to be an unlikely requirement as a number of insulins which show little ability to dimerize (for example, guinea-pig insulin and some nitrated insulins) are biologically active, albeit weakly so.[14]

As described above, the biological activity of insulin is usually directly proportional to the receptor binding as measured by the ability of the insulin to displace radioactively-labelled insulin from the receptor.[20] The activity depends on the integrity of the three-dimensional structure of insulin; no insulin with a disturbed tertiary structure is fully active.[14,29] Some of the residues involved in dimerization, in particular B24-Phe and B25-Phe, and possibly the adjacent B12, B16, and B26, probably bind the receptor initially through non-covalent interactions (see Figure 10.2). A1, A19, and A21 have also been postulated to bind the receptor; however, these can be modified without complete loss of activity and evidence is accumulating that they are on the periphery of the receptor binding region. Secondary conformational changes or degradation of insulin necessary for the biological activity remain possibilities. The importance of the tertiary structure is clearly a consequence of the wide separation in the sequence of groups important to receptor binding which must be brought together in the correct relative positions.

In the insulin receptor complex, it is probable that the receptor has a concave surface complementary to that of the hormone. The surface may have maximum dimensions of 20 × 20 Å but as residues on the periphery of the insulin receptor binding region are not essential the interaction at this point may not be close. The receptor almost certainly must have a large hydrophobic region complementary to that formed by residues B24-Phe, B25-Phe, B12-Val, and possibly part of B26-Tyr and A19-Tyr. It must also be able to satisfy the hydrogen bond donors of the main chain of residues B24–B26 and so may have a polypeptide chain capable of forming an antiparallel $\beta$-sheet interaction. On the periphery it must contain a series of charged groups which can form ionic interactions with A1-$\alpha$-amino, A4-glutamyl carboxylate, A21-$\alpha$-carboxylate, B22-guanidinium, and B13-glutamyl carboxylate. It thus involves a larger surface area and more extensive interactions than exhibited in the dimer accounting for the higher association constant.

Studies on glucagon show that almost the whole molecule is required for full biological activity although loss of the N-terminal histidine has less effect on the receptor binding than the potency. It has been suggested that receptor binding involves hydrophobic interactions and that a helical conformer is induced or stabilized at the receptor.[8] This conformer has organized hydrophobic regions which might interact with complementary regions of the receptor. Other charged interactions, i.e. with $Arg^{17}$ and $Arg^{18}$, may further increase the receptor binding, but interaction of the N-terminal residues could enhance the biological response without a proportional increase in the stability of the hormone–receptor complex. The ease of forming a helical conformer in the correct environment means that a stable tertiary structure is not required in glucagon.

## V. CONCLUSIONS

The plasma membrane has been established as the primary site of insulin and glucagon action. Both hormones are specifically bound to a site on the plasma membrane, the hormone receptor. However, the two hormones have different mechanisms for receptor recognition: insulin action requires a relatively rigid preformed conformer whereas glucagon action depends upon induction of a specific conformer at the receptor. In each case, a signal is thereby created, eventually leading to a biological response. In the case of glucagon, the effector system has been shown to be adenylate cyclase. For insulin, the mechanism for target cell activation has not been clarified. Attempts at the isolation and structural characterization of the receptor function are currently in progress, and a solution appears resolvable by currently available techniques. Studies of ligand–protein interaction in a homogeneous system should help to overcome present uncertainties about the precise binding mechanism and could yield new insight into the relationship between those molecules and the regulatory functions of biological membranes.

## VI. REFERENCES

1. Stadie, W. C., Haugaard, N., Marsh, J. B., and Hills, A. G. *Am. J. Med. Sci.*, **218**, 265, (1949).
2. Pastan, I., Roth, J., and Macchia, V. *Proc. Natl. Acad. Sci. U.S.A.*, **56**, 1802, (1966).
3. Sutherland, E. W., and Rall, T. W. *J. Biol. Chem.*, **232**, 1077, (1957).
4. Rodbell, M. *J. Biol. Chem.*, **242**, 5744, (1967).
5. Kahn, C. R. *J. Cell Biol.*, **70**, 261, (1976).
6. Catt, K. J., and Dufau, M. L. *Ann. Rev. Physiol.*, **39**, 529, (1977).
7. Tager, H. S., and Steiner, D. R. *Proc. Natl. Acad. Sci. U.S.A.*, **70**, 2321, (1973).
8. Sasaki, K., Dockerill, S., Adamiak, D. A., Tickle, I. J., and Blundell, T. L. *Nature (London)*, **257**, 751, (1975).
9. Panijpan, B., and Gratzer, W. B. *Eur. J. Biochem.*, **45**, 547, (1974).
10. Blanchard, M. H., and King, M. V. *Biochem. Biophys. Res. Commun.*, **25**, 298, (1966).
11. Berner, H., and Edelhoch, H. *J. Biol. Chem.*, **246**, 1785, (1971).
12. Chan, S. J., Keirn, P., and Steiner, D. F. *Proc. Natl. Acad. Sci. U.S.A.*, **73**, 1964, (1976).
13. Steiner, D. F., Clark, J. L., Nolan, C., Rubenstein, A. H., Margoliash, E., Atne, B., and Oyer, P. E. *Recent Prog. Horm. Res.*, **25**, (1969).
14. Blundell, T. L., Hodgkin, D. C., Dodson, G. G., and Mercola, D. A. *Adv. Protein Chem.*, **26**, 280, (1972).
15. Rodbell, M., Krans, H. M. J., Pohl, S. L., and Birnbaumer, L. *J. Biol. Chem.*, **246**, 1861, (1971).
16. Pohl, S. L., Krans, H. M. J., Birnbaumer, L, and Rodbell, M. *J. Biol. Chem.*, **246**, 2295, (1972).

17. Hruby, V. J., Wright, D. E., Lin, M. C., and Rodbell, M. *Metabolism,* **25**, suppl. 1, 1323, (1976).
18. Birnbaumer, L, and Pohl, S. L. *J. Biol. Chem.,* **248**, 2056, (1973).
19. Giorgio, N. A., Johnson, C. B., and Blecher, M. *J. Biol. Chem.,* **249**, 428, (1974).
20. Freychet, P., Brandenburg, D., and Wollmer, A. *Diabetologia*, **10**, 1, (1975).
21. Kahn, C. R., Freychet, P., Roth, J., and Neville, D. M., Jr. *J. Biol. Chem.,* **249**, 2249, (1974).
22. Terris, S., and Steiner, D. F. *J. Biol. Chem.,* **250**, 8389, (1975).
23. De Meyts, P., Van Obberghen, E., Roth, J., Wollmer, A., and Brandenburg, D. *Nature (London),* **273**, 504, (1978).
24. Pollet, R. J., Staendert, M. L., and Haase, B. A. *J. Biol. Chem.*, **252**, 5828, (1977).
25. Gliemann, J., Gammeltoft, S., and Vinten, J. *J. Biol. Chem.,* **250**, 3368, (1975).
26. Cuatrecases, P. *J. Biol. Chem.,* **247**, 1980, (1972).
27. Ginsberg, B. H., Kahn, C. R., Roth, J., and De Meyts, P. *Biochem. Biophys. Res. Commun.,* **73**, 1068, (1976).
28. Blundell, T. L., and Wood, S. P. *Nature (London),* **257**, 197, (1975).
29. Pullen, R. A., Lindsay, D. G., Wood, S. P., Tickle, I. J., Blundell, T. L., Wollmer, A., Krail, G., Brandenburg, D., Zahn, H., Gliemann, J., and Gammeltoft, S. *Nature (London),* **259**, 369, (1976).

Cellular Receptors for Hormones and Neurotransmitters
Edited by D. Schulster and A. Levitzki


CHAPTER 11

# ACTH receptors

Dennis Schulster
*Hormones Division, National Institute for Biological Standards and Control, Holly Hill, Hampstead, London NW3 6RB, U.K.*
Robert Schwyzer
*Institute of Molecular Biology and Biophysics, Swiss Federal Institute of Technology, CH-8093, Zürich-Hönggerberg, Switzerland*



## I. INTRODUCTION

### (i) ACTH molecules with biological activity

The physiologically active peptide hormone responsible for stimulating corticosteroid synthesis by the fasciculata cells of the adrenal cortex has long been recognized as corticotropin (ACTH), secreted by the anterior lobe of the pituitary. However, it has been reported by many laboratories that

ACTH (39 amino acids) also exists in various high molecular weight forms. These peptides were designated 'big' and 'intermediate' ACTH and have been described in pituitaries, tumours, and plasmas of several species including man.[1,2] (For reviews see References 3 and 4 and chapters by Nicholson and Lowry *et al.* in Reference 5). The 'big' ACTH of human plasmas and tumours may be cleaved by tryptic digestion to an immunologically ACTH-like fragment and this procedure results in the emergence of biological activity (Gewirtz *et al.*, 1974, cited in Reference 5). Two ACTH peptides of molecular weights, 24,000 and 34,000 isolated from ovine pituitary glands were shown to be susceptible to degradation by tissue enzymes (Lee and Lee, 1977, cited in Reference 5). Similar high molecular weight forms of ACTH have been found in mouse pituitary tumour cells (Eiper *et al.*, 1976, cited in Reference 5), in human plasma, and a normal human pituitary (Orth and Nicholson, 1977, see Reference 5). Labelling with [$^3$H]glucosamine and [$^3$H]mannose as well as concanavalin A-agarose binding studies suggest that 'big' ACTH and possibly 'intermediate' ACTH are glycoproteins. The corticotropin precursor from rat pituitary with molecular weight 30,000 is also believed to be the precursor of the lipotropins, $\beta$-melanotropins, and endorphins and has therefore been called 'pro-opiocortin'.[6]

### (ii) The biological activities of ACTH

The biological activities of ACTH are shown in Table 11.1. Of these, perhaps the most noteworthy are its growth-promoting and steroidogenic actions on the adrenal cortex. Thus isolated zona fasciculata-reticularis cells are stimulated about 50-fold in their production of glucocorticoids (e.g. corticosterone and cortisol) and zona glomerulosa cells about two-fold in the output of aldosterone and other corticosteroids. In most mammals, the glomerulosa cells are associated with the outer capsule of the adrenal gland, while the fasciculata and reticularis cells comprise the bulk of the adrenal cortex and are arranged in columns between the glomerulosa cells and the inner core of medullary cells.

The N-terminal part of ACTH also produces extra-adrenal effects. Its darkening effects on amphibian skin are similar to those observed with $\alpha$-MSH. The *in vitro* melanotropic activity in the frog (*Rana pipiens*) of a variety of structural analogues of ACTH and $\alpha$- and $\beta$-MSH have been evaluated (see Chapter 12; also Schwyzer, 1977, in Reference 5).

ACTH also stimulates lipolysis in fat cells with release of glycerol and free fatty acid. In terms of potency, however, it should be noted that adrenal cells are usually 10–100 times more sensitive to ACTH than either fat cells or frog skin. The median effective dose ($ED_{50}$) for ACTH stimulation of isolated adrenocortical cell steroidogenesis is about $10^{-5}$ i.u. ml$^{-1}$ ($10^{-11}$ M).

Table 11.1. Some biological properties of ACTH

| *In vivo* |
|---|
| 1. Increases the weight of adrenal glands in normal animals and maintains the adrenal weight in hypophysectomized animals |
| 2. Stimulates corticoid production as estimated in the adrenal venous blood |
| 3. Reduces ascorbic acid content of adrenal gland |
| 4. Causes eosinopenia and thymic involution |
| 5. Enhances erythropoiesis in hypophysectomized animals |
| 6. Increases metabolic rate of hypophysectomized rats |
| 7. Induces deciduoma in hypophysectomized–oophorectomized rats |
| 8. Increases weights of sex accessories in hypophysectomized–castrated rats |
| 9. Maintains muscle glycogen in hypophysectomized animals |
| 10. Enhances glycogen deposition in the liver |
| 11. Acts as a galactopoietic agent |
| 12. Antagonizes the action of growth hormone |
| 13. Causes an increase of liver fat in fasted animals |
| 14. Induces an elevation of serum-free non-esterified fatty acids |
| 15. Increases blood ketone bodies in fasted rats |
| 16. Stimulates melanophore expansion (melanin dispersion) in amphibians and reptiles |
| 17. Facilitates acquisition and retention of learned behaviour in rats and humans |
| *In vivo*, in the absence of adrenals |
| 18. Suppresses the increased capillary permeability induced by exudin |
| 19. Causes an increase in liver fat in animals maintained with corticoids |
| 20. Produces hypergranulation of the renal juxtaglomerular cells |
| 21. Influences metabolism of cortisol |
| *In vitro* |
| 22. Stimulates corticosteroid production and aldosterone production by the adrenal cortex |
| 23. Causes melanin dispersion in skins of amphibians and reptiles |
| 24. Releases non-esterified fatty acids from rat and rabbit adipose tissue |
| 25. Induces the uptake and oxidation of glucose by rat mammary tissues |
| 26. Inhibits incorporation of amino acids into adipose tissue protein |

For the stimulation of lipolysis in isolated fat cells the $ED_{50}$ value is $10^{-9}$ M. Most of the work in the ACTH-receptor field has therefore concentrated on the adrenocortical cell, although the fat cell has undoubtedly also proved to be a useful system.

Interesting actions of ACTH and its fragments or analogues have also been reported on the central nervous system and these analogues appear to facilitate the acquisition and retention of learned behaviour in the rat and in humans (de Wied, 1977, in Reference 5; Beckwith, 1976, Sandman, 1976, and Kastin, 1977, cited in Reference 4). Little is known regarding brain receptors for these peptides but undoubtedly future work will be directed towards this area.

## II. STRUCTURE–ACTIVITY RELATIONSHIPS

The study of structure–activity relationships, mainly by means of synthetic chemistry, has provided fundamental insight into the chemical organization of hormonal information in the ACTH molecule and into the specific requirements of different ACTH receptors for stimulation[3,4] (Schwyzer, 1977, in Reference 5).

ACTH from five different sources is invariably an open-chain peptide of 39 amino acid residues (Figure 11.1). The species differences reside in the C-terminal two-thirds of the molecule and the amino acid replacements are all conservative of the general biological and physicochemical properties; ACTH is chemically related to $\alpha$-melanotropin ($\alpha$-MSH), $\beta$-melanotropin ($\beta$-MSH), and the lipotropins (LPH) through a common hepta- or tetrapeptide sequence flanked by tyrosine and proline (Figure 11.1; see also Chapter 12). These hormones, plus the endorphins and enkephalins, constitute the opiocortin family and are derived biogenetically from one precursor protein, pro-opicortin.[6]

In aqueous solution, ACTH is a flexible molecule (Schwyzer, 1977, in Reference 5). It has early been postulated that the molecule in binding to the more rigid receptor, might adapt itself to the recognition site or recognition subunit of the receptor (the discriminator) and thus ensure a kinetically, thermodynamically, and functionally optimal interaction.[7] ACTH has not been crystallized.

The synthesis and testing of dozens of ACTH analogues, derivatives, and fragments has revealed the following facts about the one-dimensional organization of hormonal information. (1) Discrete (although sometimes overlapping) sequences of adjacent amino acids are responsible for the different components of the total biological action (sychnological organization into 'continuate words'). (2) Different target cells may be stimulated by different portions of the sequence (pleiotropic action). (3) Partial sequences, obtained by synthesis, can produce effects similar to those they elicit when contained in the complete molecule.

These facts can be interpreted in terms of hormone–receptor interactions by assuming that different parts of the discriminator interact with different portions of the ACTH sequence to produce different effects, and that—as a consequence of molecular flexibility—the hormone can adapt itself to discriminators of various types of target cell that may have different one- and three-dimensional requirements for ACTH binding.

The salient features of our knowledge about the molecular anatomy of

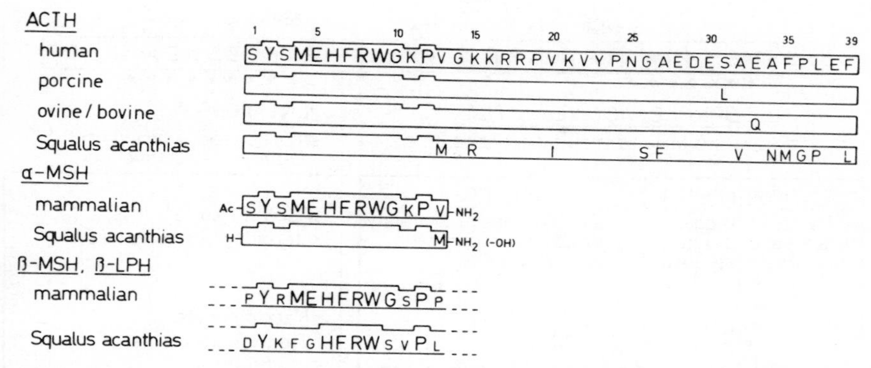


Figure 11.1. The amino acid sequence of corticotropin: species differences and chemical relationships with the melanotropins and lipotropins. For the one-letter notation for amino acid residues, see the List of Abbreviations. In the ACTH and $\alpha$-MSH series, only the amino acid replacements are shown; the molecules are otherwise identical to the first sequence. For $\beta$-LPH and $\beta$-MSH only the invariant amino acids that are homologous with the ACTH and $\alpha$-MSH sequences are shown (large capitals), as well as those in their immediate vicinity. The broken lines indicate adjacent $\beta$-MSH- and $\beta$-LPH-specific sequences. The $\beta$-LPH molecule contains (towards the C-terminus) the sequences of the endorphins and [Met$^5$]enkephalin (see Chapter 17)

ACTH in one-dimensional space are summarized in Figure 11.2. All of the known actions of ACTH are exerted by the N-terminal portion, $ACTH_{1-24}$-tetracosapeptide. Attachment of the C-terminal pentadecapeptide ($ACTH_{25-39}$) enhances the antigenicity and the *in vivo* duration of action (probably by a biochemical protection mechanism in the bloodstream). $ACTH_{1-24}$-tetracosapeptide is more potent than the native hormone *in vitro*, assayed using isolated adrenal cells.

The active site can be further subdivided into a potentiator sequence, a message sequence, and an address sequence for adrenal cortex steroidogenic receptors. The 'message 1' stimulates the steroidogenic response and the production of cyclic AMP. Because of this latter action, the receptors responsible for it are called '$\beta$-type' in analogy to $\beta$-adrenergic receptors. The same subdivision holds with respect to rat adipocytes, the address sequence contains an additional 'message' ('message 3') that activates an adipocyte $\alpha$-type receptor for $Mg^{2+}$ transport that can be blocked by the $\alpha$-blocker, phentolamine.[8]

The situation is still more complex for ACTH action on the embryologically related melanophores and central nervous system (CNS) cells. The 'address' portion, 14–24, reduces the melanophore-specific activity of $ACTH_{1-13}$-tridecapeptide ('negative' address) especially through

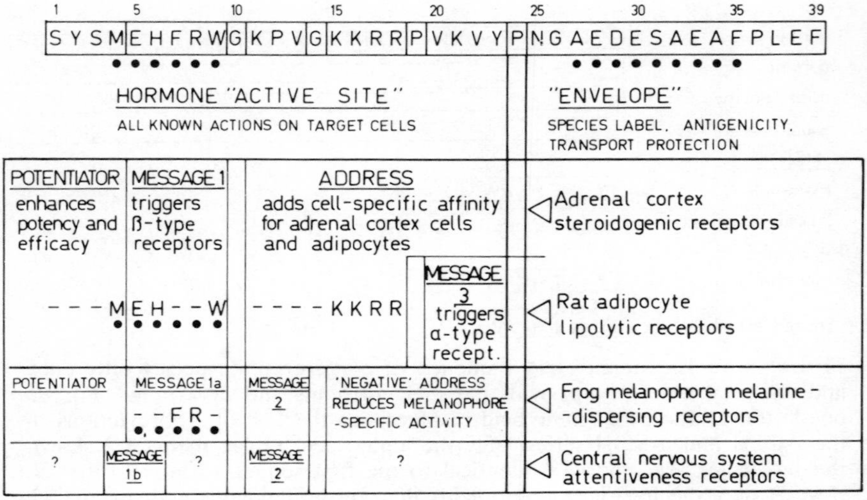

1 5 10 15 20 25 30 35 39

S Y S M E H F R W G K P V G K K R R P V K V Y P N G A E D E S A E A F P L E F

HORMONE "ACTIVE SITE"
ALL KNOWN ACTIONS ON TARGET CELLS

"ENVELOPE"
SPECIES LABEL, ANTIGENICITY, TRANSPORT PROTECTION

POTENTIATOR enhances potency and efficacy
MESSAGE 1 triggers β-type receptors
ADDRESS adds cell-specific affinity for adrenal cortex cells and adipocytes
Adrenal cortex steroidogenic receptors

- - - M E H - - W
- - - - K K R R
MESSAGE 3 triggers α-type recept.
Rat adipocyte lipolytic receptors

POTENTIATOR
MESSAGE 1a - - F R -
MESSAGE 2
'NEGATIVE' ADDRESS REDUCES MELANOPHORE-SPECIFIC ACTIVITY
Frog melanophore melanine-dispersing receptors

?
MESSAGE 1b
?
MESSAGE 2
?
Central nervous system attentiveness receptors

Figure 11.2. The molecular anatomy of ACTH. For the one-letter notation for amino acids see the List of Abbreviations. The amino acid sequence is divided into segments conveying specific information for various target cells. Solid circles (•) indicate amino acid residues with a potential for nucleating or assuming $\alpha$-helical conformation. Large capitals in the boxes indicate the residues that are most essential for producing the effect. The segments are loosely called potentiator, message, address, and envelope according to their function and with some analogy to human communication. $\beta$-Type receptors are those that—in analogy to $\beta$-adrenergic receptors—respond with enhanced cyclic AMP production (for a qualification, see Section V). The $\alpha$-type receptor of adipocytes enhances $Mg^{2+}$ transport and can be blocked by $\alpha$-adrenergic inhibitors. The analogy is strengthened by the fact that $\beta$-adrenergic agonists can also cause lipolysis and, in certain species, melanin dispersion (see also Chapter 12). Potency is the inverse median effective dose and efficacy is synonymous with intrinsic activity (ratio of the maximal effect obtained with a modified hormone divided by the maximal effect of the native hormone)

the basic sequence 15–18. However, its segment 11–13 acts as a 'message 2' for melanine dispersion—acting independently of the 'message 1'. The potentiation of the effect of 'message 1' is also caused by the N-terminal tetrapeptide. However, the melanotropic action of 'message 1' is mediated not mainly through the residues Glu-His-Trp as for steroidogenesis and lipolysis, but mainly through -Phe-Arg-. If one assumes an $\alpha$-helical conformation for 'message 1' in contact with the receptors, then the adrenal and fat cell receptors would interact with one side of the helix, the melanophore receptor with the opposite side (Schwyzer, 1977, in Reference 5). A further displacement of the essential amino acids of 'message 1' has taken place for the CNS receptors that also respond to 'message 2'. These details are presented more fully in Chapter 12.

It is chemically interesting and perhaps of functional importance, that many of these segments are divided from one another by the amino acids glycine and proline. These compare with the other amino acids by causing, respectively, an increased or decreased rotational freedom of the peptide chain about their $\alpha$-atoms: they appear to act as spacers between the 'continuate words' conveying different biological information. The potentiator–'message 1' junction is a noteworthy exception: it can be shown that the potentiator acts by producing additional binding force for the 'message 1' through direct interaction with the discriminator and/or by stabilizing an $\alpha$-helical conformation of the 'message 1' in contact with the discriminator by virtue of its additional $\alpha$-nucleating potential.[9]

## III. ACTH–RECEPTOR: LOCALIZATION AND BINDING STUDIES

The ACTH molecule has been chemically coupled to large inert polymers such as cellulose, agarose, and polyacrylamide. These complexes, comparable in size to the adrenal cell itself, could nevertheless stimulate adrenal steroidogenesis and it was established that this bioactivity was not due to cleavage of active peptides from the complex.[10] ACTH therefore acts without entering the cell and such studies support the idea that specific ACTH receptors are located on the outer cell membrane surface.

For binding studies, high specific activity ligand is necessary in order to detect receptors of high affinity. Tritiated and $^{125}$I-labelled ACTH derivatives have been used for these investigations. However, there are problems in the use of both these types of labelled ligands. Tritiated hormones can only be prepared with relatively low specific activities—insufficient to detect low numbers of high affinity receptors. Iodinated peptides are analogues, that may be expected to have properties different from those of the native hormone. The iodine atom is itself relatively large, and likely to introduce change in a small peptide that may affect its conformational properties during interaction with the receptor. Similarly, during the iodination procedure susceptible amino acids may be oxidized or otherwise chemically modified, and the bioactivity thereby affected. For rigorous assessment of the biological activity of an iodinated hormone it is essential to completely separate the products of the iodination reaction: labelled and non-labelled hormone and damaged fragments. For the larger glycoprotein and protein hormones (TSH, hCG, PTH etc.) iodination seems to have little effect on bioactivity and this approach has provided much important information on receptors for these hormones. ACTH, however, is much more susceptible to inactivation following iodination, and this has proved to be a major stumbling block to further advances in this field.

In early work using $^{125}$I-labelled ACTH, subcellular fractions from mouse adrenal tumour cells were employed.[11] Although binding and adenylate cyclase activity were studied using very different conditions, it was nevertheless found that, for several enzymically- or chemically-modified ACTH derivatives their widely differing potencies, as adenylate cyclase activators, paralleled their capabilities to inhibit binding of $^{125}$I-labelled ACTH. In these early studies it was reported that following iodination using chloramine-T, the iodinated product (<4%) could be completely freed by chromatography from residual ACTH, and that the labelled hormone (iodinated at tyrosine$^{2}$) had at least 50% of the bioactivity of the native hormone. However, it has become clear that preparation of iodinated ACTH which retains the biological properties of the hormone, is more difficult than was originally thought. Di-iodo substitution at Tyr$^{2}$ caused a reduction in potency of 97%, and at Tyr$^{23}$ a reduction of 43%, when compared with $ACTH_{1-24}$ itself.[12]

A method has been described[13] for the preparation of a biologically active monoiodo-derivative of ACTH. Following iodination using chloramine-T and purification by isoelectric focusing and CM-cellulose chromatography, biological activity was restored to the monoiodo-derivatives by treatment with cysteine (known to reduce oxidized methionine residues). However, no binding studies have been reported using this material and it is therefore difficult to evaluate.

Lactoperoxidase and $H_2O_2$ have been used for the preparation of iodinated ACTH of specific activity 1,300–2,700 Ci mmol$^{-1}$. This procedure is potentially less likely to cause oxidation of methionine and other residues, than the chloramine-T method, however, the difficulty in evaluating the biological activity of such a product lies in the problems of purification to a homogeneous monoiodinated derivative, free from non-labelled ACTH. Nevertheless, the binding of this iodinated ACTH to intact isolated adrenal cells has been examined.[14] Binding sites of both high affinity ($K_D' = 2.5 \times 10^{-10}$ M) and low affinity ($K_D' = 1 \times 10^{-8}$ M) were observed (Figure 11.3) together with a correlation between occupancy of the high affinity binding sites and stimulation of steroidogenesis.

A tritiated derivative Phe$^{2}$, Nor-valyl[$^{3}H_2$]$^{4}$$ACTH_{1-24}$ has been synthesized with specific activity 7 Ci mmol$^{-1}$. In the fat cell assay for lipolytic activity this radioactive analogue was ten times less potent than $ACTH_{1-24}$ although the maximal velocities were the same for both compounds. Binding studies on isolated fat cells[15] revealed the presence of at least two types of binding sites: high affinity sites ($K_D' = 9 \times 10^{-9}$ M) and low affinity sites ($K_D' = 1 \times 10^{-6}$ M).

A tritiated derivative of $ACTH_{1-24}$ has also been synthesized with biological properties identical to those of non-labelled ACTH.[16] However, although $^{3}$H-labelled hormones have proved invaluable for some receptor binding studies (notably vasopressin, see Chapter 14), for the adrenal with activation evident at $10^{-10}$ M ACTH, it is clear that $^{3}$H-labelled derivatives (specific activity 40 Ci mmol$^{-1}$) are insufficiently radioactive for the facile

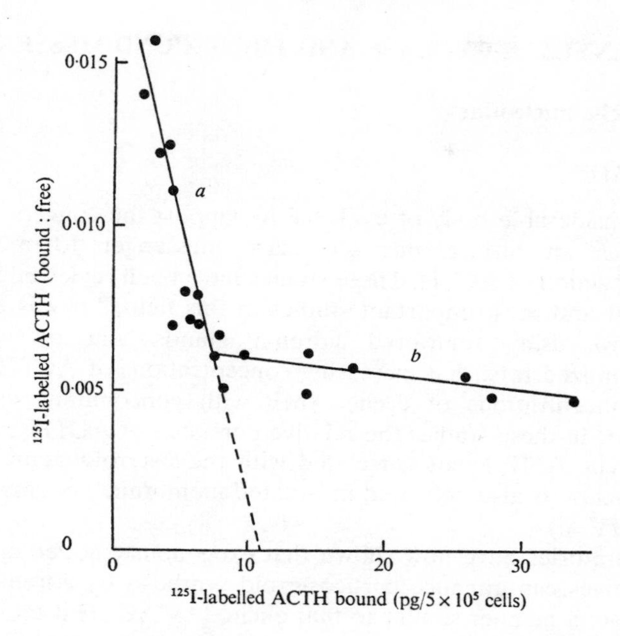


Figure 11.3. Scatchard plot of the binding data obtained with isolated rat adrenal cells. Cells were incubated for 20 min with $^{125}$I-labelled ACTH (150 $\mu$Ci $\mu g^{-1}$), then centrifuged at 2,000 rev $min^{-1}$ for 1 min, and radioactivity in the cell pellet determined. Non-specific binding was determined in the presence of 25 $\mu$g of ACTH. Combined data from five experiments is shown; each determination was in duplicate. Regression analysis by the method of least squares gave values for $K_D^a = 2.5 \times 10^{-10}$ M with 3,000 sites/cell and for $K_D^b = 1 \times 10^{-8}$ M with 30,000 sites/cell. Reproduced with permission from Reference 14

detection of such high affinity receptors. $^{125}$I-labelled derivatives are obviously of greater potential use for this, having specific activities in the region of 2,000 Ci $mmol^{-1}$.

Hydrophilic and lipophilic spin labels (electron spin resonance probes) have been used to examine the effects of $ACTH_{1-24}$ receptor-binding on protein conformation and lipid fluidity in the membrane.[17] Increased lateral mobility of the hormone–receptor complex was implicated as a consequence of hormone binding. It was suggested that crystalline patches of phospholipid, formed by aggregation of several hormone–receptor complexes on the outer plasma membrane surface, may act as the transducer to activate the adenylate cyclase on the inner membrane surface.

## IV. ADENYLATE CYCLASE AND THE SECOND MESSENGER

### (i) Role of cyclic nucleotides

#### *(a) Cyclic AMP*

There is a considerable body of evidence to support the concept that cyclic AMP acts as an intracellular secondary messenger to mediate the steroidogenic action of ACTH. These studies have been reviewed.[10,18] In one of the earliest and most important studies in this field,[19] it was established both *in vitro* using quartered adrenal glands and *in vivo* using hypophysectomized rats, that increasing concentrations of ACTH produced increasing concentrations of cyclic AMP with concomitant increases in steroid output. In these studies the relative potencies of ACTH analogues in increasing cyclic AMP levels correlated with their steroidogenic potencies. Adenylate cyclase is also activated in isolated membrane systems by ACTH (see Section IV. ii).

Many laboratories have now shown that exogenously added cyclic AMP, or its derivatives can enhance corticosteroid synthesis by adrenal tissue or isolated cells in a manner similar to that elicited by ACTH itself. Moreover, the locus of the cyclic AMP steroidogenic effect lies between cholesterol and pregnenolone, at the same point on the pathway observed for the effect of ACTH itself. It is, however, necessary to add remarkably high concentrations of exogenous cyclic AMP in order to stimulate steroidogenesis (e.g. $ED_{50}$ values of 3 mM cyclic AMP are reported for isolated adrenal cells). Thus the amount of exogenously added cyclic AMP required to produce half maximal steroidogenesis is about 100 times that elicited from the cell by a half maximal dose of ACTH. However, cyclic AMP is rapidly destroyed by phosphodiesterases within the cell, and metabolism studies have shown that it is likely that only 1% of added cyclic AMP survives to find its intracellular site of action.

Isolated adrenal cells have been examined in the superfusion system for continuous incubation, in which steroidogenic responses are apparent within 24 s of adding ACTH.[10] Cyclic AMP outputs were compared with steroidogenesis following single injections of ACTH and a striking similarity in the characteristics of the two responses was evident. For different amounts of ACTH: (1) the time to reach maximum output rate was the same for both steroidogenesis and cyclic AMP; (2) the initial stimulatory rates as well as the subsequent decay rates were very similar for the two responses; (3) the ratio of the peak heights (maximum output rates) for steroidogenesis was the same as that for cyclic AMP; and (4) the ratio of the areas under the curves (total response) for steroidogenesis equalled that for cyclic AMP. The characteristics of the dynamic responses thus provide strong support for a direct link between cyclic AMP output and steroidogenesis in ACTH stimulated adrenal cells.

Table 11.2. Dose–responses of isolated adrenal cells to ACTH

| $ACTH_{1-39}$ (M) | Cyclic AMP (ng adrenal$^{-1}$ h$^{-1}$) | Corticosterone ($\mu$g adrenal$^{-1}$ h$^{-1}$) |
|---|---|---|
| 0 | 0.28 ± 0.06 | 0.11 ± 0.05 |
| $4 \times 10^{-11}$ | 0.32 ± 0.08 N.S. | 0.13 ± 0.04 N.S. |
| $1 \times 10^{-10}$ | 0.70 ± 0.13† | 0.39 ± 0.06‡ |
| $2 \times 10^{-10}$ | 1.31 ± 0.19* | 0.65 ± 0.01* |
| $1 \times 10^{-9}$ | 3.46 ± 0.33* | 1.10 ± 0.05* |
| $2 \times 10^{-9}$ | 10.07 ± 1.39* | 1.17 ± 0.06* |
| $2 \times 10^{-8}$ | 43.06 ± 5.33* | 1.25 ± 0.07* |
| $2 \times 10^{-7}$ | 47.58 ± 7.75* | 1.12 ± 0.06* |

Effect of porcine $ACTH_{1-39}$ on cyclic AMP and corticosterone output by isolated adrenal fasciculata-reticularis cells: combined data (means ± S.E.M.) from three experiments, each comprising duplicate incubations. Statistical significances were determined by the paired Student's t-test: N.S. not significant; †$P < 0.05$ (probability significant); ‡$P < 0.01$ (significant); and *$P < 0.001$ (highly significant) compared with values in the absence of added ACTH. Data from Reference 20.

Using isolated adrenal cells and conventional static incubation procedures, the dose–response characteristics and time-course relationships have been established.[20] Increased cyclic AMP output was evident within 1 min of ACTH addition and before any discernible increase in steroidogenesis. Moreover, only those concentrations of ACTH that stimulated steroidogenesis also increased cyclic AMP in this system (Table 11.2). In depicting these data (and in many other similar adenylate cyclase-linked hormone systems) the issue has been confused by presenting the whole dose–response curve on one graph. At high, well above physiological, concentrations of ACTH there is a vast over-production of cyclic AMP (Table 11.2) and in graphs containing all these data together,[20] small increases in cyclic AMP observed at low, 'physiological' concentrations of ACTH ($3 \times 10^{-10}$ M), are not evident despite low basal outputs. It may then *appear* that there are concentrations of ACTH capable of stimulating steroidogenesis but incapable of increasing cyclic AMP output.

These data (Table 11.2) also, provide clear evidence for the presence of 'spare receptors' in the adrenal cells. Thus at an ACTH concentration (about $10^{-9}$ M) which was maximal for steroidogenesis, cyclic AMP outputs were only 9% of maximum. This correlates well with the *in vivo* data,[19] and it is clear there is an enormous reserve capacity for adrenal cell cyclic AMP production; only a small proportion of the cell's potential for cyclic AMP production need be activated in order to switch on steroidogenesis maximally.

In this context, it is of interest that examination of the binding of $^{125}$I-labelled ACTH to intact isolated adrenal cells, has also demonstrated the presence of 'spare receptors' for ACTH (see later). It was calculated that steroidogenesis was maximal when about only 12% of the total adrenal cell binding sites were filled.[14] It seems that the large amounts of cyclic AMP produced at high concentrations of ACTH ($> 10^{-9}$ M) are attributable to the filling of these 'spare receptor' sites not directly involved in steroidogenesis. Similar observations regarding the existence of 'spare receptors' have been made for other hormonal systems e.g. the catecholamine response in turkey erythrocytes and rat parotid gland (see Chapter 15, Section IX). Only a small fraction (1%) of the $\beta$-adrenergic receptors need be occupied by hormone in order to elicit the maximal biochemical response. As previously discussed (Chapter 2, Section VI. ii) the concept of 'spare or reserve receptors' may be misleading, since it is likely that all the receptors participate in the response. Mathematical considerations suggest that 'spare receptors' provide the necessary total number of receptors, such that the maximal response can be attained at very low hormone occupancy.

The effects of cholera toxin on isolated adrenal cells are also consistent with an intermediary role for cyclic AMP in the stimulation of steroidogenesis.[21] The characteristics of the dose-response curve for cholera toxin stimulated steroidogenesis were strikingly similar to those for cyclic AMP output. The maximal and half-maximal effective doses of the toxin were almost identical for both responses. Although steroid outputs are similar to those in response to ACTH, cyclic AMP values are much lower and the excessive production of cyclic AMP at high hormonal concentration (attributed to 'spare' ACTH receptors) is not evident. The mechanism of action of cholera toxin is believed to operate via the GTPase associated with adenylate cyclase (see Chapter 4), and thus bypass the cell membrane ACTH–receptor system. Thus the close relationship between cyclic AMP and steroidogenesis elicited by cholera toxin, demonstrates that small changes in cyclic AMP within the cell do correlate with stimulation of steroidogenesis, and imply that similarly small cyclic AMP changes are involved in the ACTH response. However, in one aspect these studies showed an important discrepancy between the two responses: no increase in steroidogenesis was discernible until 60 min after the addition of cholera toxin, whereas cyclic AMP production was switched on 20 min earlier—thus supporting the work of others questioning the obligatory involvement of cyclic AMP (see Section V).

Despite the multiplicity of further evidence favouring an intermediary role for cyclic AMP (including the effects of various agonists and antagonists, see Section II), the stimulatory effects of methylxanthines and the ability of ACTH to activate protein kinase in whole cells,[10] nevertheless there is now considerable evidence incompatible with an obligatory role for cyclic AMP. These data and the alternatives to the simple 'second messenger concept' are discussed in Section V.

*(b) Cyclic GMP*

Exogenous cyclic GMP stimulates steroidogenesis in rat adrenal slices and isolated adrenal cells.[22] Small changes in cyclic GMP in isolated adrenal cells have been observed in response to low concentrations of ACTH.[23] Cyclic GMP measurements have been hampered by the expense and relative insensitivity of this assay. However, using a sensitive method based on the highly specific binding of GDP to *Escherichia coli* polypeptide chain elongation factor Tu, a correlation between ACTH-induced steroidogenesis and cyclic GMP (but not cyclic AMP) has recently been reported.[24] Maximal stimulation (two-fold) of cyclic GMP levels was observed at 5 $\mu$I.U. ml$^{-1}$ ($5 \times 10^{-12}$ M) ACTH, with concomitant increases in protein kinase activity. These workers have proposed cyclic GMP as the second messenger, acting via activation of cyclic GMP-dependent protein kinase.

**(ii) Regulation of adenylate cyclase**

*(a) Guanine nucleotides*

Much of our understanding of the mode of action of guanine nucleotides in adenylate cyclase systems has been gained from investigations on the glucagon sensitive hepatic system and the catecholamine-responsive enzyme of erythrocytes and lymphocytes. The mechanism whereby GTP is thought to act synergistically with hormones, firstly to initiate and subsequently to terminate activation of adenylate cyclase, has been previously described (Chapter 4).

Much less mechanistic information has derived from investigations on the action of guanine nucleotides on adrenocortical adenylate cyclase than from those using the catecholamine- and glucagon-responsive systems. However, the latter systems have served as good models, and the adrenal system has been shown to have several features in common with them. Studies on the activation by naturally occurring nucleotide triphosphates have shown that GTP is the most potent for the rat and the bovine adrenal enzyme;[25,27] ITP was considerably less effective and the pyrimidine nucleotides, UTP and CTP, were very weak activators.

The adrenal enzyme is activated by a guanine 5′-($\beta$, $\gamma$-imido)-triphosphate (p(NH)ppG). This activation is slow and is not reversed by thorough washing or detergent solubilization of p(NH)ppG—pretreated membranes, although it may be competitively inhibited, but not reversed, by GTP.[25,26] It is therefore likely that, as in other systems, adrenal membranes contain a specific GTPase associated with the adenylate cyclase. Maximal activation of the adenylate cyclase by p(NH)ppG required 20–30 min at 30 °C, after which time the enzyme remained fully active; the time required to reach this state was not affected by varying p(NH)ppG concentrations in

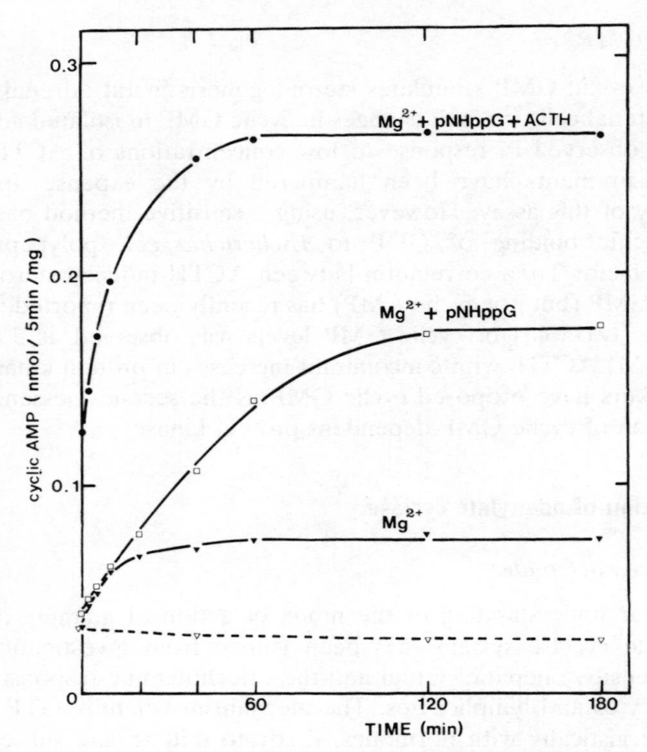


Figure 11.4. Kinetics of heterotropic activation of bovine adrenal adenylate cyclase at 20 °C.

Adrenal membranes were incubated in the presence of the effectors shown. At the times indicated the adenylate cyclase activity of aliquots was determined. Concentrations of the effectors were: $ACTH_{1-24}$, 1 $\mu$m; p(NH)ppG, 0.1 mM. $MgCl_2$ was 5 mM in both the cyclase assay and the preincubation. Note that while the final activity attained in the presence of both $ACTH_{1-24}$ and p(NH)ppG is only 55% greater than that in the presence of p(NH)ppG alone, the initial rate of activation (0–10 min) is at least 500% greater in the presence of hormone. Reproduced with permission from Reference 28

the range 0.1–1.7 mM.[26] That the rate-determining step in this activation is independent of the free p(NH)ppG concentration, implies that it involves a conformational change in the cyclase system occurring after binding of the GTP analogue.

ACTH increases the rate of activation rather than the maximal activity of adenylate cyclase in adrenal membranes incubated with p(NH)ppG

(Figure 11.4). This may be due to an ACTH-induced increase in the rate of the otherwise slow conformational change in the cyclase complex which is influenced by guanine nucleotide binding. For glucagon and the catecholamines a similar action has been proposed in the hepatic and erythrocyte systems respectively.

*(b) Divalent cations*

The magnesium ion is an important regulator of adenylate cyclase. The substrate for the catalytic subunit is $MgATP^{2-}$, and the $K_m$ for adenylate cyclase in adrenal membrane preparations is similar to that observed for the enzyme derived from many other cells, in the range 0.15–0.25 mM $MgATP^{2-}$.[25,27] However, earlier studies have shown that adenylate cyclase activity was increased by increasing $Mg^{2+}$ concentrations, even though the ATP was largely in the complexed form of $MgATP^{2-}$. $Mg^{2+}$ may either act as an allosteric activator of the enzyme or act indirectly by lowering the concentration of $HATP^{3-}$ and other non-complexed forms of ATP which might inhibit adenylate cyclase.

Examination of the $Mg^{2+}$-dependence of rat adrenal adenylate cyclase activity in the presence of ACTH and GTP[27] led to the suggestion that ACTH and GTP act to produce a state of the enzyme which is less susceptible to inhibition by $HATP^{3-}$ (but which is inhibited by high $Mg^{2+}$ concentrations). More recent studies on hepatic and bovine adrenal adenylate cyclase favour the idea that $Mg^{2+}$ itself is an activator of the enzyme. A time-dependent increase in adenylate cyclase activity follows incubation of bovine adrenal membranes with $Mg^{2+}$ (Figure 11.4), showing that in the absence of exogenously added GTP, the adrenal enzyme is activated by $Mg^{2+}$. Recent data obtained using the adrenal system are in accordance with the earlier concept that in the fat cell, ACTH raised the affinity of the cation for a $Mg^{2+}$-binding site on the enzyme.[27,28]

Thus ACTH and GTP may act in concert to activate adenylate cyclase by increasing the affinity of $Mg^{2+}$ for a specific $Mg^{2+}$-binding site.

Calcium ions also affect the activity of adenylate cyclase, and as has been found for the enzyme from most systems, $Ca^{2+}$ at millimolar concentrations inhibits the adrenocortical enzyme. However, this differs from the results obtained using isolated adrenal cells in which it was found that increased medium $Ca^{2+}$ concentrations (up to 7 mM) increased ACTH-stimulated cyclic AMP output.[29] To account for this discrepancy it was suggested that in the intact cell, adenylate cyclase resides in a compartment of low $Ca^{2+}$ concentration and is unaffected by extracellular $Ca^{2+}$ fluctuations.

Recently it has been possible to show that in the absence of ACTH, steroidogenesis in isolated rat adrenal cells could be stimulated by calcium salts when these were presented as metastable solutions favouring colloidal calcium formation.[30] The stimulating effect of external $Ca^{2+}$ on the intact cell may relate to the coupling of the ACTH receptor–adenylate cyclase system. Thus EGTA, which chelates $Ca^{2+}$ (and other cations such as $Mn^{2+}$, $Co^{2+}$, and $Sr^{2+}$) blocks adenylate cyclase activation by ACTH in a particulate preparation from adrenal tumour cells whereas EGTA did not affect the basal activity and $^{125}$I-labelled ACTH binding to a solubilized preparation of these particles.[11] EGTA also inhibited ACTH-stimulation of both bovine adrenal adenylate cyclase[25] and the ACTH effect on the multi-receptor fat cell adenylate cyclase system.[31]

In both the adrenal cortex and the fat cell, $Ca^{2+}$ is thus implicated in transmission of the signal between ACTH-receptor and the adenylate cyclase enzymic subunit.

The role of calmodulin, the $Ca^{2+}$-dependent modulator protein found in most cells so far examined (including adrenal and fat cells), has been discussed in Chapter 4. The established activation of phosphodiesterase and (in brain tissue) of adenylate cyclase by this $Ca^{2+}$-dependent protein, and its possible association with the plasma membrane imply a central role for this factor. It now seems likely that the role of $Ca^{2+}$ in transmission of the hormone receptor signal is intimately associated with the functioning of this modulator protein, and this represents an important area for future research.

### (c) *Adenosine*

By the use of adenosine analogues, two distinct binding sites have been identified, via which adenosine alters adenylate cyclase activity in plasma membranes.[32] It is proposed that these two sites may be characterized functionally as a 'P-site' which mediates inhibition and requires an intact purine ring for activity and an 'R-site' which mediates activation of adenylate cyclase and shows a requirement for an intact ribose moiety for activity. Methylxanthines such as theophylline generally antagonize the adenosine activation at the R-site, but have no effect at the P-site.

The effect of adenosine on adenylate cyclase in Y-1 adrenal tumour cells appears to be mediated via the R-sites: chloroadenosine was as potent as adenosine in stimulating adenylate cyclase and the activation was antagonized by theophylline. In these studies the stimulatory effect of adenosine on steroidogenesis by Y-1 adrenal tumour cells was also confirmed. That the Y-1 adrenal membrane may contain both P- and R-sites was indicated by the biphasic response to adenosine, with activation at low concentrations and inhibition at higher concentrations.[32]

Recent studies on fat cell membranes[33] have demonstrated a GTP-dependence for adenosine inhibition of adenylate cyclase. It was suggested that adenosine is a ligand that, acting at one cell membrane receptor-(R-site), can inhibit through a GTP-mediated process the action of another ligand (ACTH or catecholamine) that also binds to a cell membrane receptor and activates adenylate cyclase through a second GTP-mediated process.

### *(d) ACTH*

Adenylate cyclase activity in membrane preparations has generally been found to be less sensitive to ACTH than in intact cells derived from adrenals of the same species. This loss of hormonal sensitivity on disruption of the intact cell system, has led investigators to question the significance of findings on the particulate enzyme and their relevance to events in the intact cell.

It has been found[34] that in isolated bovine adrenal cells (5-Gln, 9-Phe)$ACTH_{1-24}$ was a full agonist for cyclic AMP production but 50-fold less potent than $ACTH_{1-24}$. However, in bovine adrenal membranes (5-Gln, 9-Phe) $ACTH_{1-24}$ had no ability to activate adenylate cyclase and, on a molar ratio basis, was a more effective antagonist of $ACTH_{1-24}$ in this system than in isolated cells. The physiological relevance of studies on adrenal membrane adenylate cyclase was therefore questioned.[34] It is now clear, however, that these results relate to the problem of the general insensitivity of particulate adenylate cyclase systems. Thus in these studies half maximal responses were found at 4 nM and 100 nM $ACTH_{1-24}$ for the intact cell and particulate systems respectively.

A variety of factors can affect the hormonal sensitivity of adrenal adenylate cyclase. Since it is labile, the time involved in preparing membrane fractions is important. So too may be the extent to which the preparation is contaminated with proteolytic enzymes and the cruder the adrenal membrane preparation the more likely is hormone degradation to contribute to poor hormonal responses. It has been shown, using isolated adrenal cells, that gross ACTH degradation was related to the cellular damage incurred during the cell preparation, and that increased hormone degradation correlated with increases in the hormone concentration ($ED_{50}$) giving half maximal stimulation of steroidogeneisis. This supports the contention that ACTH degradation is primarily a non-specific process independent from the receptor binding event.[28]

Taking due regard for the above difficulties, it has now been possible to obtain adrenal membranes showing half-maximal adenylate cyclase activation by 0.3–1.0 nM ACTH[28] values very close to the $ED_{50}$ value (0.3 nM) for half-maximal stimulation of cyclic AMP production in the intact cell.[35] Moreover, in these membrane preparations, activation of adenylate cyclase was observed at physiologically relevant ACTH concentrations ($<0.1$ nm).

## V. COMPARTMENT GUIDANCE CONCEPT

The simple sequence of events: Hormone → cyclic AMP → cell-specific responses that is most probably true for $\beta$-adrenergic mechanisms, might be more complicated in the case of ACTH action. A number of observations have led to this view; the most striking ones are as follows.

1. At ACTH concentrations that produce maximal adenylate cyclase stimulation in adipocyte membrane vesicles and maximal rates of lipolysis in adipocytes, the addition of phenoxazones like actinocin and actinomycin (the antibiotic property is unnecessary) immediately doubles both rates and also the number of functional discriminators.[36] If indeed—as usually postulated—only a fraction of the cyclic AMP produced by hormonal stimulation is necessary to produce maximal secondary effects (e.g. lipolysis) then it is hard to see why the production of a still greater excess of cyclic AMP should further increase the secondary effect.
2. Certain derivatives of ACTH and its fragments produce steroidogenesis and lipolysis without causing noticeable cyclic AMP accumulation over background levels.[37]
3. ACTH fragments may also stimulate membrane adenylate cyclase without causing lipolysis[38] or steroidogenesis[39] in whole cells.
4. An increase of tissue cyclic AMP in perfused cat adrenal glands was found to be insufficient to elicit steroid release.
5. Calcium ion was found to serve as a direct messenger in the physiological activation of steroidogenesis in isolated rat adrenal cortex cells, whereas cyclic AMP was claimed to be a subserving factor maintaining full steroidogenesis.
6. In isolated rat adipocytes, hyper-osmolar glucose enhances the effect of ACTH on cyclic AMP accumulation but reduces its effect on lipolysis.

(For literature references to points 4–6 see Reference 40).

The main explanations and their logical comparison with the second messenger concept are illustrated in Figure 11.5. The *second messenger* concept assumes cyclic AMP (accumulation or production) to be a necessary and sufficient condition for causing the steroidogenic and lipolytic effects. The *dual receptor* concept assumes cyclic AMP to be a sufficient but unnecessary condition for causing the effects; they could be caused via other pathways or second messengers (this could explain observations, 2, 5, and 6 above). The *compartment guidance* concept assumes cyclic AMP (newly produced or background amounts) to be a necessary but insufficient condition to evoke the effects. Other actions of ACTH (on the same or other discriminators) are necessary to guide cyclic AMP into the correct compartment for eliciting steroidogenesis and lipolysis. This other pathway cannot produce the effect independently of cyclic AMP; it has the quality of

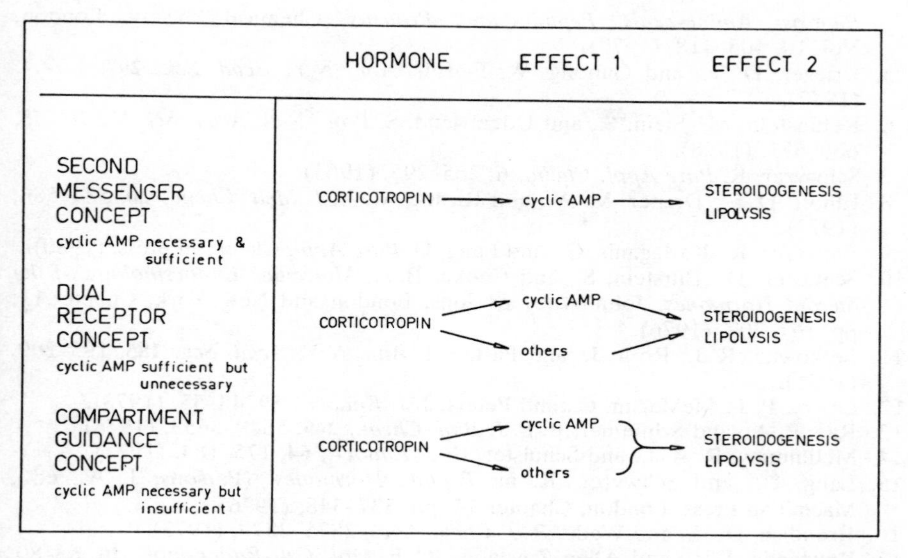


Figure 11.5. Three concepts for stimulus–effect coupling in the action of ACTH on steroidogenesis and lipolysis

inhibiting the action of ACTH on the adenylate cyclase pathway, and it works only in the presence of a sufficient external supply of $Ca^{2+}$.[40] It thus has certain characteristics of the $\alpha$-adrenergic receptor activated by the sequence 20–23 (see Section II), although this sequence appears to be unnecessary; peptides containing only the 'Message 1' (Figure 11.2) can also elicit steroidogenesis and lipolysis (Schwyzer, 1977, in Reference 5). According to the compartment guidance hypothesis, adenylate cyclase activation by ACTH would be a kind of emergency effect, becoming necessary at low background concentrations of cyclic AMP. Whether or not the concept will hold, cannot yet be decided. It does, however, suggest certain experiments that could clarify the unexplained results reported above.

## VI. REFERENCES


1. Yalow, R. S., and Berson, S. A. *Biochem. Biophys. Res. Commun.*, **44**, 439–445, (1971).
2. Coslovsky, R., and Yalow, R. S. *Biochem. Biophys. Res. Commun.*, **60**, 1351–1356, (1974).
3. Schwyzer, R. in *Specialist Periodical Reports: Amino-acids, Peptides, and Proteins*, Chemical Society, London, Vol. 9, pp. 445–462, (1978).
4. Schwyzer, R., Fauchère, J.-L., Eberle, A., Kriwaczek, V. M., Do, K., Petermann, C., Fischli, W., Oppliger, M., and Märki, W. in *Specialist Periodical*

*Reports: Amino-acids, Peptides and Proteins*, Chemical Society, London, Vol. 10, 405–418, (1979).
5. Krieger, D. T., and Ganong, W. F. Eds. *Ann. N.Y. Acad. Sci.*, **297**, 1–664. (1977).
6. Rubinstein, M., Stein, S., and Udenfriend, S. *Proc. Natl. Acad. Sci. U.S.A.*, **75**, 669–671, (1978).
7. Schwyzer, R. *Pure Appl. Chem.*, **6**, 265–295, (1963).
8. Elliott, D. E., Draper, M. W., and Rizack, M. A. *J. Med. Chem.*, **20**, 584–586, (1977).
9. Schwyzer, R., Karlaganis, G., and Lang, U. *Pure Appl. Chem.*, in press, (1980).
10. Schulster, D., Burstein, S., and Cooke, B. A. *Molecular Endocrinology of the Steroid Hormones*, John Wiley & Sons, London and New York, Chapter 11, pp. 167–207, (1976).
11. Lefkowitz, R. J., Roth, J., and Pastan, I. *Ann. N.Y. Acad. Sci.*, **185**, 195–209, (1971).
12. Lowry, P. J., McMartin, C., and Peters, J. *J. Endocr.*, **59**, 43–55, (1973).
13. Rae, P. A., and Schimmer, B. P. *J. Biol. Chem.*, **249**, 5649–5653, (1974).
14. McIlhinney, R. A. J., and Schulster, D. *J. Endocr.*, **64**, 175–184, (1975).
15. Lang, U., and Schwyzer, R. in *'Peptide Hormones'* (Parsons, J. A., ed.), Macmillan Press, London, Chapter 17, pp. 337–348, (1976).
16. Brundish, D. E., and Wade, R. *J. Chem. Soc.*, 2875–2879, (1973).
17. Rowlands, J. R., and Allen-Rowlands, C. F. *Mol. Cell Endocrinol.*, **10**, 63–80, (1978).
18. Halkerston, I. D. K. *Adv. Cyclic Nucl. Res.*, **6**, 99–136, (1975).
19. Grahame-Smith, D. G., Butcher, R. W., Ney, R. L., and Sutherland, E. W. *J. Biol. Chem.*, **242**, 5535–5541, (1967).
20. Mackie, C. M., Richardson, M. C., and Schulster, D. *FEBS Letters*, **23**, 345–348, (1972).
21. Palfreyman, J. W., and Schulster, D. *Biochem. Biophys. Acta*, **404**, 221–230, (1975).
22. Glinsmann, W. H., Hern, E. P., Linarelli, L. G., and Farese, R. V. *Endocrinology*, **85**, 711–719, (1969).
23. Sharma, R. K., Ahmed, N. K., and Shanker, G. *Eur. J. Biochem.*, **70**, 427–433, (1976).
24. Perchellet, J. P., Shanker, G., and Sharma, R. K. *Science*, 199, 311–312, (1978).
25. Glossmann, H., and Gips, H. *Naunyn-Schmiedeberg's Arch. Pharmacol.*, **289**, 77–97 (1975); **292**, 199–203, (1976).
26. Glynn, P., Cooper, D. M. F., and Schulster, D. *Biochem. Biophys. Acta*, **524**, 474–483, (1978).
27. Londos, C., and Rodbell, M. *J. Biol. Chem.*, **250**, 3459–3465, (1975).
28. Glynn, P., Cooper, D. M. F., and Schulster, D. *Mol. Cell Endocrinol.*, **13**, 99–121, (1979).
29. Sayers, G., Beall, R. J., and Seelig, S. *Science*, **175**, 1131–1133, (1972).
30. Neher, R., and Milani, A. *Mol. Cell Endocrinol.*, **9**, 243–253, (1978).
31. Yamamura, H., Lad, P. M., and Rodbell, M. *J. Biol. Chem.*, **252**, 7964–7966, (1977).
32. Londos, C., and Wolff, J. *Proc. Natl. Acad. Sci. U.S.A.*, **74**, 5482–5486, (1977).
33. Londos, C., Cooper, D. M. F., Schlegel, W., and Rodbell, M. *Proc. Natl. Acad. Sci. U.S.A.*, **75**, 5362–5366, (1978).
34. Finn, F. M., Johns, P. A., Nishi, N., and Hofmann, K. *J. Biol. Chem.*, **251**, 3576–3585, (1976).

35. Seelig, S., Lindley, B. D., and Sayers, G, *Methods Enzymol.,* **36**, 347–359, (1975).
36. Lang, U., Karlaganis, G., Vogel, R., and Schwyzer, R. *Biochemistry*, **13**, 2626–2633, (1974).
37. Sayers, G., Seelig, S., Kumar, S., Karlaganis, G., Schwyzer, R., and Fujino, M. *Proc. Soc. Exp. Biol. Med.,* **145**, 176–181, (1974).
38. Lang, U., Fauchère, J.-L., Pelican, G. M., Karlaganis, G., and Schwyzer, R. *FEBS Letters,* **66**, 246–249, (1976).
39. Bonnafous, J. C., Fauchère, J.-L., Schlegel, W., and Schwyzer, R. *FEBS Letters,* **78**, 247–250, (1977).
40. Schwyzer, R. *Bull. Schweiz. Akad. Med. Wiss.,* **34**, 263–274, (1978).

Cellular Receptors for Hormones and Neurotransmitters
Edited by D. Schulster and a. Levitzki


CHAPTER 12

# MSH receptors

Alex N. Eberle
*Institute of Molecular Biology and Biophysics, Swiss Federal Institute of Technology, CH-8093 Zürich-Hönggerberg, Switzerland*



## I. INTRODUCTION

For many vertebrates and numerous invertebrates, pigmentation and colour changes are a necessary means for protective camouflage in response to environmental threat. Such background adaptations of the skin were recognized early on, to be under the control of neural and/or hormonal mechanisms. The search for an endocrine factor regulating the movement of pigmentary organelles of poikilothermic chromatophores led to the discovery of the melanocyte-stimulating hormones (melanotropins, intermedins), a group of relatively short polypeptides secreted by the intermediate lobe of the pituitary. Numerous experiments with fish, frogs, lizards, mice, and other animals have substantiated the important role of these hormones for the physiological and morphological colour changes of many cold-blooded vertebrates as well as a small number of mammals.[1] The significance of the melanotropins in higher mammals and man, however, has hitherto been much less apparent, and their function has remained somewhat enigmatic. This may explain why MSH receptors have so far not been as widely investigated as other hormone receptors. On the other hand, recent findings of a possibly decisive involvement of MSH in neural functioning and foetal development will encourage increasing interest and effort in this field.

## II. STRUCTURE, FUNCTION, AND ASSAY OF MSH

### (i) Structural aspects

Isolation and structural analysis of the MSHs have revealed two major forms: (1) the basic $\alpha$-MSH (pI $\approx$ 10.5) containing 13 amino acid residues, an N-terminal acetyl and a C-terminal amide group, and (2) the more acidic $\beta$-MSH (pI $\approx$ 5.8) with 18 residues and both ends unprotected (Table 12.1; for review see Reference 2). Human $\beta$-MSH, originally isolated as a docosapeptide (22 residues) from human pituitaries, seems to be an artifact due to proteolysis of $\beta$-lipotropin during the extraction.[3] $\alpha$-MSH has been found to have an identical structure in all mammals so far investigated (although the precise structure of human $\alpha$-MSH is still unknown). In the dogfish *Squalus acanthias*, $\alpha$-MSH is not acetylated and valine$^{13}$ is replaced by methionine.[4] This minor modification can be explained with one single base mutation of the corresponding gene. In contrast to $\alpha$-MSH, the $\beta$-forms show considerable species variation due to a relatively large number of DNA/RNA base differences (within the mammals: 0–3; between mammals and elasmobranchs: 11–14; between the two elasmobranchs *Squalus acanthias* and *Scyliorhinus canicula*: 8). Considering the fact that about 300 million years ago the elasmobranchs diverged from the main line of evolution and 100 million years later the two dogfish species diverged from each other, this gives a mutation rate of two base changes per 100 million years. The greater variability of $\beta$-MSH during evolution and the phylogenic persistence of the $\alpha$-MSH structure may reflect a much broader physiological significance of the latter hormone than of the former.

### (ii) Biological function

The melanotropins exhibit strong effects on a variety of tissues of lower vertebrates, mammals and, to some extent, also of man. In addition to its

Table 12.1 Structure of the Melanotropins

| Sequence (residues 1 2 3 4 5 6 7 8 9 10 11 12 13 14 15 16 17 18 19 20 39) | Peptide |
|---|---|
| Ser-*Tyr* -Ser-Met-Glu-*His-Phe-Arg-Trp* -Gly-Lys-*Pro* -Val -Gly -Lys-Lys-Arg-Arg-Pro-Val· · ·Phe | ACTH |
| Ac ·Ser-*Tyr* -Ser-Met-Glu-*His-Phe-Arg-Trp* -Gly-Lys-*Pro* -Val·$NH_2$ | α-MSH (mammals) |
| Ser-*Tyr* -Ser-Met-Glu-*His-Phe-Arg-Trp* -Gly-Lys-*Pro*-Met·$NH_2$ | α-MSH (dogfish) |
| Asp-Glu -Gly -Pro-*Tyr*-Arg-Met-Glu-*His-Phe-Arg-Trp* -Gly -Ser-*Pro* -Pro -Lys-Asp | β-MSH (monkey) |
| Asp-Glu -Gly -Pro-*Tyr* -Lys-Met-Glu-*His-Phe-Arg-Trp* -Gly -Ser-*Pro* -Pro -Lys-Asp | β-MSH (pig) |
| Asp-Glu -Gly -Pro-*Tyr* -Lys-Met-Glu-*His-Phe-Arg-Trp* -Gly -Ser-*Pro*-Arg -Lys-Asp | β-MSH (horse) |
| Asp -Ser -Gly -Pro-*Tyr* -Lys-Met-Glu-*His-Phe-Arg-Trp* -Gly -Ser-*Pro* -Pro -Lys-Asp | β-MSH (ox, sheep) |
| Asp-Gly -Gly -Pro-*Tyr* -Lys-Met-Glu-*His-Phe-Arg-Trp* -Gly -Ser-*Pro* -Pro -Lys-Asp | β-MSH (camel I) |
| Asp-Gly -Gly -Pro-*Tyr* -Lys-Met-Gln-*His-Phe-Arg-Trp* -Gly -Ser-*Pro* -Pro -Lys-Asp | β-MSH (camel II) |
| Asp-Gly -Ile-Asp-*Tyr* -Lys-Met-Gly-*His-Phe-Arg-Trp* -Gly-Ala-*Pro*-Met -Asp -Lys | β-MSH (dogfish *Scyliorhinus canicula*) |
| Asp-Gly-Asp-Asp-*Tyr* -Lys-Phe-Gly-*His-Phe-Arg-Trp* -Ser-Val-*Pro*-Leu | β-MSH (dogfish *Squalus acanthias*) |
| Glu· · ·Lys-Lys-Asp -Ser -Gly -Pro-*Tyr* -Lys-Met-Glu-*His-Phe-Arg-Trp* -Gly -Ser-*Pro* -Pro -Lys-Asp-Lys-Arg -Tyr-Gly-Gly· · ·Glu | β-LPH (sheep) |
| 1 39 40 41 42 43 44 45 46 47 48 49 50 51 52 53 54 55 56 57 58 59 60 61 62 63 91 | |

Common amino acid residues in italics

'classical' function as a pigment-dispersing hormone, MSH (mainly $\alpha$-MSH) affects proliferation of melanocytes, stimulates pigment production in melanocytes and melanoma cells and increases sebum secretion in rats and lipolysis in rabbit fat cells. More important in mammals and men, however, are its actions on the central nervous system (e.g. behaviour, stimulation of dopaminergic neurons), its presence during pregnancy, and its influence on foetal growth as well as its releasing activity for somatotropic hormone (for review see Reference 5). This diversity of action of MSH raises the question as to whether there might be more than one mechanism by which this hormone elicits its biological responses in different target cells (e.g. more than one type of receptor).

### (iii) Biosynthesis

The biosynthetic pathway for $\alpha$- and $\beta$-MSH are closely related. One single protein, pro-opiocortin (MW $\approx$ 30,000), seems to be the common precursor for both adrenocorticotropin (ACTH) and lipotropin (LPH) which is the prohormone of the opioid peptides.[6] In the intermediate lobe of the pituitary, $ACTH_{1-39}$ is enzymically degraded to CLIP (corticotropin-like intermediate lobe peptide, $ACTH_{18-39}$) and $ACTH_{1-13}$ which, after acetylation and amidation, yields $\alpha$-MSH (Figure 12.1). $\beta$-MSH is formed through enzymic cleavage of $\beta$-LPH via $\gamma$-LPH; in mammals, it contains the same heptapeptide, -Met-Glu-His-Phe-Arg-Trp-Gly-, as $\alpha$-MSH. It is an open question whether this could be explained by a

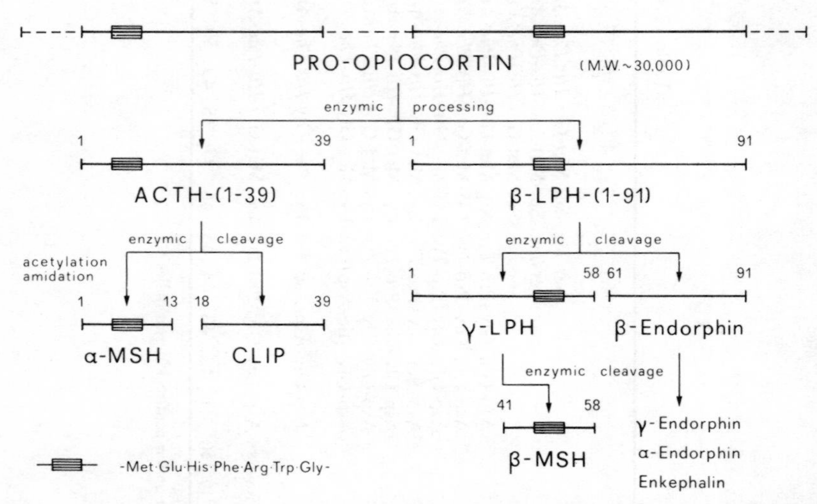


Figure 12.1. Biosynthetic pathways for ACTH, $\alpha$-MSH, $\beta$-LPH, $\gamma$-LPH, and $\beta$-MSH

duplication of just the corresponding DNA-sequence accidentally leading to a second melanotropic peptide, or whether ACTH and LPH may be derived from the same ancestral gene, the two copies of which evolved independently after duplication.

#### (iv) Bioassay and radioimmunoassay

The hormonal potency for inducing melanin migration is usually determined photometrically with an *in vitro* frog skin or lizard skin assay.[7] Native $\alpha$-MSH is the most potent melanotropic peptide and has an activity of $4 \times 10^{10}$ Units $\mathrm{mmol}^{-1}$, i.e. it is active *in vitro* at a concentration of $2 \times 10^{-11}$ M. (1 Unit (U) represents the activity of 0.04 g of a standard neurohypophyseal powder.) $\beta$-MSH is distinctly less active than $\alpha$-MSH: estimates range from $4 \times 10^{9}$ U $\mathrm{mmol}^{-1}$ (ox, horse) to $10^{10}$ U $\mathrm{mmol}^{-1}$ (monkey, pig). For routine determinations of biological fluids, the relatively expensive bioassays have been replaced by more selective and sensitive radioimmunoassays. Using $\alpha$-MSH attached uniformly by its $\mathrm{Glu}^{5}$ side chain to serum albumin as the antigen, and rabbits or goats as test animals, highly specific antisera against $\alpha$-MSH have been obtained with titres of up to $1:10^{7}$, very low or negligible cross-reactions with other hormones and detection limits of down to $< 1$ pg $\mathrm{ml}^{-1}$.[8]

## III. STIMULATION OF MELANOPHORE RECEPTORS BY $\alpha$-MSH, ITS FRAGMENTS AND ANALOGUES

As isolated functional MSH-receptor molecules are not yet available, one approach to elucidate the chemical mechanism by which the hormone acts on its target cell is by studying structure–activity relationships of a number of modified peptides. This enables one to draw conclusions about the organization of hormonal information on the one hand and about the complementary read-out devices in the receptor on the other. Up to now more than 120 $\alpha$-MSH analogues and fragments have been synthesized and biologically tested. Analysis of the melanotropic potency of 40 of these peptides and comparison with their activity on adrenal cells reveals considerable differences in structural requirements in the two systems (Figure 12.2).

The optimal chain length for maximal MSH-activity is 13 residues with both ends blocked; chain shortening or chain elongation both lead to a progressive decrease of activity (e.g. $\mathrm{ACTH}_{1-24}$ exhibits only 1% of the potency of $\alpha$-MSH). Whether the two uncharged terminal groups may simply reflect increased stability to exopeptidase action or whether they are due to a closer binding of the hormone to its receptor, remains to be established. In contrast to MSH, the ACTH-activity is strongly affected

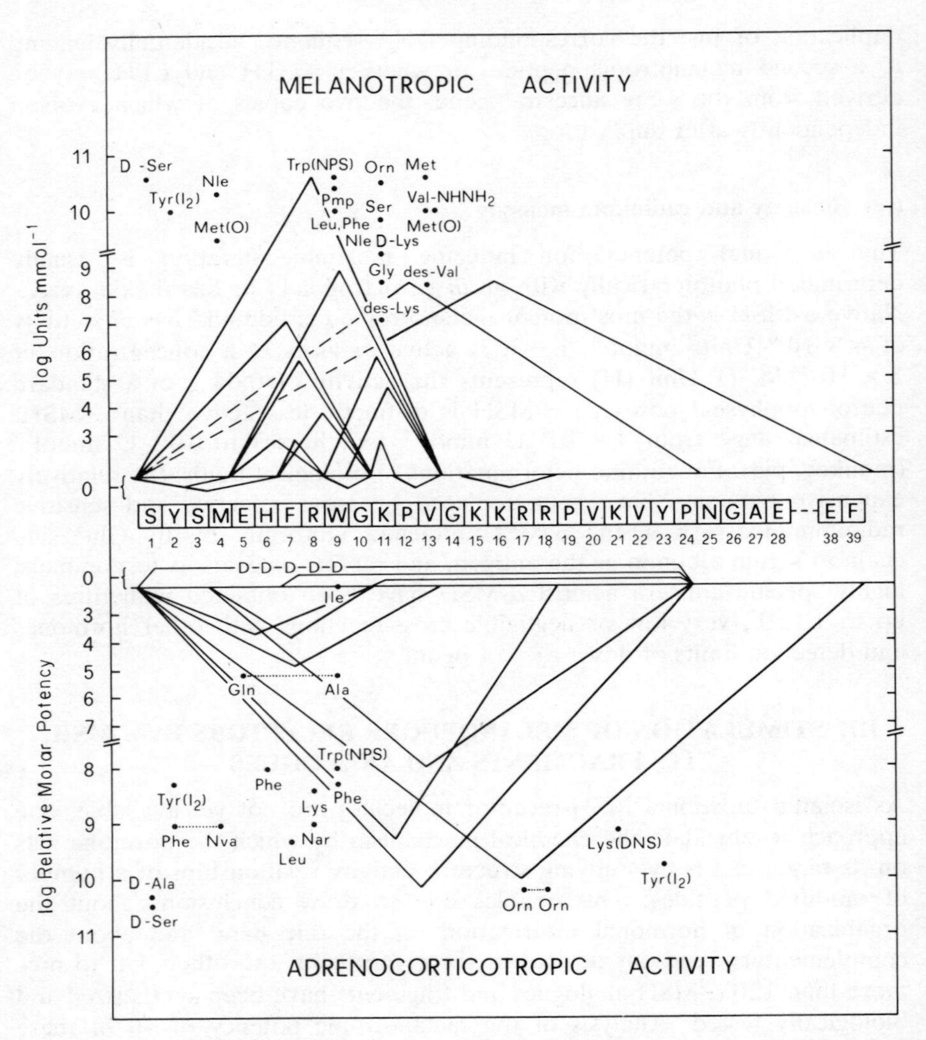


Figure 12.2. Structure-activity relationships of $\alpha$-MSH and ACTH. The tops of the triangles covering a certain partial sequence represent the activity of the corresponding peptide (melanotropic activity: Units mmol$^{-1}$; corticotropic activity: relative molar potency, $1/ED_{50}$). The symbols indicate the activity of $\alpha$-$MSH_{1-13}$ derivatives (upper part) with one single modification and of $ACTH_{1-24}$ derivatives (lower part) with one or with two (●. . . .●) modifications. $Tyr(I_2)$ = 3′, 5′-di-iodotyrosine; Met(O) = methionine-S-oxide; Pmp = *ar*-pentamethyl-phenylalanine; Nar = nor-arginine. For one letter symbols of amino acids see The List of Abbreviations

when the N-terminus is blocked or when the sequence (14–20) (part of the 'address region', see below) is lacking. This explains why, in the isolated adrenal cell system, $\alpha$-MSH has less than 0.001% of the activity shown by $ACTH_{1-24}$.

With regard to its function as a melanotropic agent, $\alpha$-MSH may be divided into three regions. The N-terminal sequence, Ac·Ser-Tyr-Ser-Met-(Glu-) is inactive and represents a potentiating sequence containing no message element to elicit the stimulus. In this region, the balance of hydrophilic (-Ser-Tyr-Ser-) and hydrophobic (Ac-, -Met-) residues seems to be an important consideration.[9] The C-terminal tripeptide, -Lys-Pro-Val·$NH_2$, exhibits melanotropic activity on its own and is 2.5 times more active than the 'classical' MSH message, His-Phe-Arg-Trp-Gly. Although the concentration necessary for melanin dispersion is much higher than that for $\alpha$-MSH ($2 \times 10^{-5}$ M and $2 \times 10^{-11}$ M respectively), nevertheless it appears that the effect is specific. Modifications of the tripeptide (especially at the lysine[11] residue) cause similar differences in potency as the same alterations of the whole tridecapeptide. Thus, the C-terminal part of $\alpha$-MSH represents an independent message sequence which consists of a hydrophilic (basic), a hydrophobic and a conformation-stabilizing element.[10] Within the 'classical' message sequence of $\alpha$- and $\beta$-MSH and of ACTH, (-Glu)-His-Phe-Arg-Trp-, $His^6$ and $Trp^9$ are particularly essential for stimulating adrenal cell and rat adipocyte receptors; they play a minor role in eliciting the melanin dispersion of melanophores. In this latter case, -Phe-Arg-, the shortest fragment with a small but definite melanotropic activity, seems to be the key portion of the message. It is embedded between two inactive dipeptides and is very sensitive to alterations.[10]

It appears that the melanophore $\alpha$-MSH discriminator contains two message-recognizing sites, one for (-Glu)-His-Phe-Arg-Trp-, and one for -Gly-Lys-Pro-Val·$NH_2$. The two sites can operate either alone or in combination to trigger melanin dispersion. In covalent combination they have a multiplicative, 'cooperative' effect. Such a phenomenon of multiple hormonal 'active sites' is not restricted to the recognition patterns of the melanophore receptor: with respect to its activity on the active avoidance behaviour in rats, $\alpha$-MSH even seems to contain three independent centres, each of which is capable of interacting with cells of the central nervous system.[11] -Met-Glu-His- constitutes one, -Phe-Arg-Trp- the second, and -Lys-Pro-Val·$NH_2$ the third centre; essential functions are attributed to $Met^4$ and $Phe^7$. Thus, structure-function relationships in $\alpha$-MSH have revealed that the elements mainly responsible for eliciting the effect in different target cells may differ from one another.[9] Most probably, different parts of one and the same molecule can interact with its discriminators with different binding constants by adopting different conformations on contact with receptor surfaces; such a hormone is said to act as a pleiotropic effector.[12]

## IV. INTERACTIONS OF MSH WITH MELANOMA CELL RECEPTORS

Cyclic AMP mimics the effects of MSH on frog dermal melanophores *in vitro* at concentrations of $10^{-2}$ M.[13] MSH increases the cyclic AMP content of pigmented dorsal *Rana pipiens* skin while having no effect on the cyclic AMP content of unpigmented abdominal skin.[14] These findings and the fact that methylxanthines darken amphibian skin have supported the hypothesis that the regulation of melanophore metabolism by MSH is mediated by the adenylate cyclase system. Because of experimental problems in preparing melanophore plasma membranes, a final proof of a selective activation of adenylate cyclase by MSH has hitherto not been possible.

An MSH-sensitive adenylate cyclase system has been found in more readily accessible membrane preparations of Cloudman mouse melanoma cells.[15] In tissue culture, these cells show a marked increase in tyrosinase activity, melanin formation, and volume (the latter only in suspension) upon treatment with $10^{-7}$ M $\beta$-MSH. The same effects have been observed with immobilized hormone ($\beta$-MSH attached to Sepharose), with $10^{-4}$ M dibutyrul cyclic AMP or—less pronounced—with cyclic AMP, and also with cholera toxin.[16] Studies with synchronized cells have shown that binding of [$^{125}$I]$\beta$-MSH to the cell membrane occurs mainly in the G2-phase of the cell cycle, whereas it is minimal during the G1- and the S-phase. This parallels the observation that the physiological responses (increase in cyclic AMP content and tyrosinase activity) are also restricted to the G2-phase.[17] FITC-(fluorescein isothiocyanate) or $^{125}$I-labelled $\beta$-MSH usually bind to the cell surface in the perinuclear area as has been demonstrated with fluorescence microscopy or autoradiography. The association constant lies around $10^8$ l mol$^{-1}$ and the number of receptors has been calculated to be approximately $10^4$ cell$^{-1}$. When the melanoma cells are exposed either to dibutyryl cyclic AMP ($10^{-3}$ M) in the presence of theophylline ($6 \times 10^{-4}$ M) or to cholera toxin (1 $\mu$g ml$^{-1}$) binding of both types of $\beta$-MSH is increased by a factor of four to five. This stimulation of MSH-binding results from an increase in the number of MSH receptors per G2 cell and, to a lesser extent, from an increase in the number of G2 cells. The affinity of the receptors for MSH is not influenced by dibutyryl cyclic AMP.[18]

Finally, one as yet unexplained observation should be mentioned: MSH inhibits growth of most wild type melanoma cells in culture whereas certain clones are not affected. There is, however, no difference in MSH-induced tyrosinase stimulation between the two cell-types.[19] This diversity could be explained by an additional mechanism of MSH-signal processing which is present only in certain types of melanoma cells. (In melanocytes, the effect of MSH is just opposite: it induces cell proliferation and increases DNA-synthesis.)

## V. MECHANISM OF MSH-ACTION

MSH elicits its effect via a multistep mechanism each part of which requires independent investigation. The starting point for such studies is certainly an analysis of the molecular anatomy and structural organization of the hormone itself, as has been presented in Section III above. The second step, namely the transfer of the information from the hormone to the receptor and the degradation of the hormone after the stimulation, remains completely obscure at the present time. The same is true for the problem of information transduction from the hormone receptor complex through the cell membrane. More data have, however, been accumulated in respect of receptor localization and of elucidating the cell specific form of the signal (Figure 12.3).

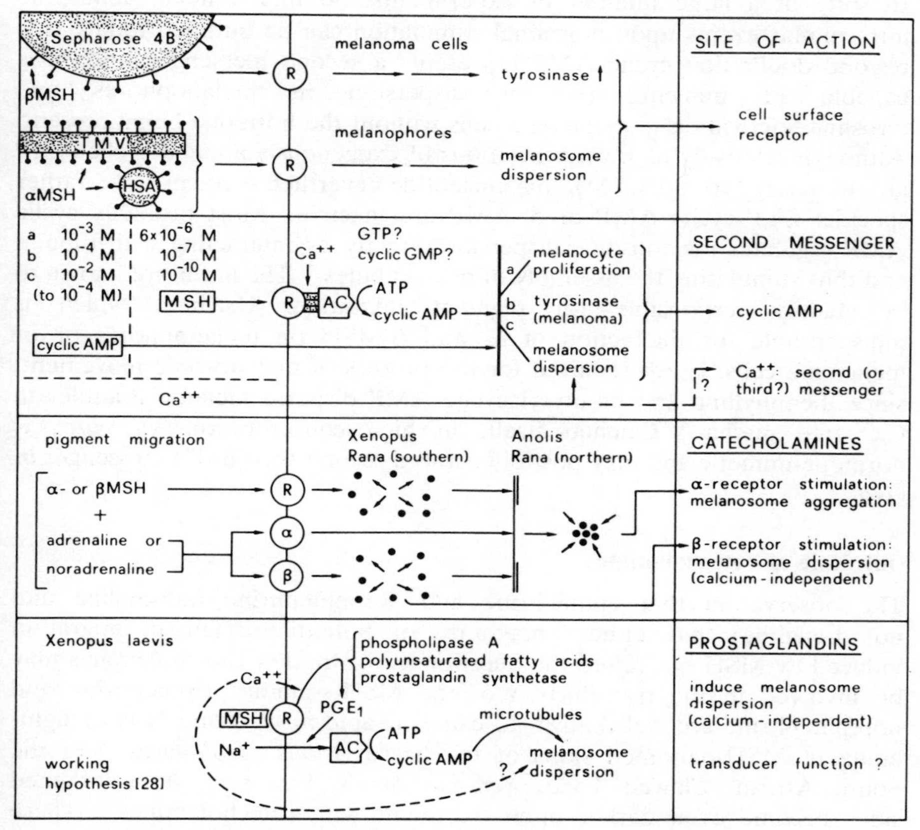


Figure 12.3. Mechanism of α-MSH-action (see text). R = receptor; AC = adenylate cyclase; $PGE_1$ = prostaglandin $E_1$

### (i) Site of action

A number of high molecular weight forms of $\alpha$-MSH and of $\alpha$-MSH fragments, in which the peptides were attached to serum albumin, thyroglobulin, ferritin, or tobacco mosaic virus, exhibited almost the same or (for TMV-linked peptides) even higher molar potency in the frog melanophore system than the free peptides.[20] These findings together with the above mentioned results obtained with $\alpha$-MSH and melanoma cells support the idea of a cell surface membrane receptor for MSH. Pinocytosis as a possible mechanism for hormonal stimulation[21] can be ruled out, because cytochalasin B does not prevent MSH from increasing intracellular levels of cyclic AMP and tyrosinase activity in melanoma cells.[22]

### (ii) Second messenger

In spite of a large number of experiments, no final scheme about the intracellular events upon hormonal stimulation can be outlined today. It is beyond doubt that cyclic AMP represents a second messenger since it is capable of inducing pigment dispersion in melanophores and tyrosinase-activation in melanoma cells without the hormone being present. Although relatively high concentrations of exogenously applied cyclic AMP are necessary ($10^{-2}$–$10^{-4}$ M), this nucleotide nevertheless seems to be rather specific: 2′:3′-cyclic AMP or 5′-AMP are inactive.[1] Most probably cyclic AMP regulates melanosome dispersion directly by interacting with tubulin and thus stimulating the assembly of microtubules.[23] The latter are known to be mainly responsible for pigment migration. Calcium, which is indispensable for the action of $\alpha$- and $\beta$-MSH on melanophores or on melanoma cells, is not required for the process of melanosome movement, since theophylline and dibutyryl cyclic AMP disperse melanin granules in $Ca^{2+}$-free media.[24] Calcium itself, in high concentrations *in vitro*, is hormone-mimetic and may possibly act as a second (or third?) messenger *in vivo*.

### (iii) Role of catecholamines

The observation that epinephrine and norepinephrine (adrenaline and noradrenaline) may either antagonize or potentiate pigment migration induced by MSH has raised the question as to whether catecholamines may be involved in the transduction of the MSH-stimulus. Epinephrine and norepinephrine are well known to cause a rapid *in vivo* and *in vitro* lightening of MSH-darkened skins of the lizard *Anolis carolinensis*.[25] In the South African Clawed Toad, *Xenopus laevis*, however, MSH-darkened skins become even darker upon treatment with catecholamines.[26] These different responses are due to a stimulation of either $\beta$-adrenergic receptors (which induce melanin dispersion) and/or $\alpha$-receptors (which are responsible for melanin aggregation). When both types of receptors are present on melanophores (e.g. in the lizard or in northern frogs), catecholamines normally aggregate the melanosomes since the $\alpha$-receptors dominate. If $\alpha$-receptors are blocked or if they do not exist on the melanophores (e.g. in *Xenopus laevis* or in southern frogs), then $\beta$-receptor stimulation increases cellular cyclic AMP and thus induces pigment dispersion.[27] This effect does not depend on calcium, in contrast to the stimulation by MSH.[28] The mechanism by which $\alpha$-receptor stimulation causes pigment aggregation is still unknown (e.g. decrease of cyclic AMP, increase of cyclic GMP?). In summary, catecholamines do not appear to affect MSH at the level of the receptor, but rather at the level of the common second messenger.

### (iv) Role of prostaglandins

$PGE_1$ exhibits melanosome dispersing activity on melanophores of *Xenopus laevis* and *Rana pipiens in vitro*.[28] Furthermore, it raises the level of cyclic AMP in dorsal frog skin and activates melanoma adenylate cyclase.[15] These effects are calcium-independent. Adenylate cyclase activation by MSH is partially blocked by 7-oxa-13-prostynoic acid (a prostaglandin antagonist) and by indomethacin (an inhibitor of prostaglandin synthetase).[15] The latter also inhibits pigment migration induced by MSH.[28] However, the facts that the concentration required for both inhibitors is rather high and that MSH does not stimulate prostaglandin synthetase in broken-cell preparations of melanoma, cause one to question a direct involvement of $PGE_1$ in mediating the MSH-stimulus.[15]

On the other hand, Van de Veerdonk has suggested that the MSH-receptor complex, as a first step, activates membranous phospholipase A.[28] This would explain the absolute calcium requirement early in the transducer chain. Newly formed polyunsaturated fatty acids would then—by the (MSH-insensitive) prostaglandin synthetase—be transformed to $PGE_1$, which would in turn cause an influx of sodium ions.[29] Since sodium influx may be coupled to pigment dispersion,[30] increased amounts of sodium could lead to a rise of intracellular cyclic AMP.[28] Although a direct experimental proof is lacking, this outline nevertheless represents an interesting working hypothesis.

## VI. FUTURE RESEARCH

Considerable insight into the mode of MSH-action has been achieved within the past few years; we are, however, still far from understanding in detailed molecular terms how the hormone stimulates melanophores, melanocytes, or melanoma cells. Even less is known about the interaction of MSH with non-pigmentatary cells such as neurons. It is just this field which will demand increased efforts in the near future. Questions about the nature of the MSH-receptor (→ receptor isolation) and about the fate of the hormone interacting with the target cell (→ degradation of MSH) are being investigated as a first priority. To this end, suitable tools such as labelled $\alpha$-MSH derivatives carrying photoaffinity, one or several different radioactive and/or fluorescence labels at defined sites, will be required.[31] Furthermore, high molecular weight forms of $\alpha$-MSH for cooperative affinity labelling[20] seem to be very promising.

Is there any practical reason for investigating MSH receptors? First of all, of course, MSH does play a role in various physiological processes (see above), and therefore, it is important to know more about the mechanism of MSH-action for a better understanding of MSH-dependent dysfunctions. A potential new application of MSH receptor binding could be envisaged in the treatment of melanoma by specific antibodies. Such antibodies could, for example, carry cytotoxins or residues containing boron atoms for neutron activation therapy,[32] and would be very promising if they were easily accessible and if there were enough antigens on the cell surface. Instead of using naturally occurring antigens, artificial ones such as MSH could be introduced and attached irreversibly to the cell membrane (e.g. by photoaffinity labelling). Furthermore, highly specific antibodies against such an antigen can be produced in large quantities.[8] Thus, in general, hormone receptors, in addition to their function as a switch for controlling intracellular metabolism, may perhaps sometimes play a role in the therapy of hormone-responsive cells.

## VII. REFERENCES


1. Bagnara, J. T. and Hadley, M. E. *Chromatophores and Color Change* Prentice-Hall Inc., Englewood Cliffs, New Jersey, (1973).
2. Geschwind, I. I. in *Structure and Control of the Melanocyte* (Della, Porta, G., and Mühlbock, O., eds.) Springer-Verlag, Berlin, pp. 28–44, (1966).
3. Scott, A. P., and Lowry, P. J. *Biochem. J.*, **139**, 593–602, (1974).
4. Bennett, H. P. J., Lowry, P. J., McMartin, C., and Scott, A. P., *Biochem. J.*, **141**, 439–444, (1974).
5. Tilders, F. J. H., Swaab, D. F., and van Wimersma Greidanus, T. B. (eds.) *Front. Hormone Res.*, Karger, Basel, Vol. 4, (1977).
6. Rubinstein, M., Stein, S., and Udenfriend, S. *Proc. Natl. Acad. Sci. U.S.A.*, **75**, 669–671, (1978).
7. Shizume, K., Lerner, A. B., and Fitzpatrick, T. B., *Endocrinology,* **54**, 553–560 (1954). Eberle, A. and Schwyzer, R., *Helv. Chim. Acta*, **58**, 1528–1535 (1975). Tilders, F. J. H., van Delft, A. M. L., and Smelik, P. G., *J. Endocrin.*, **66**, 165–175, (1975).
8. Kopp, H. G., Eberle, A., Vitins, P., Lichtensteiger, W., and Schwyzer, R. *Eur. J. Biochem.*, **75**, 417–422 (1977). Bürgisser, E., Eberle, A., and Schwyzer, R. *Experientia*, **34**, 935, (1978).

9. Eberle, A. and Schwyzer, R., in *Surface Membrane Receptors* (Bradshaw, R. A., Frazier, W. A., Merrell, R. C., Gottlieb, D. I., and Hogue-Angeletti, R. A., eds.) Plenum Publ. Corp., New York, pp. 291–304, (1976).
10. Eberle, A., and Schwyzer, R. *Clin. Endocrin.*, **5, suppl.**, 41s–48s, (1976).
11. Greven, H. M., and de Wied, D. *Front. Hormone Res.*, **4**, 140–152, (1977).
12. Schwyzer, R., and Eberle, A. *Front. Hormone Res.,* **4**, 18–25, (1977).
13. Bitensky, M. W., and Burstein, R. S. *Nature (London)*, **208**, 1282–1284, (1965).
14. Abe, K., Butcher, R. W., Nicholson, W. E., Burd, W. E., Liddle, R. A., and Liddle, G. W. *Endocrinology*, **84**, 362–368, (1969).
15. Kreiner, P. W., Gold, C. J., Keirns, J. J., Brock, W. A., and Bitensky, M. W., *Yale J. Biol. Med.*, **46**, 583–591, (1973).
16. Pawelek, J., Wong, G., Sansone, M., and Morowitz, J. *Yale J. Biol. Med.*, **46**, 430–443, (1973).
17. Varga, J. M., DiPasquale, A., Pawelek, J., McGuire, J. S., and Lerner, A. B., *Proc. Natl. Acad. Sci. U.S.A.*, **71**, 1590–1593. (1974).
18. DiPasquale, A., McGuire, J., and Varga, J. M., *Proc. Natl. Acad. Sci. U.S.A.*, **74**, 601–605, (1977).
19. Pawelek, J., Sansone, M., Koch, N., Christie, G., Halaban, R., Hendee, J., Lerner, A. B. and Varga, J. M. *Proc. Natl. Acad. Sci. U.S.A.*, **72**, 951–955, (1975).
20. Eberle, A., Kriwaczek, V. M., and Schwyzer, R., *FEBS Letters,* **80**, 246–250, (1977). Kriwaczek, V. M., Eberle, A., Müller, M., and Schwyzer, R., *Helv. Chim. Acta*, **61**, 1232–1240, (1978).
21. Varga, J. M., Moellmann, G., Fritsch, P., Godawska, E., and Lerner, A. B. *Proc. Natl. Acad. Sci. U.S.A.*, **73**, 559–562, (1976).
22. DiPasquale, A., and McGuire, J. *Exp. Cell Res.*, **102**, 264–268, (1976).
23. Steiner, M. *Nature (London)*, **272**, 834–835, (1978).
24. Vesely, D. L., and Hadley, M. E. *Pigment Cell*, **3**, 265–274, (1976).
25. Horowitz, S. B., *J. Cell. Comp. Physiol.*, **51**, 341–357, (1958).
26. Graham, J. D. P. *J. Physiol.*, **158**, 5p–6p, (1961).
27. van de Veerdonk, F. C. G., and Konijn, T. M. *Acta endocr. (Copenh)*, **64**, 364–376, (1970).
28. van de Veerdonk, F. C. G., and Brouwer, E. *Biochem. Biophys. Res. Commun.*, **52**, 130–136 (1973). van de Veerdonk, F. C. G., *Pigment Cell*, **3**, 275–283, (1976).
29. Ramwell, P. W., and Shaw, J. E. *Recent Prog. Horm. Res.*, **26**, 139–187, (1970).
30. Novales, R. R., *Physiol. Zool.*, **32**, 15–28, (1959).
31. Eberle, A., and Schwyzer, R. *Helv. Chim. Acta*, **59**, 2421–2431, (1976).
32. Leukart, O., Caviezel, M., Eberle, A., Escher, E., Tun Kyi, A., and Schwyzer, R. *Helv. Chim. Acta*, **59**, 2184–2187, (1976).

Cellular Receptors for Hormones and Neurotransmitters
Edited by D. Schulster and A. Levitzki


CHAPTER 13

# Angiotensin II receptors

Kevin J. Catt and Greti Aguilera
*Endocrinology and Reproduction Research Branch, National Institute of Child Health and Human Development, National Institutes of Health, Bethesda, Maryland 20014, U.S.A.*



## I. INTRODUCTION

The identification of specific receptor sites for angiotensin II (AII) was originally performed in adrenal gland, uterus, and kidney by binding studies with $^{125}$I-labelled angiotensin II.[1,2] In subsequent receptor studies, either tritiated or monoiodinated AII was used to identify and characterized AII receptors in a variety of target tissues, including adrenal cortex[3–6], aortic smooth muscle,[7–8] uterine smooth muscle,[9–10] renal glomeruli,[11] and brain.[12] Thus, specific binding sites for angiotensin have been identified in each of the major target tissues upon which the peptide is known to act. The preparation of tritiated[13] and monoiodinated[14–16] forms of angiotensin II suitable for radioassay and receptor binding studies has been described. While the tritiated peptide has the advantage of being chemically identical with the native octapeptide, the monoiodinated tracer retains substantial biological activity and has the advantage of much higher specific activity and greater convenience in receptor binding studies.

The properties of the AII receptors present in the several target tissues were found to be generally similar to those of other peptide hormone receptors, with high specificity and affinity for angiotensin II and related peptides, and predominant localization in the plasma membrane. In most studies, the particulate binding fractions with specificity for angiotensin II have not been subjected to highly selective purification procedures, and have frequently consisted of microsomal particles prepared by high-speed centrifugation after preliminary removal of the low-speed sediment from homogenates of the target tissue. However, analysis of such preparations by density gradient centrifugation has indicated that angiotensin receptors are associated with the plasma membrane fraction.[3,6,8,17] The localization of specific receptor sites in the plasma membrane does not imply that the peptide acts only at the cell membrane level, though evidence for intracellular actions of AII has yet to be presented. There is little doubt that the initial interaction of angiotensin with its target cells occurs via specific receptor sites in the plasma membrane, with subsequent activation of effector systems leading to increased steroidogenesis and aldosterone production.

The binding affinities of the AII receptors measured in most target tissues have been in the range $10^9$–$10^{10}$ $M^{-1}$.[1,3,7,9,11,12] These binding constants imply that half-saturation of receptors would occur at angiotensin II concentrations of 0.1–1 nM, whereas blood angiotensin II levels[18] are much lower than this (about 0.025 nM) under basal conditions. Such discrepancies between receptor binding affinity and plasma hormone concentration are not uncommonly found when hormone receptors are characterized, and may reflect the ability of many target cells to be activated by only fractional occupancy of the hormone receptor sites (see Chapter 2). Another possibility in the case of angiotensin receptors is that locally generated angiotensin could have a role in stimulating target cell responses, if tissue renins generate angiotensin II in concentrations higher than those normally present in the circulation. It is also possible that the effective binding affinities of peptide hormone receptors in the intact target tissue are higher than those measured by equilibrium binding studies performed *in vitro* with cells or tissue fractions. However, the validity of the binding sites for AII as biologically active receptors has been demonstrated by several factors: their specificity for AII derivatives, with affinities in general agreement with the biological activities of agonist and antagonists;[21] their specific tissue distribution in relevant target organs; their ability to bind angiotensin II in quantitative correlation with specific responses, such as contractile changes,[7] and aldosterone production;[19] and the ability of specific antagonists to block both peptide binding and steroidogenic responses at the target cell level.[20]

## II. LOCATION OF RECEPTORS

The exact distribution of angiotensin II receptors in target tissues has yet to be determined in detail. In the adrenal gland, angiotensin II receptors are concentrated in the zona glomerulosa,[17] though in some species binding sites are also present in the zona fasciculata. The relative binding of angiotensin II to glomerulosa and fasciculata cells of the rat is shown in Figure 13.1. In other species, including the dog and cow, angiotensin II receptors are quite abundant in the fasciculata cells, though still less concentrated than in the glomerulosa zone. The presence of vascular tissue in many target organs that

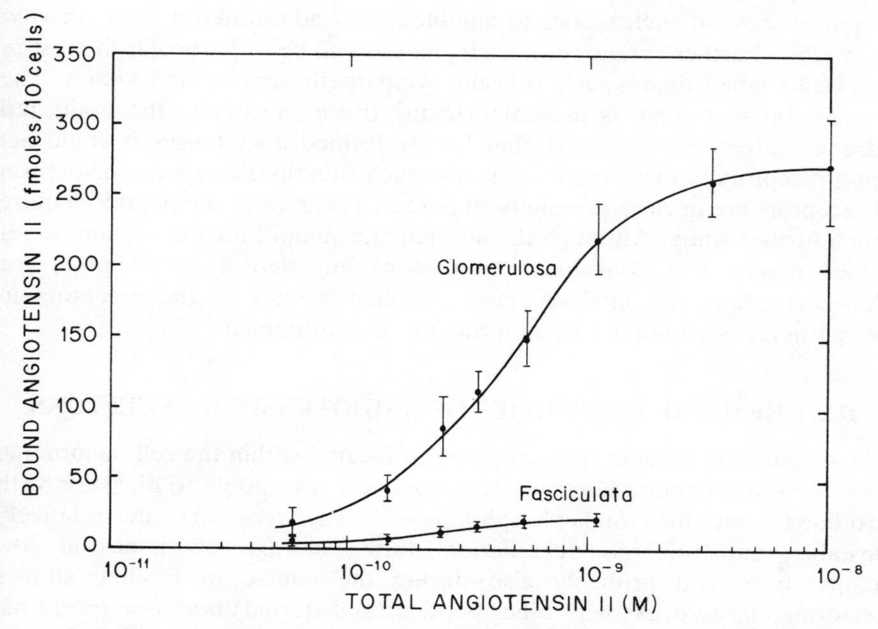


Figure 13.1. Specific binding of $^{125}$I-angiotensin II by dispersed rat adrenal glomerulosa and fasciculata cells during incubation at 24 °C

are homogenized and analysed for angiotensin II receptors could present the problem of receptor heterogeneity in the resulting homogenate. However, angiotensin II receptors in vascular tissue are quite difficult to analyse directly, and even in the well-studied receptors of the rabbit aorta, smooth muscle cells comprise only a fraction of the tissue homogenate. For this reason, vascular receptors probably represent only a minute proportion of the sites present in target organs such as the adrenal, uterus, and nervous tissue. The lack of a direct approach to analysis of vascular receptors has led to the use of uterine muscle as a more abundant source of smooth muscle receptors, and many of the properties of the uterine receptors are probably identical with those of the vascular receptors. However, a suitable preparation for direct analysis of vascular receptors would be highly desirable, to permit examination of the role of receptor modulation in determining the vascular responses to angiotensin II during alterations in electrolyte balance, renin secretion, and aldosterone secretion.

The use of homogenates and purified tissue preparations such as adrenal glomerulosa and smooth muscle from the uterus has been of value in this regard, as has the analysis of binding in brain homogenates[11] and renal glomeruli.[12] This approach has made it possible to derive detailed correlations between the concentration, affinity, and specificity of angiotensin II receptors in individual target cells, and the known responsiveness of such tissues to angiotensin II administered either *in vivo* or *in vitro*. Further extension of such analysis will be of particular interest in less well-studied tissues such as brain, sympathetic ganglia, and kidney. The recognition that renin is present in many tissues, including the brain and adrenal, raises the possibility that locally formed angiotensin II could act upon receptors in these organs. Clearly, such situations in which angiotensin II receptors are in close proximity to potential sources of the peptide require much further study. Although the adrenal and smooth muscle are important target tissues and convenient sources of angiotensin II receptors for characterization and analysis, more detailed studies of the receptors in neural tissue is an important area for future development.

## III. CHEMICAL PROPERTIES OF ANGIOTENSIN II RECEPTORS

The angiotensin binding sites are proteins located within the cell membrane, and like other peptide hormone receptors, are susceptible to digestion with proteolytic enzymes and phospholipase,[3,8] The receptors are relatively unstable, and undergo degradation during storage of tissues at low temperature, and probably also during the course of binding studies performed *in vitro*. As with other peptide (and steroid) hormone receptors, the free sites are degraded more rapidly than those occupied by the homologous peptide. The angiotensin receptors of the adrenal cortex are relatively uninfluenced by changes in divalent cation concentration, but show higher binding with increasing sodium concentration up to levels of about 100 mM. The angiotensin receptors, like those for glucagon, are also influenced by guanyl nucleotides, and show an increased dissociation rate constant and lower binding affinity in the presence of GTP and analogues.[6] The instability of angiotensin II receptors is particularly apparent during solubilization, and relatively little success has attended attempts to extract the receptors in a soluble state with non-ionic and other detergents. It would be important to perform more detailed analysis of the angiotensin II receptors in soluble form, in order to characterize the molecular properties of the receptor sites and to analyse their regulation by factors that could modify their affinity or activity within the intact cell membrane.

In the adrenal, the relationship between angiotensin II receptors and ACTH receptors remains an important question. The distribution of such receptors within the zones of the adrenal, the effects of ACTH and angiotensin II upon differentiation of the adrenal and anatomical zones, the relationship of these actions to the acute effects of these regulatory peptides upon the zona glomerulosa, as well as the relationships between peptide hormone receptors and electrolytes during control of aldosterone secretion by the zona glomerulosa, are in need of more detailed elucidation.

## IV. PREPARATION AND ANALYSIS OF PARTICULATE ADRENAL RECEPTORS FOR ANGIOTENSIN II

Specific binding sites for angiotensin II and related peptides have been demonstrated in homogenates and subcellular fractions prepared from adrenal cortex, kidney, and aortic smooth muscle. The binding sites in adrenal and kidney homogenates were identified with $^{125}$I-labelled angiotensin II; those in aortic smooth muscle homogenates were originally demonstrated with tritiated angiotensin II. Both radioiodinated and tritiated angiotensin II have been employed more recently to identify binding sites in the adrenal cortex and the renal glomerulus. In most studies, the particulate binding fraction with specificity for angiotensin II has not been subjected to highly selective purification procedures, and is usually employed as a high speed pellet prepared from homogenates of the target tissue. Where localization of the binding sites has been pursued during fractionation, the available evidence has indicated that the angiotensin II receptors are located in the plasma membrane fraction.

### (i) Angiotensin II tracers

Tritiated angiotensin II offers the advantage of high biological activity in the tracer employed for binding studies, but its relatively low specific activity is not satisfactory for the demonstration of high affinity binding sites. For this purpose, $^{125}$I-labelled angiotensin II of specific activity approximately 1,000 Ci mmol$^{-1}$ has proved to be more satisfactory, and has several advantages as tracer for binding studies in adrenal cortex preparations. Tritiated angiotensin II provides a specific activity of 30–50 Ci mmol$^{-1}$, reportedly with retention of full biological activity.[13] Radioiodinated angiotensin II tracer for binding studies is relatively easily prepared by iodination of pure synthetic angiotensin II (Beckman or Schwarz–Mann) by the chloramine-T method. The biological activity of monoiodo-angiotensin II has been variously estimated to be from 25% to 80%, while that of di-iodo-angiotensin II is considerably lower (2–24%). For this reason, it is essential to prepare the monoiodinated peptide for receptor binding studies. This is most readily achieved by labelling angiotensin II to relatively low specific activity (e.g. 1 mCi of $^{125}$I to 50 $\mu$g of angiotensin II), followed by isolation of the monoiodo-peptide by ion change chromatography upon DEAE-Sephadex. Radioiodinated monoiodo-angiotensin II prepared by ion exchange chromatography is available commercially from Schwarz–Mann, New England Nuclear, and other suppliers. The specific activity of $^{125}$I-angiotensin II prepared by this method is usually in the region of 600–1,000 $\mu$Ci $\mu$g$^{-1}$, and the presence of the monoiodo-peptide can be confirmed by pronase digestion and chromatography of the labelled amino acid residues. Monoiodinated angiotensin II is unusually stable during storage, and retains high specific activity for prolonged periods after labelling. This phenomenon is probably due to the phenomenon of 'decay catastrophe', the destruction of radioiodinated peptides which results from the intense ionization accompanying the radioactive decay of $^{125}$I. The resulting simultaneous disappearance of both radionuclide and peptide causes relatively prolonged retention of the high specific activity. After preparation, the labelled peptide should be stored as frozen aliquots at $-60$°C, and used only once for binding studies after thawing. When quantitative binding studies are to be performed, with derivation of association constants and binding site concentrations, the labelled peptide should be characterized in terms of its biological activity and its ability to bind to excess receptor sites. Suitable corrections for these factors can then be employed during the computation of binding site concentrations and affinity.

### (ii) Preparation of binding fractions

Although vascular and uterine smooth muscle have been shown to bind angiotensin II, the most abundant receptor sites are located within the adrenal cortex. In addition, adrenal glands are relatively easy to obtain and process for preparation of peptide binding fractions, and the present description will concentrate on the use of such tissues for binding studies.

Bovine or dog adrenal glands obtained immediately after slaughter are kept in ice-cold buffered saline until arrival at the laboratory. The glands are then quickly dissected to isolate the cortical layers, which are minced and homogenized in a Dounce homogenizer in 20 mM sodium bicarbonate, followed by filtration through fine nylon mesh. More than 90% of the angiotensin II binding sites in the filtered cortex homogenate are recovered in the 1,500 g sediment, and in the 20,000 g sediment of the 1,500 g supernatant solution. The concentration of receptor sites per milligram of protein is about two-fold higher in the 20,000 g fraction than in the 1,500 g sediment. When more concentrated preparations of angiotensin II binding sites are required, the initial dissection of the bovine adrenal cortex should exclude the majority of the fasciculata layer, with retention of the capsule and glomerulosa layer. For most binding-inhibition studies of angiotensin II agonists and antagonists, this additional step is not necessary, since adequate binding activity is obtained by homogenization of the whole adrenal cortex.

For preparation of rat adrenal particles, rat adrenals are dissected out immediately after sacrifice and homogenized in 20 mM sodium bicarbonate in a Dounce homogenizer, followed by filtration through fine nylon mesh. The binding activity of rat adrenals is somewhat more stable to storage than that of the bovine adrenal, and glands frozen immediately can be stored at −70 °C with moderate retention of binding activity. The filtered homogenate is diluted with ice-cold buffer, centrifuged at 20,000 g to sediment the majority of the binding particles, and the pellet is resuspended in 50 mM Tris.HCl buffer for binding studies. As with bovine glands, higher concentrations of angiotensin binding sites per milligram of homogenate protein can be obtained by selective homogenization of the zona glomerulosa layer. In the rat, this can be easily prepared by enucleating the gland through a small incision by digital pressure, to give a capsular layer containing the majority of the zona glomerulosa and the highest concentration of angiotensin II binding sites.

### (iii) Preparation of adrenal cells

Dispersion of adrenal cortex or zona glomerulosa cells for binding and metabolic studies can be satisfactorily performed by collagenase digestion of rat, rabbit, cat, or dog adrenal glands. Relatively less success has attended the use of bovine adrenal glands, despite the much greater quantities of material available from this source. This difference may depend partly upon time delays during procurement and transport of bovine adrenal glands, whereas tissue obtained directly from small animal species can be more rapidly processed in the laboratory. Although the technical aspects of cell dispersion by collagenase are relatively simple, there is a great deal of variability in the viability and biological responsiveness of cells prepared by this procedure. In general, it is necessary to employ the lowest possible concentration of collagenase to achieve cell dispersion, to select for batches of collagenase which have minimum deleterious effects upon the biological responsiveness of the cell preparation, and to standardize the procedure so that the dispersed cells are subjected to the minimum amount of handling and centrifugation once dispersion is complete.

### (iv) Binding studies

Uptake of tritiated and radioiodinated angiotensin II by particulate adrenal cortex fractions is characterized by relatively rapid kinetics, with attainment of a steady-state after incubation for 30–45 min. with concentrations of angiotensin II in the range $10^{-9}$–$10^{-11}$ M. Degradation of both angiotensin II receptors and the labelled peptides can occur rapidly at high temperatures, and for this reason most binding studies have been performed at room temperature, most commonly at 20 °C. For reproducible binding studies, it is preferable to use a temperature-regulated bath at 20 °C, rather than to rely upon the constancy of ambient temperature. At lower temperatures, the degradation of receptor sites and tracer is relatively slow, but the rate of binding of angiotensin II is markedly reduced, and much longer times are required for the establishment of an equilibrium or steady-state. When binding is performed for more prolonged periods, a decline from the steady-state value is commonly seen, probably reflecting progressive degradation of free hormone in the incubation medium and consequent dissociation of bound hormone from receptor sites. This process is probably also partly attributable to a certain degree of degradation of the peptide receptor sites. A suitable compromise for quantitative binding studies is to perform incubation at 20 °C for relatively short time periods, from 30–60 min, having ensured that the chosen time interval is adequate to permit attainment of a steady-state for all concentrations of peptide hormone employed during binding studies.

It is important to note that isolated adrenal cells constitute an excellent preparation for binding studies with angiotensin II, given the possibility that enzymic methods of cell dispersion could reduce the angiotensin II receptors by proteolysis of the cell surface. This is less likely to occur when collagenase is employed for cell dispersions, and is more prominent after trypsinization of adrenal tissues to give isolated single adrenal cells. In intact adrenal cells, it is possible to relate plasma-membrane binding to activation of steroidogenesis, and to perform binding studies at 37 °C with less risk of peptide degradation than in adrenal homogenates and subcellular particles.

After incubation, the particle- or cell-bound radioactivity is isolated by filtration of the incubation suspension through 0.45 $\mu$m Millipore filters, or glass fibre filters when cells are analysed. For binding studies with

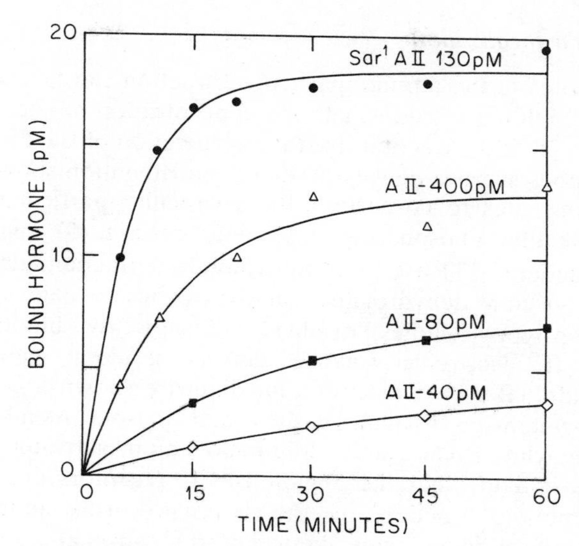


Figure 13.2. Binding kinetics of $^{125}$I-labelled angiotensin II (40–400 pM) and 130 pM Sar$^1$-angiotensin II by collagenase-dispersed dog glomerulosa cells during incubation at 37 °C

[$^3$H]angiotensin II, filters are dissolved or suspended in Biosolve or Aquasol solution prior to counting, and radioactivity determined in a liquid scintillation spectrometer. Bound [$^{125}$I]iodoangiotensin II can be similarly quantitated by liquid scintillation counting if desired, though this is more conveniently performed by counting in an automatic $\gamma$-spectrometer, with appropriate positioning of the filters in plastic counting vials. Examples of the rate of binding of $^{125}$I-labelled angiotensin II and the more potent agonist [Sar$^1$]angiotensin II by collagenase-dispersed adrenal cells are shown in Figure 13.2.

## (v) Characteristics of adrenal cortex receptor sites

### *(a) Stability*

The binding properties of bovine adrenal cortex fractions are maintained for up to 12 h at 0 °C, but freezing followed by storage in liquid nitrogen and subsequent thawing frequently causes loss of binding affinity for angiotensin II, and reduced selectivity for the octapeptide. During binding studies performed at 20 °C, relatively little degradation of receptors is detectable during preincubation of the binding fraction for periods equivalent to those subsequently required during equilibrium binding studies.

### (b) Subcellular localization

Most of the sites in the adrenal particulate fraction can be recovered in the 1,500 g and 20,000 g fractions, and the concentration of sites is about twice as high in the 20,000 g fraction. Further purification of the 20,000 g fraction by discontinuous sucrose density gradient centrifugation shows the majority of the binding sites to co-purify with a vesicular particulate fraction enriched in alkaline phosphatase, adenylate cyclase, 5′ nucleotidase, and $Na^+/K^+$-dependent ATP′ase. The mitochondrial fraction identified by the presence of succinic dehydrogenase and cytochrome oxidase activity, and electron microscopy studies, displays relatively low binding avidity for angiotensin II. Isopycnic sucrose density gradient centrifugation of particulate adrenal cortex receptors has likewise shown a close correlation between angiotensin II binding sites and plasma membrane enzymes including adenylate cyclase and sodium–potassium activated ATP′ase. It is therefore most likely that the angiotensin II receptors are located in the plasma membrane fraction of the adrenal cortex homogenate. The distribution of binding sites between the adrenal glomerulosa and fasciculata-reticularis layers has not been extensively studied, but as expected the receptor sites appear to be most dense in the glomerulosa layer, at least in the adrenals of the cow, rat, and dog.

### (c) Binding properties

Scatchard plots derived from equilibrium binding and binding-inhibition data obtained at 20 °C show the presence of high-affinity binding sites with

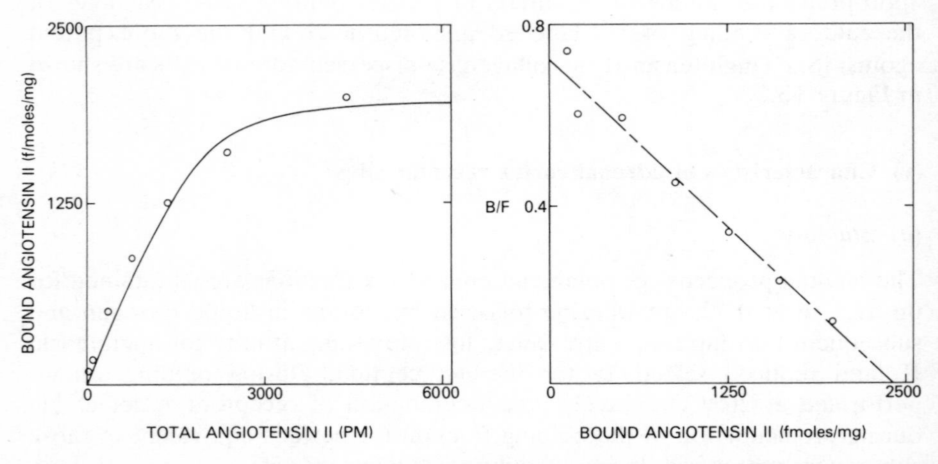


Figure 13.3. Saturation curve (a) and Scatchard plot (b) of angiotensin II binding to rat adrenal capsule particles during incubation for 45 min at 37 °C

$K_a$ of approximately $10^9\ M^{-1}$, and sometimes a second order of sites with $K_a$ of about $10^8\ M^{-1}$. In most experiments, the higher affinity sites are the predominant form observed, and the low-affinity sites are of minor quantitative significance. The binding of [$^{125}$I]iodoangiotensin II by rat adrenal capsule homogenate, expressed as both saturation and Scatchard plots, is shown in Figure 13.3. The initial binding of angiotensin II to receptors behaves as a second-order reaction with $k_a$ of $2.4 \times 10^5\ M^{-1}\ s^{-1}$. Dissociation of angiotensin II from prelabelled adrenal receptors occurs as a biphasic process, with two components of half-life 4.5 and 23 min. The dissociation of [$^{125}$I]angiotensin II previously bound to dispersed adrenal cells, following addition of an excess of the unlabelled hormone, is relatively rapid and almost complete, again indicating the surface location of specific angiotensin binding sites in the target cell. The magnitude of the association constants obtained for angiotensin II binding sites in rat adrenal cortex is consistently somewhat higher than that of the bovine preparation, being commonly between $10^9\ M^{-1}$ and $10^{10}\ M^{-1}$. In adrenal cortex particles, and sometimes in isolated adrenal cells, the binding of labelled angiotensin II is not completely reversible, and dissociation studies show an initial rapid phase followed by a prolonged phase of slow release of the tracer peptide.

Preincubation experiments have shown that relatively little receptor degradation occurs during incubation of adrenal particles at 20 °C for up to 60 min. On the other hand, rapid degradation of the free or unbound tracer angiotensin II occurs during incubation with subcellular particles from bovine or rat adrenal cortex. This process is inhibited by reduction of temperature, by addition of peptides such as glucagon and insulin, EDTA, and sulphhydryl-protecting agents such as dithiothreitol (DTT). The rate and extent of angiotensin II binding to receptor sites is also influenced by the concentration of sodium ion in the incubation medium, and maximum binding activity is obtained when binding assays are performed in the presence of 120–140 mM NaCl. Although the rate of degradation of angiotensin II peptides is reduced by addition of carrier peptides and DTT, prolonged incubation periods may still be accompanied by effects attributable to significant degradation of the labelled peptide. Elution of labelled angiotensin II from adrenal binding sites at low pH, followed by rebinding studies with fresh adrenal particles, has shown that the receptor-bound angiotensin is not subjected to significant degradation. By contrast, as stated above, the free angiotensin II in the incubation medium is relatively rapidly degraded, at a rate depending upon the temperature and protective agents present. It should also be noted that changes in the concentration of free angiotensin II during binding studies may introduce a significant problem in the computation of binding constants, since all such calculations are based upon assumptions regarding the concentrations of bound and free hormone.

*(d) Specificity of angiotensin II binding*

Apart from the non-specific protective effects of certain peptides and proteins upon tracer degradation, true competition with tracer angiotensin II for binding to adrenal receptor sites is observed only with peptides related to angiotensin II. Thus, no binding-inhibition occurs with glucagon, insulin, ACTH, and parathyroid hormone, as well as with a variety of proteins. An exception to this is the ability of the hypothalamic peptide, LHRH, to compete for binding to adrenal receptors at high concentrations, probably due to similarities of structure with the C-terminal region of angiotensin II. On the other hand, a close correlation exists between the binding-inhibition potency of a variety of angiotensin II analogues and fragments, and the biological activity of these compounds in smooth muscle and adrenal response systems. The $Val^{5-}$ and $Ile^{5-}$ forms of angiotensin are equipotent in terms of binding activity, and the C-terminal 2–8 heptapeptide formed by deletion of the N-terminal aspartic acid residue is equipotent with the intact octapeptide. Angiotensin I and the synthetic tetradecapeptide display less than 5% of the activity of angiotensin II, and the activities of the (1–7)-heptapeptide, (3–8)-hexapeptide, (4–8)-pentapeptide, and (5–8)-tetrapeptide are orders of magnitude below that of the intact

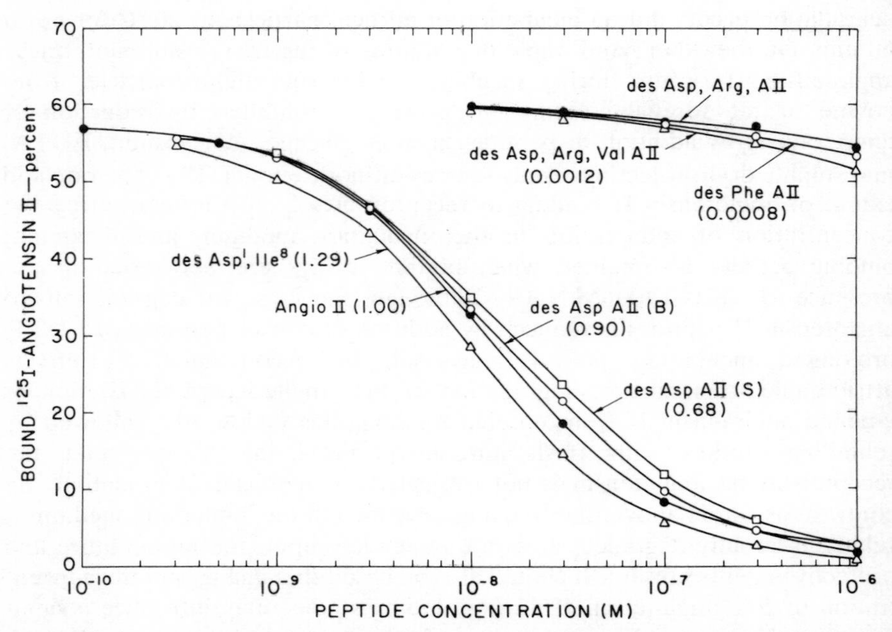


Figure 13.4. Binding-inhibition activities of angiotensin II analogues and fragments in an adrenal cortex radioligand–receptor assay

octapeptide (Figure 13.4). In addition, peptide analogues which are totally devoid of biological activity (e.g. the [Phe[3], Val[4], Tyr[8]] analogue of angiotensin II) are also devoid of binding-inhibition activity *in vitro*. By contrast, the highly potent angiotensin II agonist, [Sar[1]]angiotensin II, shows significantly increased binding-inhibition activity for adrenal cortex receptors, in keeping with its ten-fold higher activity upon smooth muscle responses and aldosterone production by adrenal cells *in vitro*.

### *(e) Radioligand–receptor assay of angiotensin II analogues*

The radioligand assay provides a precise and convenient system for evaluation and comparison of the binding-inhibition properties of angiotensin II analogues. Just as the competitive binding activities of angiotensin II agonists have been found to be proportional to their biological activities, as described above, the binding-inhibition activity of a wide range of angiotensin II antagonists has similarly been correlated with the activity of these compounds upon angiotensin II responses in smooth muscle and adrenal cortex. Such antagonists, formed by C-terminal substitution of angiotensin II with isoleucine or alanine, exhibit

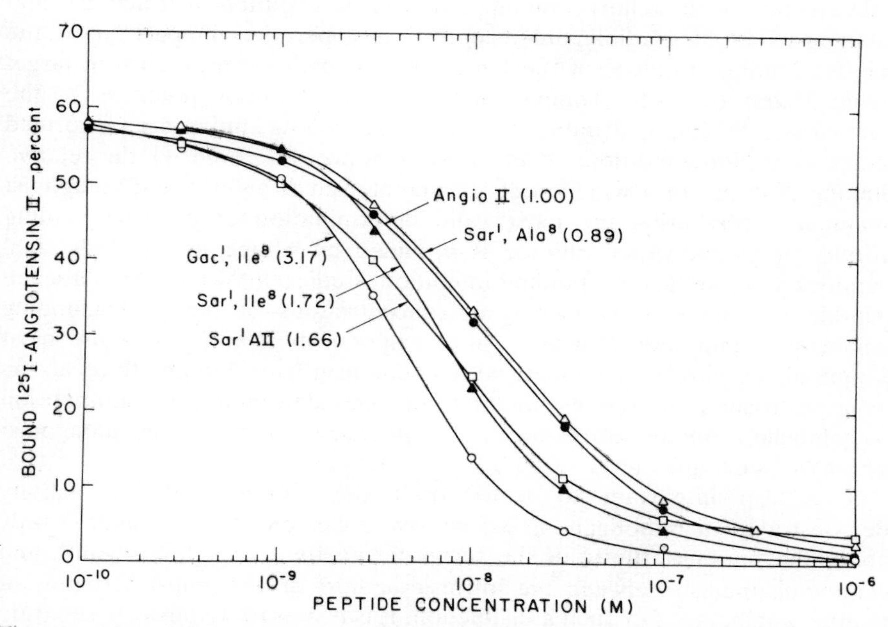


Figure 13.5. Effect of N-terminal substitution with (sarcosine[1]) and (guanidoacetic[1]) residues upon receptor binding–inhibition activities of angiotensin II and antagonist derivatives

binding-inhibition potencies in proportion to their known activities as antagonists of angiotensin II upon smooth muscle. Replacement of the aromatic phenylalanine residue, necessary for the biological activity of angiotensin II, by aliphatic residues including glycine, alanine, leucine, and isoleucine, has resulted in the formation of powerful competitive antagonists with relatively little agonist activity.

In addition to the important effect of such changes at the C-terminus upon the antagonist activity and binding potency of angiotensin II analogues, accompanying changes in the N-terminal aspartic acid with sarcosine results in a significant increase in both binding activity and antagonist activity of the resultant peptide. Replacement of aspartic acid by guanidoacetic acid has a similar though even more marked enhancing effect upon the binding activity of [Ile$^8$]angiotensin II (Figure 13.5). Such enhancement of activity appears to be related to the basicity of the N-terminal portion of the molecule, and is completely abolished by substitution of acidic residues such as succinic acid at the N-terminus of angiotensin II. The enhancement of binding-inhibition potency by N-terminal sarcosine substitution is due to increased binding affinity of the modified peptide for the angiotensin II receptor site, and may also be to some extent influenced by decreased degradation of the peptide by angiotensinase A in the adrenal cortex.

Derivation of the relative binding potencies of angiotensin II agonists and antagonists in the radioligand–receptor assay permits comparison of the relative binding affinities of these peptides for their receptor sites in target tissue. Extension of the comparison of binding-inhibition potencies to the comparison of binding affinities is valid if the binding studies are performed under saturation conditions. If this is so, it is possible to derive the relative binding affinities of a wide variety of angiotensin II agonist and antagonist analogues. Obviously, the most valid determination of intrinsic binding affinity of an individual peptide is provided by subjecting the individual peptides to homologous binding-inhibition studies, in which the labelled peptide is displaced by increasing concentrations of the corresponding unlabelled compound. However, since angiotensin II antagonists are most commonly employed to compete with angiotensin II for binding to receptors in target tissues, the relative binding potencies determined by competition with labelled angiotensin II are highly practical aspects of the antagonist activity of such analogues.

It is also important to realize that the binding-inhibition activity demonstrated in radioligand–receptor assays can provide information only about the binding affinity of the receptor site for the peptide ligand, and cannot distinguish between the intrinsic agonist or antagonist activities of peptide analogues. For such a distinction, it is necessary to perform separate determinations of agonist and antagonist activity in appropriate target cell response systems. As long as the interpretation of the binding-inhibition potency of peptide analogues is confined to the interaction between receptor and hormone, the adrenal radioligand–receptor assay provides a direct and quantitative method for evaluation of the role of binding affinity in the action of both agonists and competitive antagonists upon the response of target tissues to angiotensin II.

## V. INTERACTION OF ANGIOTENSIN II RECEPTORS WITH PEPTIDE FRAGMENTS AND ANALOGUES

There is a marked need to clarify the differential responsiveness of angiotensin II receptors in various tissues to angiotensin II agonists and antagonists. The use of particulate angiotensin II receptors in adrenal, vascular, and uterine homogenates to analyse the binding properties of agonists and antagonists has been well described,[20,21] and similar studies have been performed in other target tissues. Typical examples of binding–inhibition curves derived in adrenal particles with angiotensin II fragments and an antagonist analogue is shown in Figure 13.4 and 13.5. In particular, the differential effects of the des-Asp$^1$-heptapeptide have been analysed in detail, as this compound has been proposed to act as a mediator of the actions of angiotensin II in the adrenal gland.[22] This proposal was based upon the apparently greater activity of the heptapeptide upon the adrenal than upon smooth muscle. However, all studies suggesting that the heptapeptide is an intermediate of angiotensin II in the adrenal have been indirect in nature, and there is as yet no compelling evidence that the heptapeptide has an essential role in this regard.

In several studies, the heptapeptide has been noted to be somewhat less active than the octapeptide in stimulating aldosterone production *in vivo* and *in vitro*, rather than more active as might be expected if it acts as an essential intermediate in the action of angiotensin II. When incubated with rat glomerulosa cells *in vitro* the heptapeptide is much less active than angiotensin II, largely due to a significantly higher rate of peptide degradation in this species, with more rapid metabolism of the heptapeptide. Recent observations in dispersed adrenal cells have indicated that the octapeptide is not significantly converted to the heptapeptide during receptor binding and stimulation of aldosterone production,[23] so at least *in vitro* the heptapeptide is not an essential intermediate in the action of angiotensin II on zona glomerulosa cells. Similar findings after infusion of angiotensin II in intact rats and analysis of receptor-bound angiotensin have shown that the octapeptide can also act on steroidogenesis *in vivo* without conversion to the heptapeptide.[24]

A close correlation between receptor binding of angiotensin II and stimulation of aldosterone production has been observed in dispersed zona glomerulosa cells, as illustrated in Figure 13.6. The existence of such a

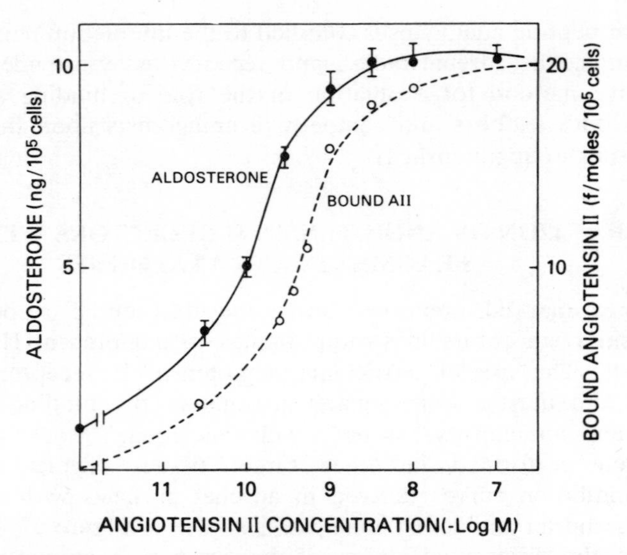


Figure 13.6. Relationship between angiotensin II binding and aldosterone production in collagenase-dispersed dog adrenal glomerulosa cells

correlation between peptide binding and activation, and the corresponding inhibition of both events by angiotensin antagonists[20,25] indicates that the binding sites for [$^{125}$I]iodoangiotensin II are indeed the biologically relevant receptors sites that modulate the actions of angiotensin upon adrenal steroidogenesis.

## VI. REGULATION OF RECEPTOR NUMBER AND AFFINITY

It is clear that the angiotensin II receptors in various target tissues are significantly modulated by alteration in electrolyte balance and in circulating angiotensin II levels. However, the regulation or control of angiotensin II receptors in adrenal, smooth muscle, and other target tissues, requires more detailed studies. Changes in electrolyte balance have been shown to alter both the affinity and the tissue concentration of the angiotensin receptor sites of the adrenal gland.[26] These changes grade into the hypertrophy and hyperplasia that occur in the adrenal during alterations in sodium ion balance, and which contribute to the enhanced adrenal responsiveness to angiotensin II during sodium deficiency. Even short periods of sodium ion deficiency cause significant changes in the receptors of the adrenal gland, and these appear to correlate with the enhanced sensitivity to angiotensin II which accompanies such sodium ion restriction.[27]

The extent to which circulating angiotensin II either maintains its receptor sites, or by analogy with several other peptide hormones, exerts negative regulation upon the target cell receptors, has yet to be determined. However, there is evidence to suggest that angiotensin II may exert both of these effects under appropriate conditions, since administration of angiotensin II has been shown to increase receptor sites in the adrenal,[28] and to decrease them in smooth muscle.[10] Furthermore, removal of endogenous angiotensin II by nephrectomy has been reported to both decrease[26] and increase[29] the adrenal receptor sites, and administration of the peptide has led to loss of adrenal receptors in nephrectomized rats.[29] These conflicting effects of angiotensin II on target-cell receptors may be dose-related, and more detailed analyses of these effects are necessary to determine the degree to which they occur during physiological regulation of the adrenal gland by angiotensin II.

The importance of these actions of angiotensin II in regulating its specific receptor sites occur *in vivo* during physiological and pathological changes in the activity of the renin–angiotensin system has recently begun to be clarified. Thus, there is evidence that the circulating level of angiotensin II is involved in the regulation of adrenal receptors in normal animals,[28] and that adrenal receptors undergo rapid changes in affinity and number when sodium ion intake is restricted and blood angiotensin levels are elevated.[27] Furthermore, the increase in adrenal angiotensin II receptors that

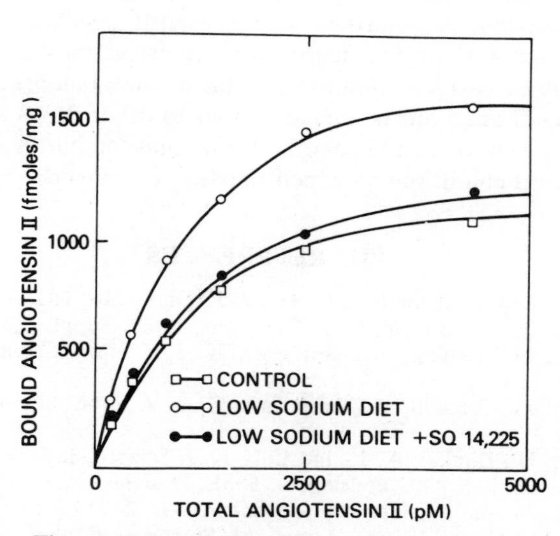


Figure 13.7. Effects of low-sodium ion diet and treatment with converting enzyme inhibitor (SQ 14,225) on angiotensin II receptor sites in rat adrenal glomerulosa cells measured by saturation analysis

accompanies sodium restriction in the rat was shown to be completely abolished by concomitant infusion of an inhibitor of angiotensin I converting enzyme (SQ 14,225), to block the formation of angiotensin II (Figure 13.7). These findings indicate that the increased receptor content and sensitivity of the adrenal in sodium-deprived animals are mediated by the rising blood angiotensin II concentration, and that the renin–angiotensin system is the major determinant of aldosterone secretion in acute sodium restriction.

In vascular tissues, prior occupancy of receptors by endogenous angiotensin II has been implicated in the diminished pressor responses to the peptide during sodium ion deficiency.[28] Also, differences between the angiotensin II receptors in adrenal and muscle have been reported, mainly from studies on the effects of angiotensin II antagonists during normal and low sodium ion intake.[29–31] During short-term sodium deficiency, adrenal and renal glomerular receptors behave similarly in exhibiting an initial increase in binding affinity[25,32] as well as a later rise in receptor numbers.[24,25,32]

More detailed information about the conformation of angiotensin II in solution and at the receptor site should be derived by binding studies employing fluorescent or other suitably labelled derivatives of angiotensin II agonists and antagonists, and by the use of affinity labelling methods to form covalently coupled hormone–receptor complexes. The relationship between the amino acid sequence of angiotensin II and the conformation of angiotensin II that confers biological activity as agonists or antagonists, also require more extensive analysis. The presently available antagonists of angiotensin II exhibit varying degrees of agonist activity in different target tissues, and under various conditions of electrolyte balance. In this regard, the angiotensin II antagonists appear similar to other drugs and antagonists, which commonly show variable agonist : antagonist activities according to the target tissues and conditions in which they are employed.

## VII. REFERENCES


1. Lin, S. Y., and Goodfriend, T. L. *Am. J. Physiol.*, **218**, 1319–1328, (1970).
2. Goodfriend, T. L., and Lin, S. Y. *Circ. Res.*, **26/27, Suppl. I**, 163–170, (1970).
3. Glossmann, H., Baukal, A., and Catt, K. J. *J. Biol. Chem.*, **249**, 825–834, (1974).
4. Glossmann, H., Baukal, A., and Catt, K. J. *J. Biol. Chem.*, **249**, 664–666, (1974).
5. Glossmann, H., Baukal, A. J., and Catt, K. J. *Science*, **185**, 281–283, (1974).
6. Catt, K. J., Baukal, A., Ketelslegers, J. M., Douglas, J., Saltman, S., Fredlund, P., and Glossmann, H. *Acta Physiol. Lat. Amer.*, **24**, 515–519, (1974).
7. Baudouin, M., Meyer, P., and Worcel, M. *Biochem. Biophys. Res. Commun.*, **42**, 434–440, (1971).
8. Devynck, M. A., and Meyer, P. *Am. J. Med.*, **61**, 758–767, (1976).
9. Rouzaire-Dubois, B., Devynck, M. A., Chevillote, E., and Meyer, P. *FEBS Letters*, **55**, 168–172, (1975).

10. Devynck, M. A., Rouzaire-Dubois, B., Chevillotte, E., and Meyer, P. *Eur. J. Pharmacol.*, **40**, 27–37, (1976).
11. Bennett, J. P., and Snyder, S. H. *J. Biol. Chem.*, **251**, 7423–7430, (1976).
12. Sraer, J. D., Sraer, J., Ardaillou, R., and Mimoune, O. *Kidney International*, **6**, 241–246, (1974).
13. Morgat, J. L., Lam Thanh, H., and Fromageot, P. *Biochim. Biophys. Acta*, **207**, 374–376, (1970).
14. Lin, S. Y., Ellis, H., Weisblum, B., and Goodfriend, T. *Biochem. Pharmacol.*, **19**, 651–662, (1970).
15. Neilsen, M. D., Jorgensen, M., and Giese, J. *Acta Endocrinologica*, **67**, 104–116, (1971).
16. Corvol, P., Rodbard, D., Drouet, J., Catt, K. J., and Menard, J. *Biochim. Biophys. Acta*, **322**, 392–400, (1973).
17. Douglas, J., Aguilera, G., Kondo, T., Catt, K. J. *Endocrinology*, **102**, 685–696, (1978).
18. Cain, M. D., Coghlan, J. P., and Catt, K. J. *Clin. Chim. Acta*, **39**, 21–34, (1972).
19. Douglas, J., Fredlund, P., Saltman, S., and Catt, K. J. *Endocrinology*, **96**, A181 (Abstr.), (1975).
20. Douglas, J., Saltman, S., Fredlund, P., Kondo, T., and Catt, K. J., *Circ. Res.*, **38, Suppl. II**, 108–112, (1976).
21. Saltman, S., Baukal, A., Waters, S., Bumpus, F. M., and Catt, K. J. *Endocrinology*, **97**, 275–282, (1975).
22. Goodfriend, T. L., and Peach, M. J. *Circ. Res.*, **36/37, Suppl. I**, 38–48, (1975).
23. Douglas, J., Bartley, P., Kondo, T., and Catt, K. J. *Endocrinology*, **102**, 1921–1923, (1978).
24. Aguilera, G., Capponi, A., Baukal, A., Fujita, K., Hauger, R., and Catt, K. J., *Endocrinology*, **104**, 1279–1285, (1979).
25. Saltman, S., Fredlund, P., and Catt, K. J. *Endocrinology*, **98**, 894–903, (1976).
26. Douglas, J., and Catt, K. J. *J. Clin. Invest.*, **58**, 834–843, (1976).
27. Aguilera, G., Hauger, R., and Catt, K. J. *Proc. Natl. Acad. Sci. U.S.A.*, **75**, 975–979, (1978).
28. Hauger, R. L., Aguilera, G., and Catt, K. J. *Nature (London)*, **271**, 176–177, (1978).
29. Pernollet, M. G., Devynck, M. A., Matthews, P. G., and Meyer, P. *Eur. J. Pharmacol.*, **43**, 361–372, (1977).
30. Thurston, H., and Laragh, J. H. *Circ. Res.*, **36**, 113–117, (1975).
31. Williams, G. H., McDonnel, L. M., Raux, M. C., and Hollenberg, N. K. *Circ. Res.*, **34**, 384–390, (1974).
32. Steele, J. M., and Lowenstein, J. *Circ. Res.*, **35**, 592–600, (1974).
33. Williams, G. H., Hollenberg, N. K., and Braley, L. M. *Endocrinology*, **98**, 1343–1350, (1976).
34. Beaufils, M., Sraer, J., Lepreux, C., and Ardaillou, R. *Am. J. Physiol.*, **22230**, 1187–1193, (1976).

Cellular Receptors for Hormones and Neurotransmitters
Edited by D. Schulster and A. Levitzki


CHAPTER 14

# Oxytocin and vasopressin receptors

Serge Jard
*Collège de France, Laboratoire de Physiologie Cellulaire, 11 Place Marcellin, Berthelot 75231, Paris Cedex 05, France*



## I. INTRODUCTION

Oxytocin and vasopressins are nonapeptides (Figure 14.1). They are synthesized by hypothalamic secretory neurons and liberated by nerve terminals from these neurons in contact with blood capillaries from the hypophysis neural lobe. Oxytocin and vasopressin are produced by different neurons and liberated under the influence of different stimuli. They are transported along the axons and secreted out of the cells in a complexed form with a carrier protein, neurophysin. Oxytocin and vasopressin are associated with distinct neurophysins that are closely related structurally. The peptide-neurophysin complex rapidly dissociates when liberated in the blood. The half-life of oxytocin and vasopressin in the blood is very short (one to a few minutes depending on the mammalian

OXYTOCIN (a)

Cys-Tyr-Ile-Gln-Asn-Cys-Pro-Leu-Gly ($NH_2$)

ARGININE-VASOPRESSIN (b)

Cys-Tyr-Phe-Gln-Asn-Cys-Pro-Arg-Gly ($NH_2$)

LYSINE-VASOPRESSIN (c)

Cys-Tyr-Phe-Gln-Asn-Cys-Pro-Lys-Gly ($NH_2$)

Figure 14.1. Primary structure of oxytocin and vasopressins. (a) Oxytocin was the first biologically active peptide obtained by complete synthesis in 1953 by Du Vigneaud *et al.*[11] (b) Present in most mammalian species including man. (c) Present in the pig and other members of the hog family. A very large number of oxytocin and vasopressin structural analogues have been prepared

species considered). They are rapidly eliminated by urinary excretion and destroyed by enzymic hydrolysis within the liver and kidneys.

In solution in dimethylsulphoxide oxytocin and vasopressin have a rather compact structure (Figure 14.2).

Oxytocin and vasopressin elicit a large variety of physiological and pharmacological effects in mammals (for review see Reference 3). These effects are usually classified into two categories: oxytocic effects (oxytocin more potent than vasopressin) and vasopressic effects (vasopressin more potent than oxytocin). The main oxytocic effects are: enhancement of uterine myometrium contractility (uterotonic effect), enhancement of uterine contractility of myoepithelial cells from the mammary gland (galactobolic effect), and increase in glucose oxidation by adipose cells (insulin-like effect). The main vasopressin effects are: reduction in urinary flow-rate with concomitant increase in urinary osmotic pressure (antidiuretic effect), increase in muscular tone of peripheral blood vessels with subsequent increase in systemic blood pressure (vasopressor effect), and increase in glycogen breakdown by the liver (glycogenolytic effect). In addition, vasopressin is active in stimulating the release of adrenocorticotropin (ACTH) by the hypophysis glandular lobe; fragments of the oxytocin molecule stimulate the release of melanocyte stimulating hormone (MSH) by hypophysis intermediary lobe. Oxytocin and vasopressin exert opposite effects on consolidation of memorization processes when administered in brain ventricles.

Among all these effects of neurophyseal peptides, the uterotonic and galactobolic effects of oxytocin, and the antidiuretic effect of vasopressin

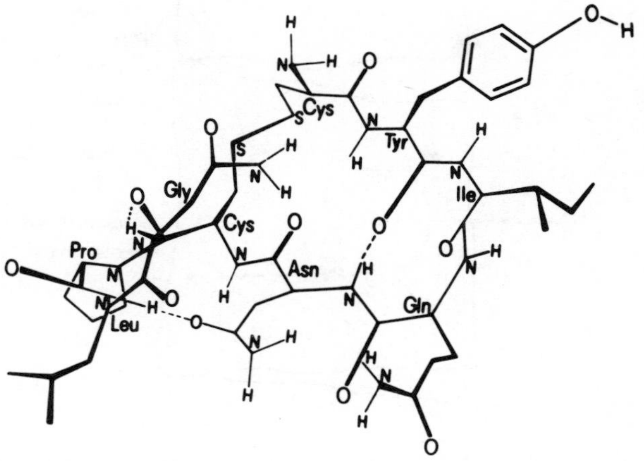

Figure 14.2. Three dimensional structure of oxytocin molecule in dimethylsulphoxide. Molecule contains two $\beta$-turns: one involving the sequence tyrosyl-isoleucyl-glutamyl-asparagyl is closed by a hydrogen bond between the NH of aspargine and the C–O of tyrosine. The second involving the peptide sequence cysteinyl-prolyl-leucyl-glycylamide is stabilized by a hydrogen bond between the C–O of cysteine and the NH of glycinamide. Reproduced with permission from Walter and Glickson, *Proc. Natl. Acad. Sci. U.S.A.*, **70**, 1199–1203, (1973)

can be elicited by the administration of low doses of hormone which are comparable to the amounts of hormone secreted by the neurohypophysis under physiological conditions. Furthermore, the more powerful stimuli responsible for an increased secretion of oxytocin and vasopressin are clearly related to the effects of these hormones on the uterus, the mammary gland and the kidney: increases in oxytocin secretion in response to suckling or to distension of uterine cervix occurring during parturition: increase in vasopressin secretion in response to an increase in concentration or a decrease in volume of extracellular fluids. The other biological effects of oxytocin and vasopressin (with the possible exception of their effects on the central nervous system) are elicited by the administration of hormone doses which are much higher than (or at the upper limit of) the hormonal secretion rates measurable under physiological conditions. The physiological significance of these effects is not yet clearly understood.

It has been convincingly established that the antidiuretic effects of vasopressin on the mammalian kidney is mediated by cyclic AMP. A vasopressin-sensitive adenylate cyclase has been found in epithelial cells

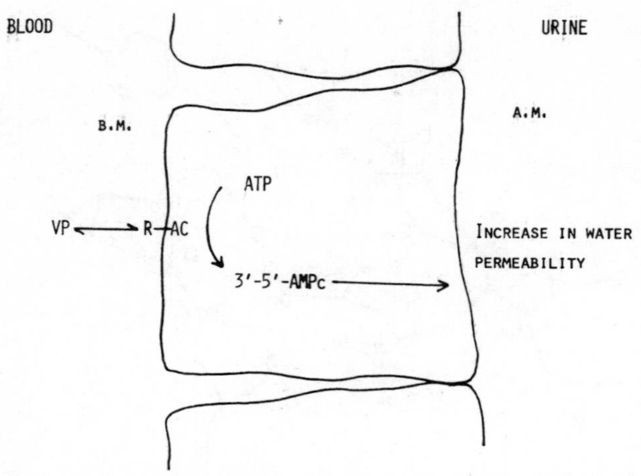


Figure 14.3. Mechanism of vasopressin action on epithelial cells from the amphibian urinary bladder and collecting ducts from the mammalian kidney. Vasopressin (VP) interacts with a specific receptor (R) located at the external surface of the basal plasma membrane (BM). The receptor triggers an increased production of 3′-5′ cyclic AMP by adenylate cyclase (AC). Cyclic AMP is responsible for an increase in water permeability of the apical plasma membrane (AM)

from two discrete segments of the mammalian nephron: the collecting duct and the ascending limb of the loop of Henle. Exogenous cyclic AMP mimics the vasopressin effect on water permeability of isolated collecting ducts as well as homologous effects of vasopressin which can be demonstrated on epithelial cells from the amphibian skin and bladder (for review see Reference 4). As diagrammatically shown in Figure 14.3, the vasopressin-sensitive adenylate cyclase is located within the basal membrane of the target cell facing the blood. Cyclic AMP induces an increase in water permeability of the luminal cell membrane in contact with the tubular fluid. Indirect evidence has been provided suggesting that membrane permeability change might be a consequence of the dephosphorylation of a specific membrane protein (protein D). This dephosphorylation being indirectly controlled by a cyclic AMP dependent-protein kinase.

There is no experimental evidence indicating that cyclic AMP might be directly involved either in the vasopressor and glycogenolytic effects of vasopressin or in the uterotonic, galactobolic, an insulin-like effects of oxytocin. The possibility that these effects might be mediated through either an increased permeability of the cell membrane to $Ca^{2+}$ ions or mobilization of sequestrated intracellular calcium has been repeatedly suggested.

## II. CHARACTERIZATION OF THE VASOPRESSIN RECEPTOR

### (i) Preparation of biological material and labelled ligands

The only vasopressin receptors characterized so far are the renal receptors from three species (pig, ox, and rat). The biological material used for receptor characterization consisted of partially purified membrane fractions prepared from homogenates of the medullo-papillary portion of the kidney. These fractions were enriched in vasopressin-sensitive adenylate cyclase activity and did not inactivate vasopressin. Characterization of the vasopressin receptor was made possible by the preparation of ([$^3$H]tyrosyl)-lysine-vasopressin and ([$^3$H]tyrosyl)-arginine-vasopressin of specific activities in the range 10–40 Ci mmol$^{-1}$. The tritiated peptides were prepared by catalytic dehalogenation of their respective di-iodotyrosyl derivatives in the presence of pure tritium.[5] Radioiodinated vasopressin which can be prepared much more easily than tritiated vasopressin could not be used for receptor characterization since it is essentially inactive biologically.

### (ii) Binding* of tritiated vasopressin to kidney membranes

Binding of tritiated vasopressin to renal medulla membranes is time-dependent and reversible as illustrated in Figure 14.4 in the case of [$^3$H]lysine-vasopressin binding to pig kidney membranes. The rate of hormone–receptor complex formation decreased when temperature was lowered. The time needed to reach equilibrium value increased when the initial hormonal concentration in the membrane incubation medium was lowered. The time-course of hormone–receptor complex formation and dissociation can be accounted for by the following reaction scheme:

$$R + H \underset{k_{-1}}{\overset{k_1}{\rightleftharpoons}} RH$$

involving a reversible binding of vasopressin molecules (H) to a homogenous population of independent receptor sites (R). The dose-dependency for hormonal binding measured under steady-state conditions is also compatible with such a model; the dose-binding curve is

*In the following description, only the specific component of binding will be considered (see Section I, Chapter 2).

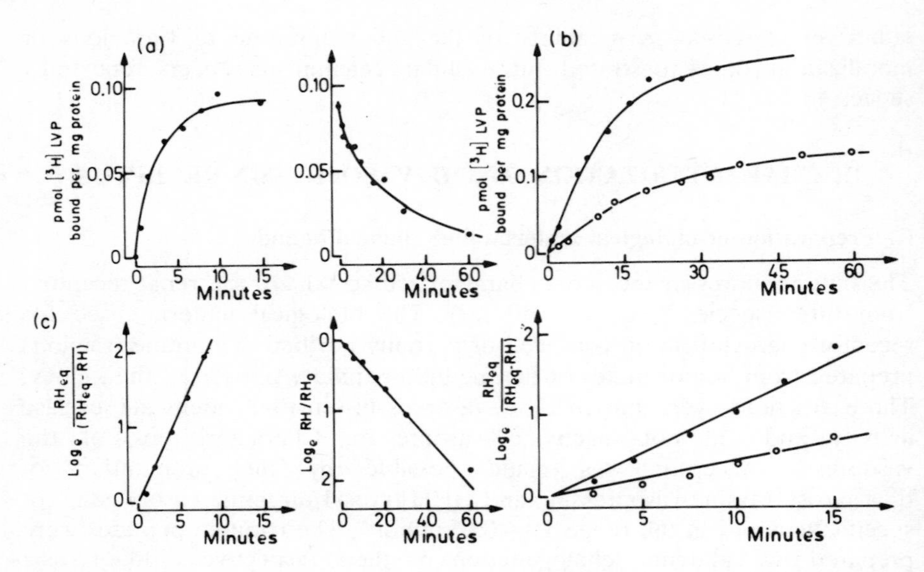


Figure 14.4 Time-courses of the association and dissociation of [$^3$H]vasopressin with membrane receptors from the pig kidney. The experiment illustrated in (a) was performed at 30 °C in the presence of $10^{-8}$ M [$^3$H]lysine-vasopressin ([$^3$H]LVP). Reversal of binding was induced by a 20-fold dilution of the membrane incubation medium. The experiment illustrated in (b) was performed at 15 °C using two different [$^3$H]LVP concentrations: $2 \times 10^{-8}$ M (•) and $5 \times 10^{-9}$ M (○). The two experiments were performed using different membrane preparations. The semilogarithmic plots of the association and dissociation curves are shown in (c). Modified from Bockaert *et al.*[6]

of an almost pure Michaelian type (Table 14.1). Binding of tritiated vasopressin could be inhibited by unlabelled vasopressin and a large series of structural analogues. These studies failed to reveal any heterogeneity in the population of receptor sites with respect to both their accessibility for the different active peptides and their affinity for a given peptide. Characteristics of renal vasopressin receptors from the three mammalian species so far studied are given in Figure 14.5.

Table 14.1. Characteristics of renal vasopressin receptors from three mammalian species

| Species | Maximal binding capacity (pmol/mg protein) | Dissociation constant (nM) |
|---|---|---|
| Pig | 1.0 | 20 |
| Ox* | 3.6 | 4 |
| Rat | 0.2 | 4 |

*Data obtained with [$^3$H]arginine-vasopressin.[8]

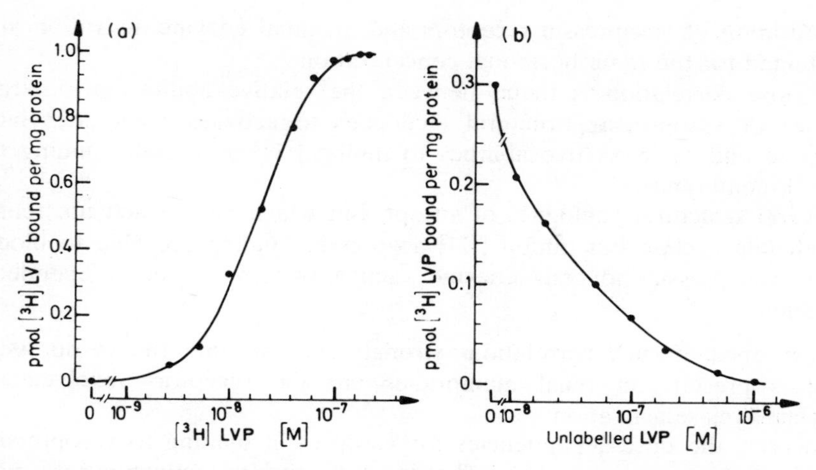


Figure 14.5. Dose-dependency of [$^3$H]vasopressin binding to pig kidney membranes. (a) Specific [$^3$H]lysine-vasopressin binding ([$^3$H]LVP) was measured under steady-state conditions for the indicated concentrations of labelled hormone. (b) Effect of increasing concentrations of unlabelled hormone on the binding of a constant amount of labelled hormone ($10^{-8}$ M). Modified from Roy *et al.*[7]

The vasopressin receptor from the pig kidney could be extracted from the membrane in a soluble and active form under the influence of non-ionic detergents. Determination of the hydrodynamic parameters of the soluble material indicated that the hormone is associated with highly assymetrical protein: Stokes radius, 57 Å, frictional ratio, 1.8, and molecular weight, 83,000 (G. Guillon, unpublished results).

### (iii) Relation of vasopressin binding to activation of renal adenylate cyclase

Numerous correlations have been established between vasopressin binding to renal membrane and adenylate cyclase activation.

1. Intrarenal distributions of vasopressin binding capacity and vasopressin-sensitive enzyme activity are superimposable.
2. Vasopressin receptors and enzyme responsiveness to hormonal stimulation appear simultaneously during the post-natal phase of kidney development in the rat.
3. Parallel modifications in number of vasopressin receptors and magnitude of maximal enzyme stimulation by the hormone can be demonstrated under several experimental situations in the rat (acute or chronic administration of exogenous vasopressin or chronic changes in the rate of secretion of endogenous vasopressin).

4. Saturation of vasopressin receptors and maximal enzyme activation are obtained for the same hormonal concentration.
5. A close correlation is found between the relative abilities of a large series of vasopressin structural analogues to activate renal adenylate cyclase and their relative abilities to inhibit [$^3$H]vasopressin binding to renal membranes.
6. Several structural analogues of vasopressin which do not activate renal adenylate cyclase but inhibit [$^3$H]vasopressin binding are able to block the vasopressin-induced enzyme activation in a dose-dependent manner.*

The existence of such correlations strongly suggests that the vasopressin receptors present on renal membranes are the receptors involved in adenylate cyclase activation.

However, the dose-dependencies for vasopressin binding to vasopressin receptors and vasopressin-induced adenylate cyclase activation are not superimposable.[4,6,8] The magnitude of adenylate cyclase activation is a saturable function of receptor occupancy (non-linear coupling). The existence of a non-linear coupling is also apparent when the time-courses of vasopressin binding and adenylate cyclase activation are compared. The enzyme activation increases more rapidly with time than the hormonal binding.[8] One can account for this apparent paradox by assuming that the first hormone–receptor complexes formed are more efficient in activating adenylate cyclase than the hormone complexes formed later on. This can be achieved if the receptor can move laterally *vis à vis* the enzyme and activate several adenylate cyclase molecules. The probability for a given hormone receptor complex to do so decreases as the number of non-activated adenylate cyclase molecules decreases.

### (iv) Relation of vasopressin binding to final antidiuretic response

There is substantial experimental evidence indicating that the final antidiuretic response of the kidney is the direct consequence of an enhanced cyclic AMP production within vasopressin target cells. However the dose-dependency for vasopressin-induced antidiuresis measured in the intact animal is strikingly different from the dose-dependency for vasopressin-induced adenylate cyclase activation measured *in vitro* on membrane fractions prepared from kidney homogenates (Figure 14.6). A likely explanation for such a discrepancy is to assume that only a small fraction of cyclic AMP which can be generated by maximally stimulated

*Among vasopressin analogues behaving like competitive inhibitors are: *N*-[pivarloyl]-[2-*O*-methyltyrosine]-oxytocin ($K_d = 7 \times 10^{-6}$ M), *N*[carbamoyl]-[2-*O*-methyltyrosine]-oxytocin ($K_d = 2 \times 10^{-5}$ M), and 1[β-mercapto-β-β-cyclopentamethylene propionic acid]-4 [valine]-8[D-arginine]-vasopressin ($K_d = 1.2 \times 10^{-8}$ M).

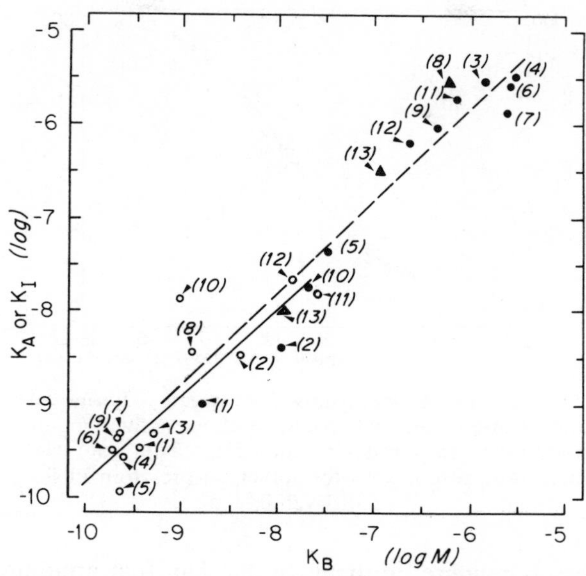


Figure 14.6. The relationship between the apparent affinities of a set of vasopressin analogues for specific binding and adenylate cyclase activation. Points on the graph refer to different vasopressin structural analogues.[10] Peptide concentrations leading to half maximal enzyme activation $K_A$ or inhibition constants for competitive blockers of vasopressin-induced enzyme activation ($K_I$) are plotted as a function of peptide concentrations leading to half saturation of vasopressin receptors from pig (dotted line) and rat renal membranes

vasopressin-sensitive adenylate cyclase is sufficient to induce a maximal change in the permeability of the vasopressin target cells. In other words, a large 'receptor reserve' for vasopressin might exist in the mammalian kidney. From available experimental data one can estimate that occupancy of less than 5% of the total number of vasopressin receptors present in the kidney is sufficient to induce maximal antidiuresis (Figure 14.7). Such an hypothesis provides a simple explanation for the observation that vasopressin analogues which behaved like partial agonists of low intrinsic activity when tested for their ability to activate renal adenylate cyclase behaved like full agonists when tested for their ability to induce an antidiuretic response *in vivo*.

Furthermore, direct comparison of the antidiuretic activities of vasopressin structural analogues with their corresponding activities on renal

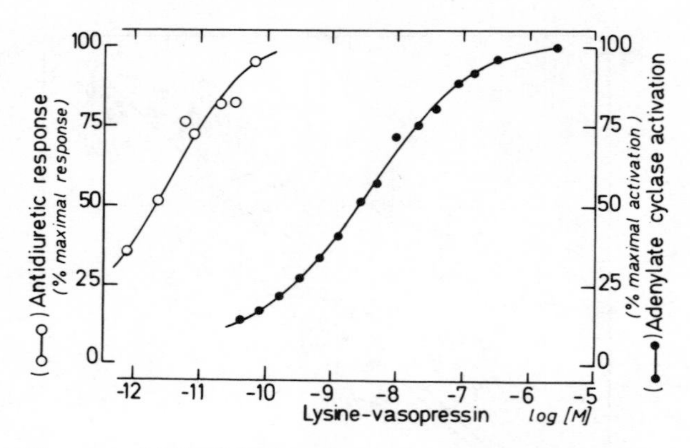


Figure 14.7. Relationships between hormonal concentration and adenylate cyclase activation or antidiuretic response in the rat. Data on the antidiuretic response in the intact rat are from Pliska and Rychlik[9]

adenylate cyclase is rendered difficult by the fact that antidiuretic potency is markedly influenced by the metabolic stability of these peptides when injected into intact animals. Making the simple assumption that the potential antidiuretic potency of a given vasopressin structural analogue is proportional to: (1) its affinity for the vasopressin receptor; (2) the magnitude of maximal adenylate cyclase activation which can be elicited by this peptide; and (3) its half-life in the intact animal. A good correlation was found between calculated and experimentally determined antidiuretic activities of a series of vasopressin structural analogues.[10]

### (v) Modulations in the number and properties of renal vasopressin receptors

Responsiveness of renal adenylate cyclase to vasopressin stimulation can be modified in response to adrenal steroids deprivation, glucocorticoids administration,[11] or acute and chronic changes in vasopressin concentration in the blood.[12] In the rat, adrenalectomy is accompanied by a progressive reduction in the magnitude of maximal adenylate cyclase activation by vasopressin. A normal response is restored by *in vivo* administration of dexamethasone, a potent artificial glucocorticoid. Glucocorticoid deficiency or administration does not significantly affect either the number of renal vasopressin receptors or their affinity for the hormone. It was suggested from these observations that glucocorticoids might control the efficiency of receptor–enzyme coupling. In the rat, experimental modifications in vasopressin blood levels induce a biphasic, concentration-dependent change in the responsiveness of renal adenylate cyclase to vasopressin stimulation. Within the physiological range of variation, the higher the blood vasopressin concentration, then the higher is the enzyme sensitivity to hormonal stimulation. Maximal sensitivity is obtained for blood vasopressin levels which correspond to those eliciting a maximal antidiuretic response, but are much lower than those needed to ensure complete saturation of renal vasopressin receptors (existence of a large receptor reserve). When blood vasopressin concentration is raised to saturating levels for the renal receptors, a marked reduction in enzyme responsiveness occurs with a concomitant reduction in receptor number.

## III. OXYTOCIN RECEPTORS

The fact that the first biochemical events triggered by binding of oxytocin to its receptors are still unknown rendered complete characterization of these receptors difficult.

### (i) Oxytocin receptors from the mammary gland and uterus

Tissue uptake of radioactivity from [$^3$H]oxytocin could be demonstrated using pieces of mammary gland from the lactating rat and segments of uterine horns taken from oestrogen-treated rats.[3] Radioactivity uptake by these two structures is linear with time up to at least one hour. The time-course of tissue uptake is strikingly different from the time-course of the biological response; the uterotonic effect of oxytocin is fully developed within 1–2 min following hormonal treatment; similarly maximum contraction of mammary strips induced by oxytocin occurs within less than 2 min. However, several arguments have been presented favouring the conclusion that uptake is in some way related to the final effect of oxytocin. Thus the ability of various oxytocin analogues to inhibit [$^3$H]oxytocin uptake was proportional to their uterotonic or galactobolic activities (Figure 14.8); the uptake is target organ specific; within a target organ uptake is target cell specific (myoepithelial cells from the mammary gland and smooth muscle cells from the oviduct).

A class of high affinity sites for tritiated oxytocin ($K_d = 1.8 \times 10^{-9}$ M) could be demonstrated on dispersed cells obtained from rat mammary tissue and $20{,}000 \times g$ particles from rat mammary gland and sow myometrium.[14,15] During fractionation of $20{,}000 \times g$ particles, by sucrose gradient centrifugation, there was a general relationship between the distribution of binding and the activity of 5′-nucleotidase, a putative marker enzyme for plasma cell membranes.

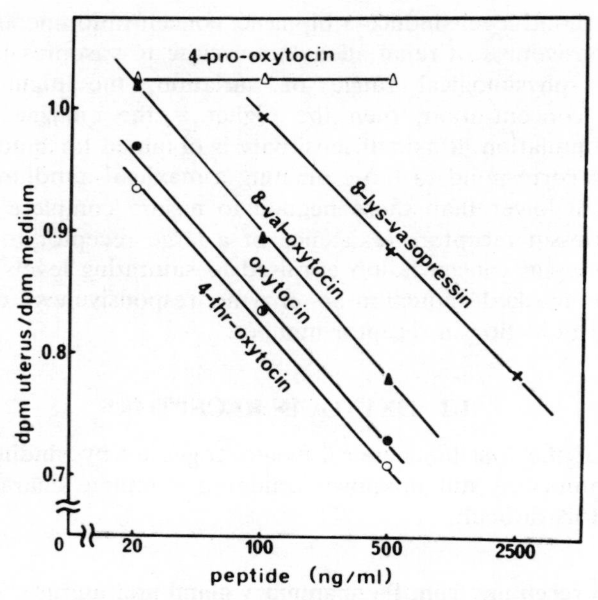


Figure 14.8. Effect of unlabelled oxytocin and analogues on the uptake of radioactivity from [$^3$H]oxytocin by uterine segments. The relative uterotonic potencies of the analogues tested are: oxytocin, 1; 4-Pro-oxytocin, $<0.01$; 8-Val-oxytocin, 0.6; and 4-Thr-oxytocin, 2. Reproduced with permission from Soloff *et al*. *Endocrinology,* **92**, 104–107, (1973)

### (ii) Oxytocin receptors from isolated rat fat cells

Under conditions where oxytocin elicits activation of glucose oxidation by isolated rat fat cells, a specific uptake of [$^3$H]oxytocin by these cells can be demonstrated.[16] Two classes of binding sites with different affinities and capacities are present on intact rat fat cells or ghosts from these cells (Figure 14.9). The sites of the first type exhibit a rather high affinity, but low capacity, for oxytocin (5 nM, 30,000 sites per cell). Dose-dependent binding of oxytocin to these sites and oxytocin-induced stimulation of glucose oxidation occur in the same range of hormonal concentration. There is a good correlation between the relative capacities of different oxytocin analogues to compete with oxytocin for binding to the high affinity sites and their relative abilities to increase glucose oxidation. Therefore, it is likely that this class of oxytocin binding sites are the oxytocin receptors involved in the insulin-like action of the hormone on rat

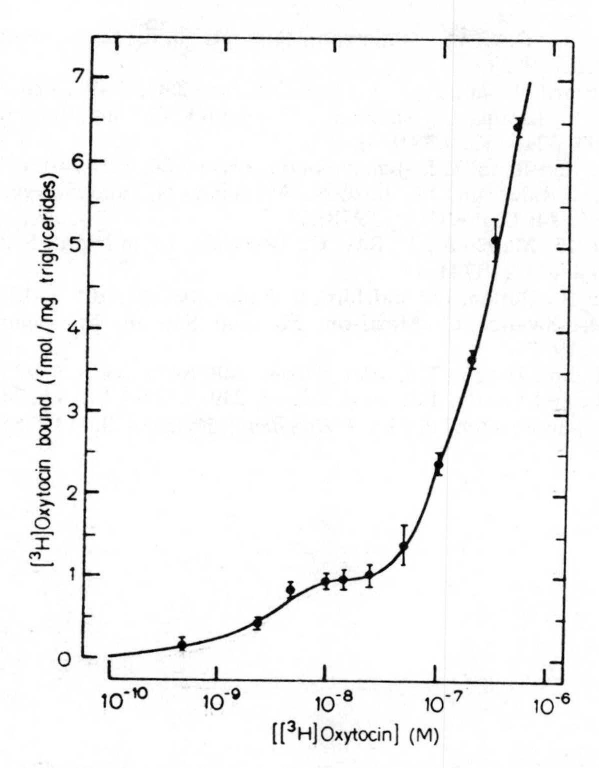


Figure 14.9. Specific binding of [$^3$H]oxytocin to isolated rat fat cells. Reproduced with permission from Bonne and Cohen, *Eur. J. Biochem.*, **56**, 295–303, (1975)

fat cells. The eventual physiological role of sites of the second category which has low affinity and high capacity for oxytocin is not yet established.

## IV. REFERENCES

1. Du Vigneaud, V., Ressler, C., Swan, J. M., Roberts, C. W., Katsoyannis, P. G., and Gordon, S. *J. Amer. Chem. Soc.*, **25**, 4879–4880, (1953).
2. Walter R., and Glickson, J. D. *Proc. Natl. Acad. Sci. U.S.A.*, **70**, 1199–1203, (1973).
3. Berde, B. (ed.) *Neurohypophysial Hormones and Similar Polypeptides: Handbook of Experimental Pharmacology*, Springer-Verlag, Berlin, Vol. XXIII. (1968).
4. Jard, S., and Bockaert, J. *Physiol. Rev.*, **55**, 489–536, (1975).
5. Pradelles, P., Morgat, J. L., Fromageot, P., Camier, M., Bonne, D., Cohen, P., Bockaert, J., and Jard, S. *FEBS Letters*, **26**, 189–192, (1972).

6. Bockaert, J., Roy, C., Rajerison, R., and Jard, S. *J. Biol. Chem.,* **248**, 5992–5931, (1973).
7. Roy, C., Barth, T., and Jard, S. *J. Biol. Chem.,* **250**, 3149–3156, (1975).
8. Hechter, O., Terada, S., Nakahara, T., Flouret, G., and Bergman, N. *J. Biol. Chem.,* **253**, 3219–3229, (1978).
9. Pliska, V., and Rychlik, I. *Acta endocrin. (Kbh),* **54**, 129–140, (1967).
10. Butlen, D., Rajerison, R., Jard, S., Manning, M., and Sawyer, W. H. *Mol. Pharmacol.,* **14**, 1006–1017, (1978).
11. Rajerison, R., Marchetti, J., Roy, C., Bockaert, J., and Jard, S. *J. Biol. Chem.,* **249**, 6390–6400, (1974).
12. Rajerison, R., Butlen, D., and Jard, S. *Endocrinology,* **101**, 1–12, (1977).
13. Soloff, M., Swartz, T., Morrison, M., and Saffran, M. *Endocrinology,* **92**, 104–107, (1973).
14. Soloff, M., and Swartz, T. *J. Biol. Chem.,* **248**, 6471–6478, (1973).
15. Soloff, M., and Swartz, T. *J. Biol. Chem.,* **249**, 1376–1381, (1974).
16. Bonne, D., and Cohen, P. *Eur. J. Biochem.,* **56**, 295–303, (1975).

Cellular Receptors for Hormones and Neurotransmitters
Edited by D. Schulster and A. Levitzki


CHAPTER 15

# Catecholamine receptors

Alexander Levitzki
*Department of Biological Chemistry, The Institute of Life Sciences, Hebrew University of Jerusalem, Jerusalem, Israel*



## I. CATECHOLAMINES AS HORMONES AND NEUROTRANSMITTERS

Catecholamines elicit a multitude of biochemical, physiological and pharmacological effects in a large variety of tissues. The wide spectrum of activities induced at the different target cells by catecholamines is brought about by the interaction of these compounds with specific catecholamine receptors. The active catecholamine can either function as a hormone at different organs or as a neurotransmitter at a postsynaptic membrane. Catecholamine receptors appear in the periphery as well as in the central nervous system (CNS), and belong to a number of classes and subclasses. The different types of catecholamine receptors are distinguished from each other in ligand specificity as well as in the nature of the biochemical response elicited subsequent to catecholamine binding to its specific receptor.

Epinephrine (adrenaline) and norepinephrine (noradrenaline) are secreted by the adrenal medulla, and thus function as hormones at peripheral organs with specific adrenoreceptors. Catecholamines also function as neurotransmitters released from catecholaminergic nerve terminals in peripheral tissues as well as in the CNS. The catecholamine released at a catecholaminergic synapse interacts with specific postsynaptic catecholamine receptors. These receptors include the $\beta$-adrenergic receptors, $\alpha$-adrenergic receptors, and dopamine receptors. Dopamine receptors appear mainly in the CNS but were recently found also in the retina and the kidney.

## II. THE CLASSIFICATION OF CATECHOLAMINE RECEPTORS

Catecholamine receptors are classified according to their ligand specificity and not according to the physiological or pharmacological activity resulting from receptor occupancy. A brief summary of the main catecholamine receptors is given in Table 15.1. The most intensive biochemical studies were carried out on the $\beta$-adrenergic receptors, whereas the $\alpha$-adrenergic receptors and dopamine receptors are less well characterized from the biochemical point of view. It is now fundamentally established that three main types of catecholamine receptors can be recognized: the $\beta$-adrenergic receptors, the $\alpha$-adrenergic receptors, and the dopamine receptors. The distinction between $\alpha$-adrenergic receptors and $\beta$-adrenergic receptors is based on the criteria which were defined in the classical studies of Ahlquist.[1,2]

The characterization of adrenoreceptors[3] is based on a two-fold procedure: (1) the relative potency (potency ratio) of a series of adrenergic agonists for eliciting the specific response and (2) the potency of an antagonist for blocking the response to a given agonist. Based on these two principles, the general definitions of $\alpha$- and $\beta$-receptors are given by the following statements:

**$\beta$-Receptors**: a $\beta$-receptor is one which mediates a response pharmacologically characterized by: (1) a relative potency series: isoprenaline $>$ epinephrine (adrenaline) $\geqslant$ norepinephrine (noradrenaline) $\gg$ phenylephrine—and (2) a susceptibility to specific blockage by pindolol, propranolol, or alprenolol at relatively low concentrations.

K

Table 15.1. The main classes of catecholamine receptors

| Type of receptor | Ligand specificity | Specific blockers | Second messenger |
|---|---|---|---|
| $\alpha$* | Norepinephrine > epinephrine > phenylephrine > isoproterenol. | Phentolamine, ergotamine, phenoxybenzamine, dibenamine | $Ca^{2+}$ Phospholipid effect |
| $\beta$† | Isoproterenol > epinephrine ⩾ norepinephrine ≫ phenylephrine. | Dichloroisoproterenol, propranolol, alprenolol, pindolol | cyclic AMP |
| Dopamine‡ | Dopamine > norepinephrine | Haloperidol, chloropromazine, some $\alpha$-blocking agents | cyclic AMP |

*On the basis of pharmacological experiments, it is now suggested that there are two types of $\alpha$-receptors: $\alpha_1$ which are postsynaptic and $\alpha_2$ which are presynaptic residing on the nerve terminal and controlling the release of catecholamines from the nerve terminal (see text).

†Two classes of $\beta$-receptors have been identified: $\beta_1$ and $\beta_2$. The distinction between the two is based on differences in the ligand specificity of the two receptors (see text).

‡A secondary biochemical response, other than cyclic AMP formation, has also been postulated recently. Thus, it may be that there is more than one class of dopamine receptor.

Comment: $\alpha$-Adrenergic receptors and $\beta$-adrenergic receptors are specific for the R stereoisomers (or the L-stereoisomers, if the optical rotatory power is considered).

**$\alpha$-Receptors**: an $\alpha$-receptor is one which mediates a response pharmacologically characterized by: (1) a relative potency series: norepinephrine $>$ epinephrine $>$ phenylephrine $\gg$ isoprenaline—and (2) a susceptibility to specific blockage by phentolamine, dibenamine, or phenoxybenzamine at low concentrations.

A summary of agonist specificity of $\alpha$- and $\beta$-receptors is given in Figure 15.1, and a summary of the chemical formulas of $\alpha$- and $\beta$-antagonists is given in Figure 15.2. The responses characteristic of $\alpha$- and $\beta$-adrenergic receptors have been further divided into subclasses. For instance, $\beta$-adrenergic responses can be divided into two subclasses: $\beta_1$-receptors and $\beta_2$-receptors. Epinephrine and norepinephrine are equipotent activators of $\beta_1$-receptors which are preferentially inhibited by practolol. Epinephrine, however, is more potent than norepinephrine in activating $\beta_2$-receptors which are inhibited by butoxamine. The $\beta$-antagonists propranolol, alprenolol, and pindolol inhibit $\beta_1$- and $\beta_2$-receptors with a similar potency.

Both types of receptors are stereospecific for the R-stereoisomer of either the agonist or the antagonist. As is seen in Table 15.1 and Figure 15.3, dopamine receptors are blocked by still another type of blocker, namely, by compounds of the phenothiazine family, such as chlorpromazine, or butyrophenones, such as haloperidol. It should, however, be emphasized that lysergic acid derivatives interact with dopamine receptors as well as with $\alpha$-receptors. Thus, ergocryptine was found to be an $\alpha$-blocker and D-LSD (lysergic acid diethylamide) was found to bind to dopamine receptors.

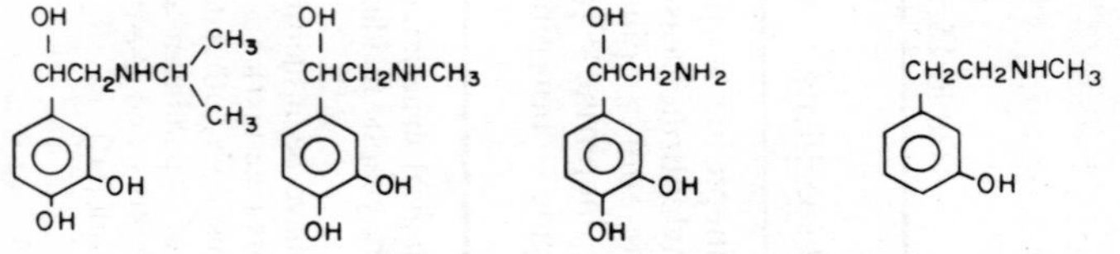

(−)ISOPRENALINE (−)ADRENALINE (−)NORADRENALINE (−)PHENYLEPHRINE

$\alpha$ 1 $<$ 2 $<$ 3 $<$ 4

$\beta$ 1 $>$ 2 $\gg$ 3 $\gg$ 4

Figure 15.1. The main agonists for $\alpha$- and $\beta$-receptors. Isoprenaline and phenylephrine are both synthetic compounds and are not found in living tissues

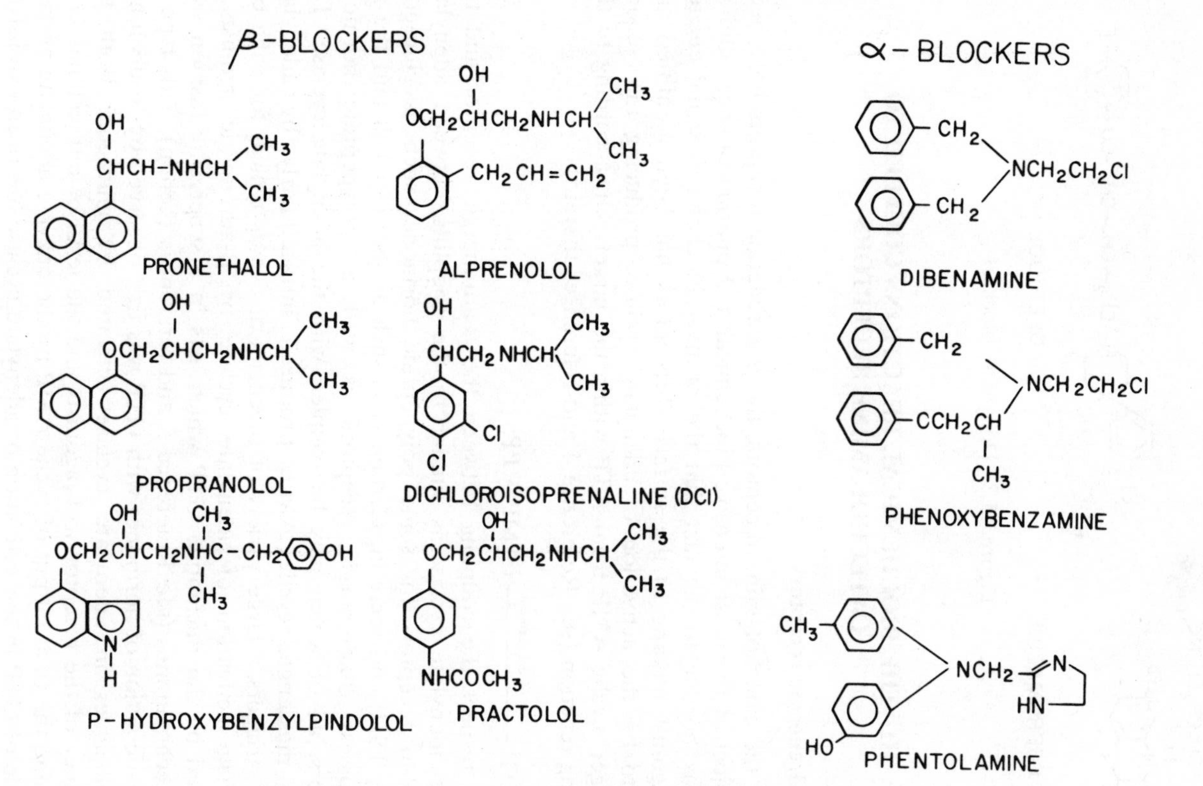


Figure 15.2. The structure of some $\alpha$- and $\beta$-adrenergic blockers

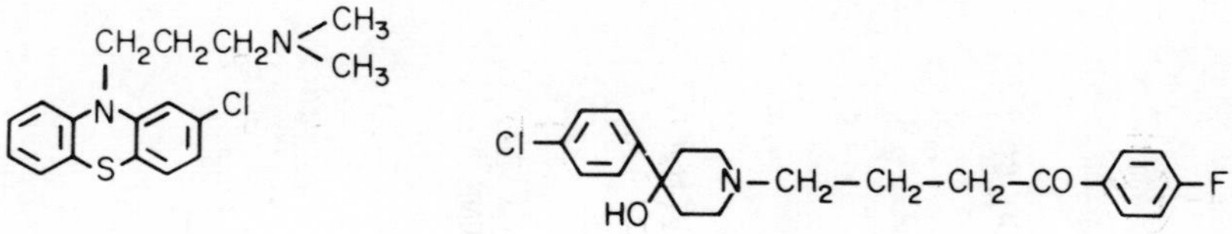

CHLORPROMAZINE

HALOPERIDOL

Figure 15.3. Dopamine blockers

## III. THE BIOCHEMICAL REACTIONS COUPLED TO CATECHOLAMINE RECEPTORS

### (i) β-Adrenergic receptors

Among the catecholamine receptors, the β-adrenergic receptors have been the subject of the most intensive biochemical and pharmacological studies. This may be due to the fact that the primary biochemical signal elicited upon agonist binding to the surface β-receptors has been identified and found to be the activation of adenylate cyclase, producing the 'second messenger' cyclic AMP from ATP within the target cell, according to the following reaction (see Reference 4 and references therein):

$$\text{ATP} \xrightarrow{l\text{ (R)-catecholamine}} \text{cAMP} + \text{PP}_i \qquad (1)$$

In this respect the coupling between the β-adrenergic receptor and the enzyme adenylate cyclase is similar to the coupling between adenylate cyclase to receptors for some polypeptide hormones such as glucagon, ACTH, TSH, and secretin. Certain cells such as the liver cell and the fat cell possess β-adrenergic receptors as well as polypeptide hormone receptors, all of which may be coupled with the adenylate cyclase. The second messenger cyclic AMP, produced intracellularly by adenylate cyclase, triggers a large variety of biochemical events typical for each cell type. The activation of adenylate cyclase by β-adrenergic agonists is mediated by the nucleotide GTP which acts in a synergistic fashion with the catecholamines (see Reference 5 and references therein). This type of synergistic action of hormones with GTP was first observed by Rodbell and his colleagues in adenylate cyclase activated by glucagon. Both the occupancy of the receptor with agonists and the level of intracellular GTP determine the final output of cyclic AMP by the enzyme adenylate cyclase. The role of GTP in the activation of adenylate cyclase has been extensively studied using non-hydrolysable analogues of GTP such as p(NH)ppG and GTPγS.[5] It is not as yet clear whether the activity of the enzyme is absolutely dependent on GTP and does not exhibit activity in its absence or possesses some basal activity in the presence of the $\beta$-adrenergic agonist alone. Recently it was found that turkey erythrocytes possess a specific $\beta$-adrenergic receptor dependent GTPase.[6]

$$\text{GTP} \xrightarrow{l\ \text{(R)-catecholamine}} \text{GDP} + \text{P}_i \tag{2}$$

It was suggested that the GTPase activity is closely associated with the adenylate cyclase and determines the steady-state level of GTP, and thus the level of adenylate cyclase activity. A general model for the interrelationship between GTP and $\beta$-agonists was recently proposed[5,7] and can be summarized by the following scheme:

$$\text{R}\cdot\text{E} \underset{}{\overset{\text{GTP}}{\rightleftharpoons}} \text{R}\cdot\text{E}\cdot\text{GTP} \overset{\text{H}}{\rightleftharpoons} \text{HR}\cdot\text{E}\cdot\text{GTP} \xrightarrow{k_3} \text{H}\cdot\text{R}'\cdot\text{E}'\cdot\text{GTP} \xrightarrow{k_4} \text{HR}\cdot\text{E} + \text{GDP} + \text{P}_i \tag{3}$$

where R is the receptor and E the enzyme, adenylate cyclase coupled to it. When both GTP and the agonist H are bound to their respective regulatory sites, the enzyme is converted to its active form E′. The E′ state exhibits GTPase activity and is responsible for the termination of the hormonal signal once the bound GTP is hydrolysed and the enzyme reverts to its inactive state E. The adenylate cyclase catalytic moiety and the GTPase catalytic moiety reside on two different subunits, both of which are coupled and associated with the third unit, namely the $\beta$-receptor, R. This state of affairs ensures both that a finite level of E′ will be achieved and that the hormone induced signal will be eventually terminated. The values of $k_3$ and $k_4$ were measured independently, and thus the value of E′ at the steady-state could be calculated.[5,7]

An alternative explanation is that the receptor and the enzyme are permanently uncoupled from each other and that the enzyme becomes activated during the formation of the transient complex H·R·E:

$$\text{HR} + \text{E}\cdot\text{GTP} \underset{k_2}{\overset{k_1}{\rightleftharpoons}} \text{HR}\cdot\text{E}\cdot\text{GTP} \xrightarrow{k_3} \text{HR} + \text{E}'\cdot\text{GTP} \xrightarrow{k_4} \text{E} + \text{EDP} + \text{P}_i \tag{4}$$

The activated enzyme E′ is converted back to E concomitantly with the hydrolysis of GTP at the regulatory site. Evidence favouring the latter mechanism has recently been obtained for the case of turkey erythrocyte $\beta$-receptor dependent adenylate cyclase.[7] Since GTP and p(NH)ppG function as activators in other hormone sensitive adenylate cyclase systems from different tissues,[8] it seems that the interrelations between hormones and GTP described in equations (3) and (4) are a general phenomenon.

These studies are in line with the fact that the $\beta$-receptor and the enzyme reside on separate macromolecules, both of which can diffuse slowly in the membrane matrix.[9] Elegant experiments performed recently by Schramm and his colleagues (see Reference 9 and references therein) have demonstrated this fact (see Chapter 4). In these experiments turkey erythrocytes, in which the catalytic activity of adenylate cyclase had been inactivated by *N*-ethylmaleimide or by heat, served to contribute the $\beta$-adrenergic receptor. Friend erythroleukaemic cells (F-cells) which possess no $\beta$-adrenergic receptors served to contribute the enzyme adenylate cyclase. The erythrocytes in which the enzyme had been inactivated were fused with the F-cells by *Sendai* virus. The cell ghosts of the fused preparation demonstrated *l*(R)-isoproterenol dependent adenylate cyclase. These experiments therefore reveal that the $\beta$-adrenergic receptor of the turkey erythrocytes must have become functionally coupled to the adenylate cyclase of the mouse F-cells. Activation by isoproterenol was demonstrable within a few minutes after fusion, and inhibitors of protein synthesis had no effect. Thus, coupling must have occurred between the pre-existing components. These fusion experiments demonstrate that the receptor has become permanently attached to the active adenylate cyclase, thus conferring hormone sensitivity to it. The rate of formation of the newly formed enzyme–receptor complex is slow, but once formed, it is rather stable.

Recent experiments on clones of lymphoma cells[10] very strongly suggest that the $\beta$-adrenergic receptor, the adenylate cyclase and the GTP regulatory unit are the products of separate genes.

### (ii) $\alpha$-Adrenergic receptors

It has been demonstrated that the primary event occurring upon occupation of the $\alpha$-adrenergic receptor by an $\alpha$-agonist in the parotid gland is the influx of $Ca^{2+}$ which functions as the 'second messenger'.[11] Furthermore, the specific $Ca^{2+}$ ionophore A-23187, when incorporated into the cell membrane, can substitute for the $\alpha$-adrenergic ligand and bypass the receptor-dependent mechanism. The influx of $Ca^{2+}$ as the primary event in the salivary gland (rat parotid) causes the efflux of $K^+$ ions with water. The efflux of potassium has also been recognized as an $\alpha$-adrenergic effect in guinea-pig liver and in adipose tissue.

The stimulation of the pineal gland with *l*(R)-epinephrine via the $\alpha$-adrenergic receptor was found to be dependent on the presence of $Ca^{2+}$ in the incubation medium and results in a seven-fold increase of cyclic GMP level. It seems from these studies that the influx of $Ca^{2+}$ is the first event induced by the $\alpha$-agonist. The formation of cyclic GMP seems to be the result of $Ca^{2+}$ influx. This is not surprising as guanylate cyclase is a $Ca^{2+}$ dependent enzyme. As indicated in Section II, $\alpha$-adrenergic receptors are involved in a variety of physiological activities, and it remains to be seen whether in each case $Ca^{2+}$ functions as the 'second messenger'. Occasionally it has been claimed that in the central nervous system, $\alpha$- as well as $\beta$-receptors are coupled to the enzyme adenylate cyclase (see Reference 11 and references therein). These observations are based on the experimental finding that $\alpha$-adrenergic blockers inhibit the formation of cyclic AMP, as was found in rat cerebral cortical tissue.

Another biochemical response elicited by the activation of $\alpha$-adrenergic receptors is the incorporation of inorganic $^{32}P_i$ into phosphatidylinositol in slices of the parotid gland.[11] This biochemical event is shown to be unrelated to the $K^+$ efflux and water secretion also induced by $\alpha$-receptor activation. Interestingly enough, the divalent cation ionophore A-23187 which introduces $Ca^{2+}$ into the cell, thus causing $K^+$ release,[11] has no significant effect on the incorporation of $^{32}P_i$ into phosphatidylinositol. Conversely, the $\alpha$-receptor induced phospholipid effect[12] is maximal in the absence of $Ca^{2+}$ in the medium when there is no $K^+$ release from the cell. In summary, it can be concluded that $\alpha$-receptor activation leads to two parallel and independent biochemical events in the rat parotid gland: (1) increase in membrane permeability towards extracellular $Ca^{2+}$ which enters the cell and causes $K^+$ release and (2) the same interaction with the $\alpha$-receptor results in the increased incorporation of $^{32}P_i$ into acidic phospholipids. This latter response is $Ca^{2+}$ independent and, in fact. maximally stimulated in its absence.

It should be noted that the phospholipid effect was shown also to be induced by the activation of the acetylcholine muscarinic receptor in the same preparation of the parotid gland. Such phospholipid effects were shown to be induced by other receptors as well (see Chapter 19).

Recently a new class of $\alpha$-receptors which are linked to adenylate cyclase were discovered. Agonist occupancy leads to enzyme *inhibition*.[13,14] This inhibition, like the stimulation, seems to depend on the presence of GTP.

### (a) *The relationship between $\alpha$- and $\beta$-receptors*

Almost every organ, tissue, or cell which possesses an $\alpha$-adrenergic receptor also possesses a $\beta$-adrenergic receptor. These two receptors elicit opposite physiological effects in the target organ (Table 15.2). Thus, it is possible that the final response of the organ in question depends on the relative activity of the two receptors. Since $Ca^{2+}$ functions as the second messenger of $\alpha$-receptor action and inhibits $\beta$-receptor dependent adenylate cyclase (at least in some cases), it may provide the link between $\alpha$- and $\beta$-receptors in systems which possess both types of adrenergic receptors. In this context it is interesting to note that in the rat parotid gland which possesses both $\alpha$- and $\beta$-receptors, the $\alpha$-blocker phentolamine

Table 15.2. Some physiological actions of adrenergic receptors

| System or tissue | Action | Receptor |
|---|---|---|
| Cardiovascular system, heart | Increased force of contraction | $\beta$ |
| | Increased rate | |
| Blood vessels | Constriction | $\alpha$ |
| | Dilation | $\beta$ |
| Respiratory system, tracheal and bronchial smooth muscle | Relaxation | $\beta$ |
| Iris (radial muscle), smooth muscle, | Pupil dilated | $\alpha$ |
| uterus | Contraction | $\alpha$ |
| Spleen | Relaxation | $\beta$ |
| | Contraction | $\alpha$ |
| Bladder | Contraction | $\alpha$ |
| | Relaxation | $\beta$ |
| Skeletal muscle | Changes in twitch tension | $\beta$ |
| | Increased release of acetylcholine | $\alpha$ |
| | Increased glycogenolysis | $\beta$ |
| Adipose tissue | Increased lipolysis | $\beta$ |
| Parotid gland | Enzyme secretion | $\beta$ |
| Parotid gland | Water and $K^+$ secretion | $\alpha$ |

slows down the fall in the level of cyclic AMP subsequent to epinephrine stimulation, as compared to a system in which the $\alpha$-blocker is absent. This effect, however, may be due to secondary biochemical events other than the direct effect of $Ca^{2+}$ on the level of adenylate cyclase activity. For example, $Ca^{2+}$ is known to activate cyclic AMP phosphodiesterase and thus an increase in intracellular $Ca^{2+}$ may result, not only in the inhibition of adenylate cyclase but also in the depletion of the cyclic AMP pool. Very recently (Reference 14 and references therein) it has been shown in some cell-free preparations that $\alpha$-agonists in the presence of GTP and $Na^+$ ions inhibit directly hormone stimulated adenylate cyclase. For example $\alpha$-agonists were found to induce adenylate cyclase inhibition in fat cells, human platelets and neuroblastoma–glioma hybrid cells. At this point, it can only be stated that the interaction of $\alpha$-receptors with $\beta$-receptors is still not fully understood in biochemical terms and requires further detailed investigation.

*(b) Is there an interconversion between $\alpha$- and $\beta$-receptors?*

Some reports in the literature have suggested that $\alpha$-receptors and $\beta$-receptors are two allosteric configurations of the same macromolecule. The pharmacological experiments upon which this hypothesis is based were performed on frog heart, where it has been claimed that $\alpha$-receptors prevail at low temperatures and transform into β-receptors at high temperatures. It was claimed that stimulation of cardiac rate and contractibility by catecholamines has the properties of a classical β-adrenergic response when experiments are performed at warm temperatures (25–37 °C) and of α-adrenergic response when experiments are performed at cold temperatures (5–15 °C). Other workers have examined this hypothesis by looking at the adenylate cyclase activity in these tissues over a wide range of temperatures. It was found that adenylate cyclase is stimulated exclusively by β-adrenergic ligands. Furthermore, the adrenergic inhibitors affecting cyclase at a wide range of temperatures were always of the β-type where α-blockers had no effect. In conclusion, the possibility of α-receptor to β-receptor interconversion still remains in view of the pharmacological experiments, although it does not now find any support from direct biochemical experiments.

### (iii) Dopamine receptors

Dopamine, like epinephrine and norepinephrine, is a neurotransmitter in the central nervous system. In some cases such as the caudate nucleus it was demonstrated that the dopamine receptor is coupled to adenylate cyclase.[15] Dopamine sensitive adenylate cyclase was also demonstrated in neuroblastoma. Dopamine receptors are distinct from the β-adrenergic receptor in their ligand specificity and response to specific blockers (Figures 15.3). Thus, dopamine receptors coupled to adenylate cyclase respond to dopamine better than to norepinephrine (Table 15.1), whereas in β-receptors the situation is reversed. β-Adrenergic blockers such as propranolol do not affect dopamine-dependent adenylate cyclase, whereas tricyclic antidepressants, such as chloropromazine, and butyrophenones, such as haloperidol, act as specific blockers of the dopamine-dependent adenylate cyclase (Table 15.1) having no effect on β-adrenergic receptor dependent adenylate cyclase. Whether dopamine receptors are coupled to other biochemical signals such as ion channels is not yet clear.

## IV. RADIOASSAY OF CATECHOLAMINE RECEPTORS

Until recently a reliable assay for catecholamine receptors was not available. The use of [$^3$H]catecholamines to monitor β-adrenergic receptors as well as other catecholamine receptors proved to be unreliable.[16,17] The failure to detect catecholamine receptors using radioactively labelled catecholamines stems from the fact that the receptor concentration accessible experimentally is far below the catecholamine receptor dissociation constant. Furthermore, catecholamines bind to many non-receptor components in the membrane preparations studies, and thus the signal to noise ratio is further lowered. In 1974 radioactively labelled β-adrenergic blockers were found to reliably monitor β-adrenergic receptors. The first ligand to be introduced was [$^3$H]DL-propranolol[17] and shortly thereafter, and independently, [$^{125}$I]hydroxybenzylpinolol[18,19] and [$^3$H]alprenolol[20,21] were introduced as specific ligands for the radioassay of β-adrenergic receptors. The use of these radioactively labelled ligands has since become a routine procedure to monitor β-adrenergic receptors in a variety of tissues.

The binding of these compounds is stereospecific for the R(–)stereoisomer(l) and all of them are displaced from the β-receptor by R(–)catecholamines(l) and not by S(+)catecholamines(d). The dissociation constants found for these β-blockers using binding experiments closely match the inhibitor constants found from their competition with catecholamines in the adenylate cyclase reaction. Recently two α-blockers, [$^3$H]dihydroergocryptine[22] and $^3$H-WB-4101[23] were introduced to monitor the α-adrenergic receptors *in vitro*. Dopamine receptors can be monitored using radioactively labelled blockers, such as [$^3$H]haloperidol, [$^3$H]chlorpromazine, and [$^3$H]spiroperidol. Radioactively labelled blockers can, in principle, be used to also monitor detergent solubilized receptors, provided that the receptor does not denature in the process of solubilization. Such attempts have been reported recently.[24] Since [$^{125}$H]hydroxybenzylpindolol exhibits extremely high affinity towards β-adrenergic receptors, it can therefore be used to monitor β-receptors in intact cells.

## V. FLUORESCENT ANTAGONISTS FOR MAPPING ADRENERGIC RECEPTORS *IN VIVO*

Recently two fluorescent β-blockers were synthesized,[25] 9-AAP and DAPN (Figure 15.4). These compounds were shown to bind in a stereospecific manner to β-adrenergic receptors *in vivo* upon their injection into rats and mice. Both peripheral β-receptors as well as β-receptors in the central nervous system bind these fluorescent antagonists and become visible in the fluorescence microscope (see Reference 25).

The use of specific fluorescent β-blockers to probe β-adrenergic receptors is complementary to the formaldehyde method of Falck and Hillarp and of the glyoxylic acid method of Lindvall and Björklund for the mapping of catecholaminergic fibres. These methods, however, do not discriminate between the different types of catecholaminergic neurons as all catecholamines condense with formaldehyde and glyoxylic acid to yield a fluorescent derivative. In addition, serotonin also yields a fluorescent derivative upon condensation with formaldehyde or glyoxylic acid, and thus serotoninergic pathways become visible in the fluorescence microscope too.

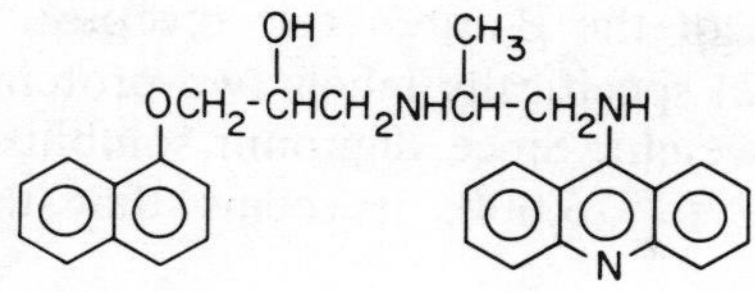

9-AAP

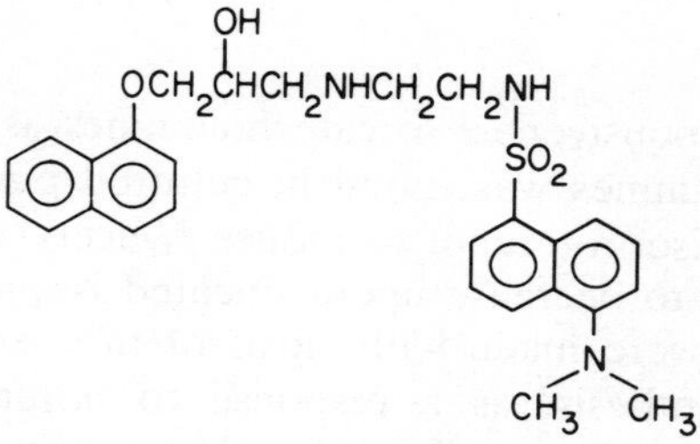

DAPN

Figure 15.4. Fluorescent $\beta$-blockers 9-AAP = 9-aminoacridino-propranolol and DAPN = Dansyl analogue of propranolol

## VI. AFFINITY LABELLING OF THE $\beta$-ADRENERGIC RECEPTOR

Affinity labelling of the $\beta$-adrenergic receptor has recently been achieved by using a reversible $\beta$-blocker to which the reactive group bromoacetyl was attached.[26] The compound *N*-(2-hydroxy-3-naphthyl-oxypropyl)-*N'*-bromoacetyl-ethylenediamine (Figure 15.1) has been shown to inhibit irreversibly the epinephrine-dependent adenylate cyclase activity without damaging the NaF dependent activity in turkey erythrocyte membranes. Furthermore, propranolol and *l*(R)-epinephrine offer protection against the affinity labelling reaction. Similarly, the compound was shown to inhibit irreversibly the hormone stimulated activity in a whole turkey red blood cell. The loss of epinephrine-dependent activity occurs concomitantly with the loss of the specific binding of [$^{3}$H]propranolol and [$^{125}$I]hydroxybenzylpindolol. More recently, the $^{3}$H-affinity label was used

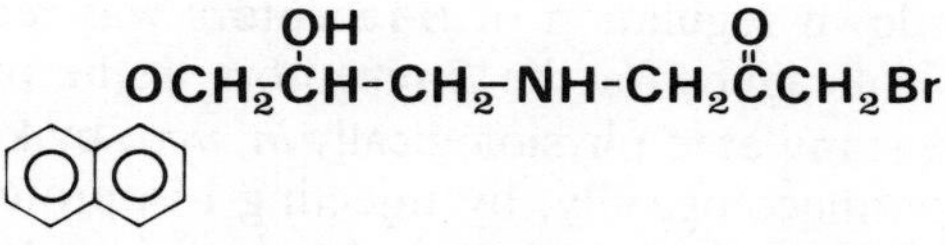

Figure 15.5. $\beta$-Adrenergic affinity label

to identify the subunits of the β-adrenergic receptor.[27] It was found that the tritiated affinity label specifically labels two protein subunits of 37,000 and 41,000 molecular weight. Since digitonin solubilized β-receptor has a molecular weight close to 75,000, it seems that the structure of the receptor is oligomeric.

## VII. SELF-REGULATION OF β-ADRENERGIC RECEPTORS

### (i) Desensitization

A decrease in the responsiveness to catecholamines as a result of repeated exposure to catecholamines was found in cultured pineal organs.[28] In that system the ability of isoproterenol to induce *N*-acetyl-serotonin-transferase activity was observed to decrease upon repeated isoproterenol stimulation. Similar observations were made with slices of the cerebral cortex. In this case, cyclic AMP synthesis, as a response to norepinephrine, was also decreased upon repeated stimulation. Catecholamine induced refractoriness was also observed *in vitro* in frog erythrocytes (see Reference 21 and references therein), human leucocytes, and macrophages. Similar effects were demonstrated in cultured fibroblasts and in glioma cells.

The availability of direct means for probing the β-adrenergic receptors led to the discovery that the concentration of a ligand can regulate the concentration or the binding properties of the receptors of the target cell. In the case of β-receptors, it has been known for some time that β-adrenergic agonists can induce functional desensitization (tachyphylaxis, tolerance) of target tissue *in vivo* and *in vitro*. Using $^{125}$I-hydroxybenzyl-pindolol, Perkins and his colleagues[21] have shown that prolonged exposure of human astrocytoma cells to β-adrenergic catecholamines leads to a decrease of 80% to 90% in the number of β-receptor binding sites without a change in affinity towards the β-antagonists. The order of potency of the catecholamines according to other studies,[21] is isoproterenol > epinephrine > norepinephrine. Various investigators also found that the β-blocker propranolol inhibits this action of agonists but does not by itself cause any decrease in the number of receptors. Perkins and his colleagues[21] also noted that the loss of the cyclase response to β-agonists is fast and precedes the loss of β-receptors from the membrane. The rapid loss of responsiveness of the cyclase system to agonist is fast and reversible whereas the loss of receptors is not. Reappearance of receptors on the membrane requires protein synthesis. Similar down regulation of β-receptors was recently reported by Kebabian and his colleagues for the β-receptors in the pineal gland. When the receptors were stimulated physiologically *in vivo*, by keeping the animals in the dark, or pharmacologically, by injecting l-isoproterenol, a rapid fall in the number of [$^3$H]alprenolol binding sites was found. Exposing the rats to light, thus decreasing the sympathetic activity, resulted in the increase of the number of β-receptors as measured by [$^3$H]alprenolol binding. These investigators also reported that the number of β-receptors on the pineal gland normally varies with a circadian periodicity which is inversely related to the cycle of neurotransmitter release.

Down regulation is an efficient mechanism to regulate receptor response, especially when the number of receptors is such that only fractional occupancy of these receptors results in maximal response. Under these circumstances, namely in the presence of 'spare receptors', a decrease in the number of receptors does not cause a decrease in the potential maximal response but shifts the dose–response curve to higher agonist concentrations. The reduction in the number of β-receptors can, in principle, account for the functional desensitization of target tissues to repeated agonist stimulation. However, it should be stressed that other mechanisms such as the reduction in receptor affinity may also be responsible for desensitization.

The molecular basis for receptor desensitization which now seems to emerge is as follows.[21] The first event which occurs upon agonist binding is the loss of coupling between the cyclase and the receptor. This phase is relatively rapid and occurs within minutes after exposure to the ligand. At a later stage, if the receptors remain occupied by the agonist the latter internalize and disappear from the membrane. Whether membrane bound cytoskeletal elements are involved in receptor to enzyme coupling and in the events leading up to desensitization is not yet clear. However, what is clear from the study of β-receptor desensitization in S49 lymphoma cells, is that cyclic AMP dependent phosphorylation is not involved. This assertion is made on the basis of the finding that mutant S49 cells, lacking protein kinase but containing β-receptor dependent adenylate cyclase, undergo desensitization. On the other hand, other mutant S49 cells possessing the β-adrenergic receptor but devoid of adenylate cyclase, do not undergo desensitization. Thus, it is clear that the process of desensitization involves processes which depend on an actively functioning enzyme.

De Vellis and Brooker reported that the generation of cyclic AMP in response to catecholamines in the 2B subclone of RC6G rat glioma cells, previously exposed to norepinephrine and refractory to further norepinephrine addition, is substantially increased by addition of RNA inhibitors and protein synthesis. They conclude that their findings support the hypothesis that protein biosynthesis is important in the development of catecholamine refractoriness.

### (ii) Supersensitivity

Supersensitization to β-adrenergic agonists is observed in different systems, and is manifested in both increased cyclic AMP synthesis and an increased number of β-receptors. It was found that after adrenalectomy of the rat, the responsiveness of rat liver adenylate cyclase to catecholamines is enhanced three- to five-fold. This increase in adenylate cyclase activity is accompanied by a three- to five-fold increase in the number of $\beta$-adrenergic receptors, as revealed by direct binding studies using [$^{125}$I]-hydroxybenzylpindolol (see Reference 29 and references therein). These changes were reversed by the administration of cortisone. It was suggested by these investigators that this increase in $\beta$-receptors and adenylate cyclase activity may be a compensatory response to the impairment in gluconeogenesis and glycogenolysis which occurs subsequent to adrenalectomy. Supersensitization of the catecholamine responsiveness in mammalian brain occurs also as a result of intraventricular injection of 6-hydroxydopamine which causes the destruction of catecholamine nerve terminals. Using [$^{125}$I]iodohydroxybenzylpindolol, an increase in the number of $\beta$-receptors in the cerebral cortex, subsequent to treatment with 6-hydroxydopamine, could be measured (see Reference 29 and references therein). Parallel to the increase in the number of $\beta$-receptors, an increase in the *l*(R)-isoproterenol dependent adenylate cyclase was detected.

## VIII. CATECHOLAMINES AS NEUROTRANSMITTERS

$\beta$-Adrenergic receptors, $\alpha$-adrenergic receptors, and dopamine receptors are involved in catecholaminergic innervation. Some of these activities are due to adrenergic stimulation where the catecholamine functions as a neurotransmitter released from presynaptic vesicles within the adrenergic synapse, acting at a postsynaptic adrenoreceptor. While there is much biochemical data concerning the action of adrenergic agonists as hormones at $\alpha$- and $\beta$-receptors, little is known about their mechanism of action as neurotransmitters. It seems, however, that $\beta$-receptor activity at $\beta$-adrenergic synapses involves the activation of adenylate cyclase, and $\alpha$-receptor activity involves changes in ion permeability where the first event is the influx of $Ca^{2+}$. For example, certain cells in the central nervous system such as the Purkinje cells of the cerebellum are innervated by noradrenergic neurons. The neurotransmitter acting at the Purkinje cells is epinephrine. It was found that epinephrine transmitter is responsible for the sustained depression of spontaneous firing of the rat cerebellar Purkinje cells. It was suggested that this norepinephrine action is mediated by cyclic AMP. Indeed, phosphodiesterase inhibitors such as papaverine were found to depress the firing of Purkinje cells. It seems, therefore, that intracellular cyclic AMP levels influences noradrenergic neurotransmission. This conclusion actually classifies these noradrenergic receptors to belong to the $\beta$-type.

The presence of $\beta$-adrenoreceptors in cerebral tissue has been confirmed by electrophysiological techniques and confirmed by direct measurements of [$^3$H]propranolol binding, [$^3$H]alprenolol binding, and

[$^{125}$I]hydroxybenzylpindolol binding (see Reference 29 and references therein). Another example where norepinephrine acts as a neurotransmitter is the rat pineal gland. In this case the norepinephrine neurotransmitter is released from the sympathetic nerves and stimulates adenylate cyclase within the pineal gland. The pineal gland is another example where the noradrenaline neurotransmitter acts on a postsynaptic membrane receptor of the $\beta$-type.

Catecholamines also function as neurotransmitters in the central nervous system by interacting with $\alpha$-adrenergic receptors (see Section II). The postsynaptic receptor is of the $\alpha_1$ type, whereas the presynaptic autoreceptors is of the $\alpha_2$ type.[30] It seems that both receptors function by changing the ion permeability (probably $Ca^{2+}$) (see Reference 30 and references therein).

The $\alpha_2$-autoreceptor is probably different from the postsynaptic $\alpha_1$-receptor found in peripheral tissues as well as in the central nervous system. The presynaptic $\alpha_2$ receptors probably control the release of catecholamines. Thus the released catecholamine is a feedback inhibitor of its own release (see Reference 30 and references therein).

## IX. SPARE RECEPTOR CONTROL OF CATECHOLAMINE ACTION

The level of the circulating catecholamine ($1.0 \times 10^{-9}$ M to $1.0 \times 10^{-8}$ M) is far below its dissociation constant to the receptor ($5 \times 10^{-7}$ M to $5.0 \times 10^{-6}$ M). Therefore, only a small fraction of the receptors become occupied at physiological levels of the hormone. At the catecholaminergic synapse, however, the catecholamine concentration can rise to $1–2 \times 10^{-4}$ M and thus saturate the receptor, at least temporarily. A direct comparison can be made between the dependence of the adenylate cyclase reaction on hormone concentration and the dependence of the hormone mediated effect on hormone concentration (Table 15.3). This can be done only in cases where the two sets of data are available. It can be seen from Table 15.3 that 50% of the biochemical response is achieved at hormone concentrations two orders of magnitude below the hormone concentrations required for 50% saturation of the adenylate cyclase reaction. It appears, therefore, that only a small fraction of the $\beta$-adrenergic receptors must be occupied by the hormone in order to saturate the subsequent biochemical response. As both biochemical signals quoted in Table 15.3 are cyclic AMP-dependent, one can calculate[31] the level of cyclic AMP attained in these systems at different receptor occupancies. It can be calculated[31] that occupancy of 0.016% of the $\beta$-adrenergic receptors in the systems described in Table 15.3 is sufficient to produce $2 \times 10^{-8}$ M to $2 \times 10^{-7}$ M cyclic AMP within 1 min. This concentration of cyclic AMP is sufficient to saturate protein kinase and protein kinase dependent processes such as the process depicted in Table 15.3. From this example it is quite clear that

Table 15.3. The $[H]_{0.5}$ values for Hormone in cyclic AMP processes

| The cyclic AMP-mediated biochemical process measured | $[H]_{0.5}$ for the hormone mediated effect (M) | $[H]_{0.5}$ for the hormone in the adenylate cyclase reaction (M) |
|---|---|---|
| Epinephrine-stimulated $Na^+$ outflux in turkey erythrocytes | $1 \times 10^{-8}$ | $6 \times 10^{-6}$ |
| Epinephrine-stimulated $\alpha$-amylase secretion by the rat parotid gland | $2 \times 10^{-7}$ | $1 \times 10^{-5}$ |

only a small fraction of the receptors becomes occupied upon hormone stimulation, and therefore most of the receptors are 'spare receptors'. These receptors are actually not 'spare', as the necessary small number of receptors that must become occupied in order to produce the biochemical signal depends on the total number of receptors. The fraction of receptors saturated will provide the level of 'second messenger' necessary to saturate the cyclic AMP dependent biochemical processes. It seems, therefore, that the terminology 'spare receptors' is inadequate since all the receptors are an integral part in generating the response.[31] (See Chapter 2, Section VI.)

There is a possible advantage to a mechanism in which the agonist has low affinity for the receptor; in this case the extent of the signal elicited becomes dependent on the *total* number of receptors per cell.[31] Such a mechanism allows for a discriminatory action of the same hormone on different tissues possessing the same receptor. Namely, whether or not different tissues will respond to the same hormone concentration depends on the total number of receptors per cell in each of the target tissues.

Two approaches can be used to analyse the existence of 'spare receptors'. One is a quantitative comparison between receptor occupancy and the dose–receptor curve. This approach can be used when quantitative measurements of receptor occupancy can be performed. Another approach which was originally used by Nickersen in 1956[32] is to block irreversibly a fraction of the receptors and determine whether full response can still be obtained. Using this technique, Nickersen was able to show that the blocking of 99% of the histamine receptors by an irreversible blocker shifts the dose–response curve with respect to the histamine agonist to higher concentrations but does not reduce the maximal response.

## X. REFERENCES


1. Ahlquist, R. P. *Am. J. Physiol.,* **153**, 586–598, (1948).
2. Ahlquist, R. P. *Ann. N.Y. Acad. Sci.*, **139**, 549–552, (1967).
3. Furchgott, R. F. in *Handbook of Experimental Pharmacology* (Blaschko, H., and Muscholl, E., eds.) Springer-Verlag, Berlin, Vol. 33, pp. 283–335, (1972).
4. Robison, G. A., Butcher, R. W., and Sutherland, E. W. *Ann. Rev. Biochem.,* **37**, 149–174, (1968).
5. Levitzki, A. *Biochem. Biophys. Res. Commun.*, **74**, 1154–1159, (1977) and Levitzki, A., and Helmreich, E. J. M. *FEBS Lett.* **101**, 213–219, (1979).
6. Cassel, D., and Selinger, Z. *Biochim. Biophys. Acta*, **452**, 538–551, (1976).
7. Tolkovsky, A. M., and Levitzki, A. *Biochemistry (USA),* **17**, 3795–3810, (1978).
8. Londos, C., Salomon, Y., Lin, M. C., Harwood, J. P., Schramm, M., Wolff, Y., and Rodbell, M. *Proc. Natl. Acad. Sci. U.S.A.,* **71**, 3087–3090, (1974).
9. Schramm, M., Orly, J., Eimerl, S., and Korner, M. *Nature (London),* **268**, 310–313. (1977); and Schulster, D., Orly, J., Seidel, G., and Schramm, M. *J. Biol. Chem*., **253**, 1201–1206, (1978).
10. Insel, P. A., Maguire, M. E., Gilman, A. G., Bourne, H. R., Coffino, P., and

Melbron, K. L. *Mol. Pharmacol.,* **12**, 1062–1069, (1976); and Ross, E. M., Haga, T., Howlett, A. C., Schwarzmeier, J., Schleifer, L. S., and Gilman, A. G. *Adv. Cyclic Nucl. Res.,* **9**, 1–17, (1978).
11. Schramm, M., and Selinger, Z. *J. Cycl. Nucl. Res.,* **1**, 181–192, (1975).
12. Oron, Y., Löwe, M., and Selinger, Z. *Mol. Pharmacol.,* **11**, 79–86, (1975).
13. Daly, J. in *Handbook of Psychopharmacology* (Iversen, L. L., Iversen, S. D., and Snyder, S. H., eds.), Plenum Press, New York, Vol. 5, pp. 47–130, (1975).
14. Sabol, S. A., and Nirenberg, M. *J. Biol. Chem.*, **254**, 1913–1920, (1979), and refs. therein.
15. Kebabian, J. W., Petzold, G. L., and Greengard, P. *Proc. Natl. Acad. Sci. U.S.A.,* **63**, 2145–2149, (1972).
16. Cuatrecasas, P., Tell, G. P. E., Sica, V., Parikh, I., and Change, K. J. *Nature (London),* **247**, 92–97, (1974).
17. Levitzki, A., Atlas, D., and Steer, M. L. *Proc. Natl. Acad. Sci. U.S.A.,* **71**, 2773–2776, (1974).
18. Aurbach, G. D., Fedak, S. A., Woodward, C. J., Palmer, J. S., Hauser, D., and Troxier, F. *Science,* **186**, 1223–1225, (1974).
19. Brown, E. M., Rodbard, D., Fedak, S. A., Woodward, C. J., and Aurbach, G. D. *J. Biol. Chem.,* **251**, 1239–1246, (1976).
20. Lefkowitz, R. J., Mukherjee, C., Civerston, M., and Caron, M. G. *Biochem. Biophys. Res. Commun.*, **60**, 703–709, (1974).
21. Ying-Fu, Su, Harden, K. T., and Perkins, J. P. *J. Biol. Chem.*, **254**, 38–41, (1979); and Mukherjee, C., Caron, M. G., and Lefkowitz, R. J. *Proc. Natl. Acad. Sci. U.S.A.*, **72**, 1945–1949, (1975).
22. Williams, L. T., Mullkien, D., and Lefkowitz, R. J. *J. Biol. Chem.,* **251**, 6915–6923, (1976).
23. Greenberg, D. A., U'Prichard, D. C., and Snyder, S. H. *Life Sci.,* **19**, 69–76, (1976).
24. Vauquelin, G., Geynet, P., Hanoune, J., and Strosberg, D. *Proc. Natl. Acad. Sci.,* **74**, 3719–3714, (1977).
25. Atlas, D., and Levitzki, A. *Proc. Natl. Acad.Sci. U.S.A.,* **74**, 5290–5294, (1977).
26. Atlas, D., Steer, M. L., and Levitzki, A. *Proc. Natl. Acad. Sci. U.S.A.,* **73**, 1921–1925, (1976).
27. Atlas, D., and Levitzki, A. *Nature (London),* **272**, 5651–5652, (1978).
28. Deguchi, T., and Axelrod, J. *Proc. Natl. Acad. Sci. U.S.A.,* **69**, 2208–2211, (1972).
29. Wolfe, B. B., Harden, T. K., and Milinoff, P. B. *Ann. Rev. Pharmacol. Toxicol.,* **17**, 575–604, (1977).
30. Langer, S. Z. *Br. J. Pharmacol.,* **60**, 481–497, (1977).
31. Levitzki, A. in *Receptors and Recognition*, (Cuatrecasas, P., and Greaves, M. F., eds.), Chapman and Hall, London, Vol. 2, Ser. A. pp. 199–299, (1976).
32. Nickersen, M. *Nature (London)*, **178**, 697–698, (1956).

Cellular Receptors for Hormones and Neurotransmitters
Edited by D. Schulster and A. Levitzki


CHAPTER 16

# Histamine receptors

Jack Peter Green and Lindsay B. Hough
*Department of Pharmacology, Mount Sinai School of Medicine of The City University of New York, New York, N.Y. 10029, U.S.A.*



## I. INTRODUCTION AND PHYSIOLOGICAL EFFECTS

The widespread distribution of histamine throughout the animal and plant kingdoms and its powerful effects on tissues have provoked steady effort to find a role for it in physiological and pathological mechanisms. In most animals, histamine causes massive vasodilation to produce a profound fall in blood pressure and bronchoconstriction to produce asphyxia.[1] These catastrophic effects so captivated scientists that most early work was directed to learn the role of histamine in disease. These efforts were succinctly summarized by Carl Dragstedt in 1947 in a parody of Kipling's poem 'Gunga Din':

Trauma, burns, and inflammation,
Headache, shock, and constipation
Show the fingerprints of some malicious fiend.
And the one that gets accused
Is that amine so abused—
Beta-imidazolethylamine.

Although histamine cannot be said to cause all of this, its role in several pathologic states is clear. Of these perhaps the best understood is the immediate hypersensitivity reaction, which results from the interaction of soluble antigen with tissue antibody.[2] After the antibody (IgE) is bound to tissue mast cells and blood basophils, exposure to the antigen initiates a series of events leading to the expulsion of intracellular secretory granules containing histamine and other biologically active substances. Differences among species in their mast cell contents as well as their varied sensitivity to histamine account for differences in the importance of histamine in anaphylaxis. In man, anaphylaxis is initiated by antigen contact and occurs as severe urticaria, itching, and laryngeal oedema. That these symptoms are only partly attributable to histamine is shown by the inability of histamine antagonists alone to prevent or annul these symptoms. Other less severe conditions resulting from antigen antibody reaction such as allergic rhinitis and dermatoses are more responsive to antihistamine therapy. In some species, anaphylaxis includes hypotension and other symptoms of shock, which are likely due to histamine release. Endotoxins cause shock in most species by releasing histamine from mast cells.[3] Histamine is also released from mast cells by many basic compounds[4] to produce anaphylactoid symptoms. Among these compounds are diverse drugs such as radiocontrast media, plasma expanders, neuromuscular blocking agents, sedatives, and anaesthetics.[5]

Trauma, venoms, detergents, and cytolytic processes also release histamine from mast cells. This effect often initiates the acute inflammatory reaction characterized by swelling, redness, heat, and pain. At least part of the reaction is due to histamine, which dilates arterioles and venules and increases vascular permeability. By dilating small vessels, histamine increases blood flow to the injured area and facilitates the physiological processes that cope with injury. In this regard histamine may serve homeostasis. It is likely that under normal circumstances, only small amounts of histamine are released by tissues, producing highly localized responses. Certainly the slow turnover of histamine in peritoneal mast cells of rat, $t_{1/2} = 4$ days,[6] suggests that histamine is normally released in small amounts. Although histamine contributes substantially to the vascular components of the inflammatory reaction, the subsequent cellular events of inflammation, i.e. the infiltration of neutrophils and macrophages, are due to substances other than histamine. However, histamine can influence these cellular events, inhibiting the cytolytic activity of T-lymphocytes as well as inhibiting its own antigen evoked release.[7] It has also been shown that stimuli, e.g. burns, endotoxin, induce histamine synthesis in cells other than mast cells, which are unidentified.[8]

Histamine affects contraction of muscle other than vascular muscle.[1] So sensitive is the guinea-pig ileum that it provides a means to assay histamine.

Bronchiolar smooth muscle responds to histamine but with marked species differences. A very small parenteral dose of histamine kills guinea-pigs in minutes by intense bronchoconstriction. The dog bronchus is relatively insensitive. Most other animals such as man are intermediate in their bronchiolar sensitivity. The sheep bronchus actually relaxes in response to histamine. Histamine also relaxes the rat uterus. On heart of some species, histamine increases both force and rate of contraction and depresses atrioventricular conduction velocity.[9]

It is now clear that histamine functions in normal gastric acid secretion in some species. The histamine content of gastric mucosa is substantial in all animals possessing true stomachs with acid secreting cells, and it is low in the gastrointestinal tracts of stomachless vertebrates.[10] The histamine containing cells are in the basal parts of the oxyntic gland. In all species examined, histamine acts directly on the acid-secreting cells to stimulate acid secretion. In the rat, histamine is the final common mediator for gastric acid secretion: the turnover of gastric histamine stores is increased by all stimuli that result in acid secretion, e.g. eating, administration of the peptide hormone gastrin, stimulation of the vagus nerve.[6] In other species, histamine very likely functions in normal acid secretion but evidence to date does not allow the inference that it is the final common mediator of acid secretion in all species.[11] If histamine is shown to function in normal acid secretion, another homeostatic role of histamine can be suggested. For although gastric acid makes little contribution to digestive processes, it reduces microbial invasion by its bactericidal action. Increasing the pH of gastric juice enhances susceptibility to invasion by bacteria[12] and perhaps to yeast as well.[13] The role of endogenous histamine in peptic ulcer is not clear, although repeated injections of histamine or histamine releasing agents produce ulcers in animals. It is certain, however, that specific antagonists that prevent the action of histamine on acid secretion are remarkably effective in treating duodenal ulcer, including that associated with hypersecretion of gastrin.[11]

Histamine affects peripheral nervous tissue. One of the most intriguing effects is the flare, an area of redness, that follows intradermal injection of histamine. The response is due to an axon reflex that occurs after stimulation of the afferent fibres by histamine.[14] Histamine has also been shown to depolarize sympathetic ganglia, as well as to potentiate or block ganglionic transmission, the response depending on the dose studied. Histamine may also be released (from cells still unidentified) after the vasoconstriction that results from stimulation of adrenergic fibres;[15] other puzzling relationships between histamine and the autonomic nervous system have been reported.[16]

Histamine also affects the brain.[14,17] Several suspected functions of histamine may be related to the hypothalamus, since its injection into this region produces hypertension, tachycardia, hypothermia, and increased water intake. Some other effects found after intraventricular injections of histamine may also be attributed to the hypothalamus. These are increases in the plasma levels of antidiuretic hormone, prolactin, and luteinizing hormone. Microinjections of histamine into the lateral hypothalamic nucleus inhibits ipsilateral but not contralateral intracranial self-stimulation. Other actions reported after intraventricular histamine are anorexia, catalepsy, and emesis; the last response has been localized to the chemoreceptor trigger zone of the area postrema.

Histamine may prove to be a mediator in the nervous system.[14,17] Studies of mammalian brain have shown that histamine meets most criteria for having transmitter function. It has non-uniform distribution with highest concentration in the hypothalamus. Subcellular fractions of brain containing nerve-endings are rich in histamine. Brain contains the specific enzymes for synthesis and metabolism of histamine. After destroying fibres in the medial forebrain bundle or the afferent fibres to the hippocampus, the activity of the specific histidine decarboxylase falls in brain regions distal to the lesion. Histamine is released from brain slices by potassium ions in a process that is dependent on ionic calcium. Histamine turns over rapidly; unlike other aromatic biogenic amines, it is not taken up by presynaptic terminals but is instead metabolized to methylhistamine which is then oxidatively deaminated by monoamine oxidase B. Neurons respond to histamine, e.g. it decreases the firing rate of cerebral cortical and brainstem neurons and increases the firing rate of hypothalamic neurons. Especially provocative is the observation that electrical stimulation of afferent fibres to the cerebral cortex or hippocampus reduces the firing rate of the neurons, and part of this effect of electrical stimulation can be blocked by histamine antagonists. As described below, histamine increases adenylate and guanylate cyclase activities in brain and in other neural tissues. Brain contains high affinity binding sites for histamine antagonists. The major deficiency in attributing a transmitter role to histamine in the vertebrate nervous system is that its precise histochemical localization has not been shown because of methodological problems. In *Aplysia*, histamine is localized to specific nerve cells, but its role as a transmitter is unproven.[18]

## II. CLASSIFICATION OF HISTAMINE RECEPTORS

Histamine receptors have been classified mainly by pharmacological procedures, and, more recently, by biochemical procedures. The principles are the same whatever the method. One measures the relative potencies of a series of agonists in modifying an activity (e.g. strength of contraction, activation of adenylate cyclase) on a tissue (e.g. stomach). The slopes of the dose–response curve for all agonists must not differ. The relative agonist potencies of the compounds are usually expressed as $ED_{50}$ values, which is the concentration of agonist giving 50% of the maximum response. If the $ED_{50}$ values of a series of agonists relative to a standard e.g. histamine, on the different tissues do not differ, the functions that were measured in the two tissues are postulated to be subserved by the same receptor. For example, the relative activities of analogues of histamine in stimulating contraction of the guinea-pig ileum and of the rat stomach were very similar, suggesting that the receptors mediating these events belonged to one category, called the $H_1$-receptor.[19,20] When the same analogues were tested on other preparations responsive to histamine, a different set of relative activities were obtained implying that a receptor (or receptors) different from the $H_1$-receptor was associated with these specific events in these specific tissues. That they comprise another homogeneous population, called the $H_2$-receptor, was suggested when the relative activities of histamine and analogues were shown to be very similar in stimulating gastric acid secretion in the rat, in inhibiting contraction of the rat uterus, and in increasing the rate of the guinea-pig atrium (table 16.1). Classification of receptors on the basis of relative agonist activities demands not only that the slope of the dose–response curves for each agonist be the same but that the time of onset of each agonist be similar. To avoid ambiguity in classification, all agonists should be full agonists, producing the same level of maximum response, differing only in affinity; for a parital agonist also acts as an antagonist, e.g. the partial agonist 2-mercaptohistamine also antagonizes the effect of histamine on acid secretion.[19].

A discrepancy between the relative activities of an agonist on different tissues may imply heterogeneity of receptors, but it is itself not persuasive evidence. The agonist may be metabolized or taken up more efficiently by one tissue than another. Part or even all the agonist activity could be due to an indirect effect. For the agonist may release endogenous histamine from its stores in some but not all tissues, a mechanism that accounts for the effectiveness of 3-pyrazolylethylamine in stimulating gastric acid secretion[21] rather than a direct effect on the $H_2$-receptor.[22] A substance may produce its effect by acting directly on the histamine receptor (e.g. to stimulate acid secretion) or through a different mediator, e.g. $\beta$-chloroethylimidazole stimulates gastric acid secretion in rat by a means other than acting on $H_2$-receptors, probably through releasing acetylcholine, but on the rat uterus it appears to act directly on the $H_2$-receptor.[19] Further, an agonist can stimulate two different histamine receptors that mediate opposing effects. Stimulation of $H_1$-receptors, at least in dog, reduces acid secretion; stimulation of $H_2$-receptors enhances acid secretion. In the dog, dimaprit produces greater maximal acid secretion than does histamine, probably because dimaprit has virtually no $H_1$-agonist activity, i.e. less than 0.0001% that of histamine.[23]

Table 16.1. Relative activities of some agonists on histamine receptors[19,20,22,23]

| Substance | Structure | $H_1$ Receptors Guinea-pig ileum | $H_1$ Receptors Rat stomach (contraction) | $H_2$ Receptors Guinea-pig atrium | $H_2$ Receptors Rat uterus | $H_2$ Receptors Rat stomach (acid secretion) |
|---|---|---|---|---|---|---|
| Histamine | HN, N ring —$CH_2$—$CH_2$—$NH_2$ | 100 | 100 | 100 | 100 | 100 |
| Thiazolylethylamine | S, N ring —$CH_2$—$CH_2$—$NH_2$ | 26.0 | | | | 0.3 |
| 2-Methylhistamine | HN, N ring ($CH_3$) —$CH_2$—$CH_2$—$NH_2$ | 16.5 | 18.6 | 4.4 | 2.1 | 2.0 |
| Triazolylethylamine | N, N, N–H ring —$CH_2$—$CH_2$—$NH_2$ | 12.7 | 10.5 | 6.8 | 9.5 | 13.7 |

| | | | | | | |
|---|---|---|---|---|---|---|
| Pyridyl-2-ethylamine | (2-pyridyl)$-CH_2-CH_2-NH_2$ | 3.0 | | | 0.6 | 0.7 |
| 4-Methylhistamine | 4-$CH_3$-imidazol-5-yl$-CH_2-CH_2-NH_2$ | 0.23 | 0.3 | 43.0 | 25.3 | 38.9 |
| Pyridyl-4-ethylamine | (4-pyridyl)$-CH_2-CH_2-NH_2$ | 0.01 | | | 0.3 | 0.7 |
| Dimaprit | $H_2N-C(=NH)-S-(CH_2)_3-N(CH_3)_2$ | <0.0001 | | 70.7 | 17.5 | 19.5 |

Relative agonist activities, measured with these cautions in mind, appear in Table 16.1. All the compounds listed except triazolylethylamine have a clearly discriminative activity on the $H_1$-receptor (e.g. the guinea-pig ileum) and the histamine receptors in other tissues, namely, the atrium, uterus, and the receptor linked to acid secretion. That these last three receptors comprise a homogeneous group, the $H_2$-receptor, is not entirely convincing from these studies with agonists. For example, the relative activity of 4-methylhistamine on the uterus is far less than on the atrium and on acid secretion. And dimaprit has greater relative activity on atrium than on the uterus or on acid secretion. There is no explanation for these discrepancies. However, the use of compounds that competitively antagonize the effects of histamine and these other agonists at the $H_2$-receptor unambiguously established that the histamine receptors in these three tissues comprise a homogeneous population.[20,22,23]

Competitive antagonists are essential for the unambiguous classification of histamine receptors. In fact our understanding of the roles of all agonists rests most securely on studies of agents that are able to reduce, in a discriminative way, the effects of agonists. Parallel displacement of the agonist dose–response curve by the antagonist shows that the agonist and antagonist are acting on the same receptor. From the $ED_{50}$ values of the agonist in the presence and absence of antagonist, one can estimate the apparent dissociation constant of the antagonist, $K_B$, which is the concentration of antagonist required to give a dose ratio of two, i.e. the concentration of antagonist that requires doubling the concentration of agonist to obtain the same effect as obtained in the absence of antagonist. The $K_B$ value is distinct for each competitive antagonist at each receptor. This value should be determined using several concentrations of antagonist. Antagonism may be expressed by measuring the dose-ratios (DR) of agonist needed to produce equal responses in the presence and absence of different concentrations of antagonists (B). Simple competitive antagonism results in a straight line with slope of 1 when log (DR − 1) is plotted against log B; in this Schild plot[19] the intercept with the abscissa is referred to as $pA_2$.[19] $K_B$ and $pA_2$ values have been extensively used in the classification of histamine receptors. For example, burimamide competitively blocked the action of histamine on the guinea-pig atrium and rat uterus, and its $pA_2$ values on these preparations (5.11 and 5.18) did not significantly differ. This agreement suggests that the receptors are homogeneous. Burimamide, in high concentrations, also antagonized the effect of histamine on the guinea-pig ileum, but the antagonism was non-competitive as shown by the Schild plot with slope of 1.32.[20] As noted above, the different relative activities of the agonist dimaprit on the guinea-pig atrium and rat uterus (Table 16.1) could suggest that the two receptors differ. As $H_2$-antagonists give nearly identical $pA_2$ values on these preparations, there is no reason to

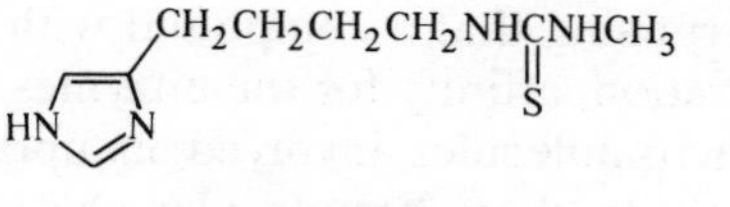

Burimamide

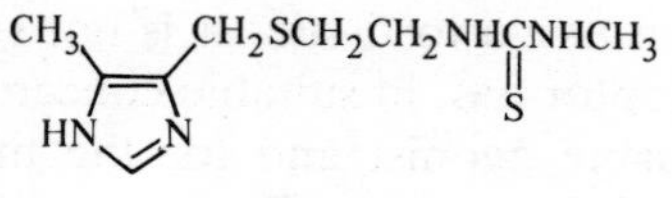

Metiamide

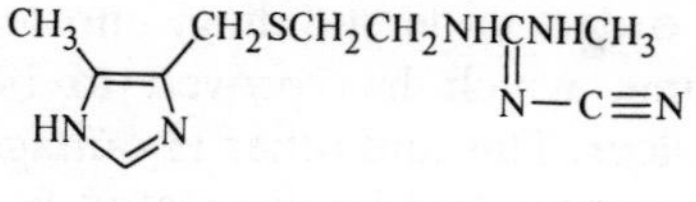

Cimetidine

invoke another receptor. Further, metiamide, an $H_2$-antagonist, has very nearly the same $pA_2$ values on these preparations even when tested with dimaprit as agonist.[23] The antagonists[24] that have been used most to classify and to characterize the $H_2$-receptor are burimamide, metiamide, and cimetidine. The $H_1$-antagonists most commonly used in receptor classification are diphenhydramine and mepyramine (i.e. pyrilamine).

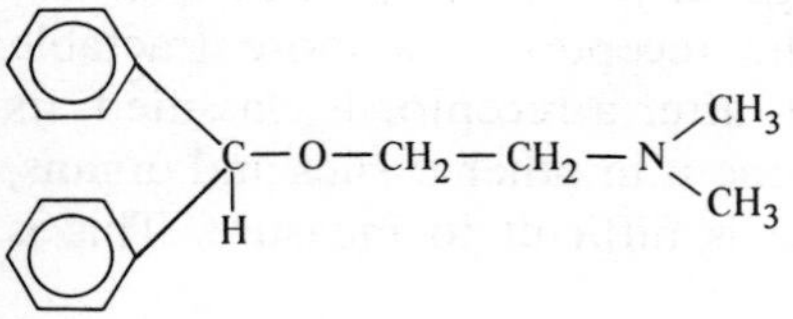

Diphenhydramine

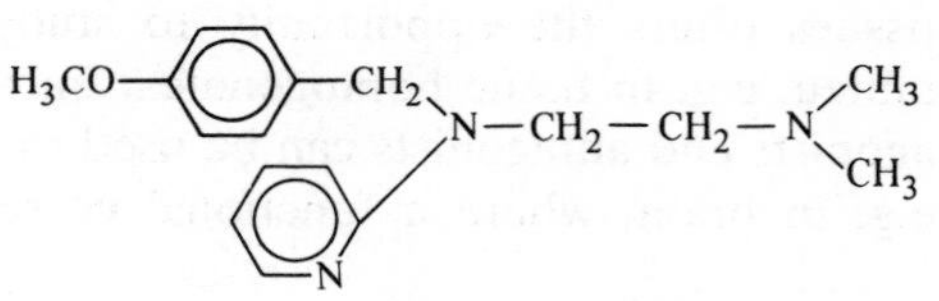

Mepyramine

It is rigorous to estimate the $pA_2$ value of an antagonist with different agonists since the displacement of the dose–response curves by the antagonist is not dependent on the affinity of the agonist. The $pA_2$ value of burimamide on the $H_2$-receptor, for example, was indistinguishable whether histamine, 4-methylhistamine, or 2-methylhistamine was used as agonist.[20] Analogously, the $pA_2$ values of mepyramine on the $H_1$-receptor did not differ when measured with six different agonists having very different relative activities.[19]

Studies of more than one antagonist are needed to classify a receptor. In one study in which 12 different compounds were examined for their antagonism of histamine on the rat atrium and uterus, three compounds gave $pA_2$ values that differed between the two tissues.[25] The reasons for the discrepancies are not known, but they are certainly due to effects other than blocking the $H_2$-receptor. Among the adventitious effects of histamine antagonists are effects on uptake and release of histamine[26,27] and effects on the enzymes that synthesize[28] and metabolize histamine.[29] All these processes vary from tissue to tissue. It is not surprising that a compound with affinity for a receptor has, in suitable concentration, affinity for the enzymes that act on the same agonist and for the macromolecules involved in uptake and binding of the agonist, for all are likely to share a molecular characteristic that allows them to react with the same compound.

A judicious and mindful use of these methods, this 'taxonomy of receptors',[30] has led to order, understanding, and fecundity. It yielded new drugs such as cimetidine, which has proved to be very effective in the treatment of duodenal ulcer. This and other $H_2$-antagonists were then used as tools to reveal the $H_2$-receptor and its associated functions in many tissues[9] and the action of drugs on this receptor.[31,32] Knowing that a receptor for a mediator is homogeneous in different tissues provides prediction of the effects of the antagonist. Showing that the receptor for a mediator in different tissues is homogenous provides convenience; after it was established that the receptor associated with rat gastric secretion was indistiguishable from that associated with guinea-pig atrial rate, the search for antagonists of histamine at the $H_2$-receptor was facilitated, for effects on the latter are more accurately and rapidly measured. Of more pervasive importance, the demonstration of a homogeneity of a receptor in different tissues offers the opportunity to study the receptor in a more tractable milieu, e.g. in tissue homogenates. Further, after a receptor is classified, its agonists and antagonists can be used to define it in other tissues and organs, e.g. in brain, where a functional correlate is difficult to measure. This is

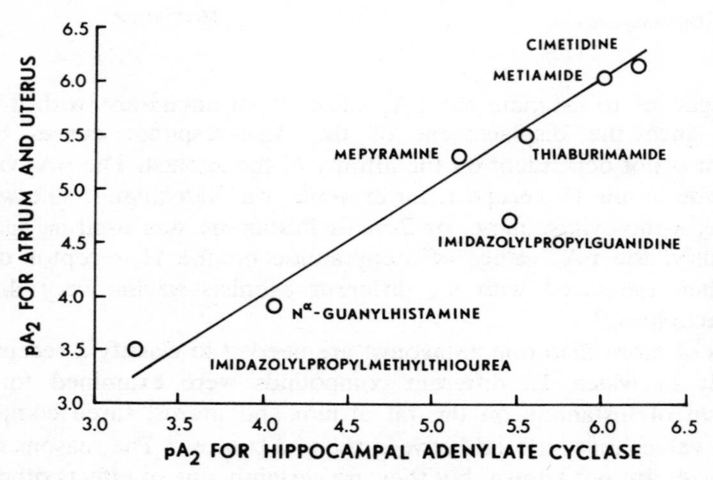


Figure 16.1. The $pA_2$ values of histamine antagonists on known $H_2$ receptors and on the histamine receptor linked to adenylate cyclase in homogenates of the guinea-pig hippocampus

exemplified in the definition of the histamine receptor linked to adenylate cyclase in homogenates of guinea-pig hippocampus.[31] The relative activities of four histamine agonists in stimulating cyclase activity did not differ from those in stimulating known $H_2$-receptors. Stimulation of the cyclase by histamine was blocked by seven different histamine antagonists, all of which gave Schild plots with slopes that did not differ from unity, indicating competitive antagonism. The Schild plot was independent of the agonist. The $pA_2$ values of all antagonists were not significantly different from those measured on the $H_2$-receptor in rat atrium and uterus (Figure 16.1). These observations imply that it is the $H_2$-receptor that is linked to adenylate cyclase in these brain homogenates.[31]

Although the procedures used to define receptors have been used for over 50 years, there appears to be a surprising innocence of them, revealed by work on all receptors and regardless whether pharmacological, physiological, or biochemical systems are studied. One recurrent error is to regard a selective agonist as a specific agonist. Dimaprit has an extraordinary selectivity for the $H_2$-receptor (0.0001% of the activity of histamine at the $H_1$-receptor[23]) and has proved useful, in conjunction with other substances, to define receptors.[31] But such selectivity is rare. Two agonists, 2- and 4-methylhistamine, are commonly abused in this way. It has been shown[20] that 4-methylhistamine has about 43% of the activity of histamine on the $H_2$-receptor, but only 0.2% of the activity of histamine on the $H_1$-receptor; the 2-methyl isomer exhibits 16.5% of the activity of histamine on the $H_1$-receptor but only 4.4% of it on the $H_2$-receptor. The high $H_2 : H_1$ ratio for 4-methylhistamine and the low ratio for 2-methylhistamine have tempted investigators to impute specificity. For example, the responses to 2- and 4-methylhistamines, at $10^{-4}$ M, have led to the inference that the effect is mediated by a specific receptor. The real test of the appropriateness of these drugs in classifying receptors must be assessed from their $ED_{50}$ values (Table 16.2). From these values, one can calculate the activation of each receptor by any concentration of agonist from equation (1):

$$\text{Effect (\% of maximum)} = \frac{e[D]}{ED_{50} + [D]} \qquad (1)$$

where e is the intrinsic activity, and [D] is the agonist concentration. At $10^{-4}$ M, 2-methylhistamine produces nearly the maximum response (i.e. 97%) on $H_1$-receptors, but also 80% of the maximum on the $H_2$-receptor. Clearly, no specificity has been evinced. 4-Methylhistamine and thiazolylethylamine are somewhat more selective, but still far from specific. At $10^{-4}$ M of 4-methylhistamine the response is 29% of the maximum on the $H_1$-receptor, as well as 97% on the $H_2$-receptor. In classifying receptors

Table 16.2. Relative activities and $ED_{50}$ values of some histamine agonists. Their chemical structures appear in Table 16.1

| Agonist | Relative activity* $H_1$ | Relative activity* $H_2$ | $ED_{50}$† ($\mu$M) $H_1$ | $ED_{50}$† ($\mu$M) $H_2$ |
|---|---|---|---|---|
| Histamine | 100.0 | 100.0 | 0.5 | 1.1* |
| 4-Methylhistamine | 0.2 | 43.0 | 250.0 | 2.6 |
| 2-Methylhistamine | 16.5 | 4.4 | 3.0 | 25.0 |
| Thiazolylethylamine | 26.0 | 0.3 | 1.9 | 366.7 |

*From the guinea-pig ileum ($H_1$), the guinea-pig atrium ($H_2$) and rat gastric acid secretion ($H_2$)[20,22]

†Calculated for the 2- and 4-methyl derivatives: $ED_{50}$ histamine $\times \frac{100}{\text{relative activity}}$

with these or other agonists, their $ED_{50}$ values relative to histamine must be measured and compared with the values on known histamine receptors.

A more common error is to ascribe specificity to the antagonist at all concentrations, as though the pharmacological categorization of the chemical endowed it with the capacity to react with only one macromolecule. Burimamide, an $H_2$-antagonist, not only releases catecholamines but it blocks the $\alpha$-noradrenergic receptor with a $pA_2$ value of 4.7,[33] which is close to its $pA_2$ value for the $H_2$-receptor, 5.1.[20] If a function—say, hyperpolarization of neural tissue—is blocked by burimamide, one cannot know that the effect is due to blockade of an $H_2$ or of an $\alpha$-adrenergic receptor. Cimetidine is a much more selective $H_2$-antagonist, but at high concentrations it blocks (non-competitively) the $\beta$-adrenergic receptor in heart and uterus and both the $H_1$ and muscarinic receptors in ileum.[34] Analogously, most $H_1$-antagonists have $K_B$ values of $10^{-8} - 10^{-10}$ M for the $H_1$-receptor, but in some work designed to define the histamine receptor, they have been commonly used at concentrations of $10^{-6}$ M and $10^{-4}$ M and occasionally at $10^{-2}$ M. Testing an antagonist at a concentration greater than 100 times the $K_B$ is almost certain to block more than the specific receptor. Non-specificity of many antagonists at high doses is well known. The $H_1$-antagonists in suitable concentrations are anticholinergic,[35] local anaesthetic, and at concentrations of $10^{-6}$ M and greater, they block the $H_2$-receptor linked to adenylate cyclase in homogenates of brain and ventricular myocardium of the guinea-pig.[31] The $H_2$-antagonists have lower affinity for their receptor than have the $H_1$-antagonists for the $H_1$-receptor. No selective $H_2$-antagonist, e.g. cimetidine, has a $pA_2$ value higher than 6.1 ($K_B = 8 \times 10^{-7}$ M). Clearly, a higher concentration of $H_2$-antagonist, e.g. cimetidine, is needed to block the $H_2$-receptor, than the concentration of $H_1$-antagonist needed to block the

$H_1$-receptor. Yet in some work designed to define a receptor, equal concentrations of each type of antagonist are used, e.g. $10^{-6}$ M, which is near the $K_B$ of the $H_2$-antagonist but 1,000 times the $K_B$ of the $H_1$-antagonist. Not surprisingly, at these concentrations the $H_2$-antagonist has no or slight effect while the $H_1$-antagonist blocks histamine, prompting the (false) inference that the activity is linked to the $H_1$-receptor. Confusion can also result from the use of only one concentration of agonist, e.g. histamine. As the antagonism is competitive and surmountable by increasing doses of agonist, one dose of agonist is unrevealing, for if its concentration is too high, it will surmount the effect of the antagonist, as shown by a glance at any dose–response curve.

## III. DISTRIBUTION OF HISTAMINE RECEPTORS

The distribution of $H_1$- and $H_2$-receptors in various tissues has been tabulated,[9] but the criteria for classification have been met only for some actions of histamine. The $H_1$-receptor mediates contraction of the guinea-pig ileum and of the bronchioles of many species. The $H_2$-receptor mediates the increased rate and force of contraction in the atrium of most species. Relaxation of the rat uterus and stimulation of gastric acid secretion in all species examined are also subserved by the $H_2$-receptor. The $H_2$-receptor on the surface of mast cells and basophils inhibits histamine release. The fall in blood pressure produced by histamine is mediated by both $H_1$-, and $H_2$-receptors. Brain contains both $H_1$- and $H_2$-receptors and both appear to be involved in the behavioural effects of histamine. Histamine decreases the firing rate of most neurons by acting on the $H_2$-receptor.

## IV. HIGH AFFINITY BINDING TO HISTAMINE RECEPTORS

Only recently have there been studies of high affinity binding of histamine. Binding of the $H_1$-antagonist, mepyramine, to the ileum yielded a $K_B$ value very nearly the same as that in blocking histamine induced contractions of the ileum.[36] Three other $H_1$-antagonists blocked the binding of mepyramine with $K_B$ values very similar to those in blocking contraction. Analogous work on rat brain membranes[37] gave $K_B$ values of 12 $H_1$-antagonists that were higher than those found on guinea-pig ileum, but the rank order of the $K_B$ values were the same. Highest levels of binding were found in the hypothalamus in rat brain, in the parietal cortex of calf brain[37] and in the cerebellum of guinea-pig brain.[37,38] The distribution of histamine binding sites did not parallel the distribution of histamine or its synthesizing enzymes. Binding was reduced after rat membranes were treated with trypsin, $\alpha$-chymotrypsin, phospholipases A and C, and Triton X-100, 0.01%.[37]

The low affinity ($K_B = 10^{-6}$ M) of the selective $H_2$-antagonists for the $H_2$-receptor may preclude their use for labelling the receptor. As amitriptyline, a tricyclic antidepressant drug, competitively antagonizes histamine at the $H_2$-receptor,[32] this receptor was labelled in brain membranes with isotopic amitriptyline in concert with other antagonists to mask sites other than the $H_2$-receptors. Analogous experiments were begun with amitriptyline to label the $H_1$-receptors as well.[39]

## V. COUPLING OF THE HISTAMINE RECEPTOR TO ADENYLATE AND GUANYLATE CYCLASES

In all tissues examined, the activation of the $H_2$-receptor has been shown to be coupled to activation of adenylate cyclase.[40] In guinea-pig brain slices, there is evidence that both the $H_1$- and $H_2$-receptors are linked to adenylate cyclase; the effects of $H_1$-antagonists on the dose–response curve to histamine suggests that activation of the $H_2$-receptor may be necessary before the $H_1$-receptor can be activated.[41] Adenosine, which enhances the effect of histamine (and other amines) on adenylate cyclase activity, appears to affect the $H_1$-receptor rather than the $H_2$-receptor in guinea-pig hippocampal slices. A study of the dose–response curves to an $H_1$-agonist and an $H_2$-agonist in the presence and absence of adenosine showed that only for the $H_1$ stimulation was there a clear decrease in the $ED_{50}$ value of adenosine (25 $\mu$M as opposed to 250 $\mu$M). The use of $H_1$- and $H_2$-antagonists also suggested that the effect of adenosine is mediated mainly by the $H_1$-receptor.[42] In mouse neuroblastoma cells, the $H_1$-receptor is linked to guanylate cyclase.[43]

Although the $H_2$-receptor is linked to adenylate cyclase in all tissues studied, it is not certain that activation of the cyclase is causally related to the function that is activated. Gastric acid secretion in amphibians is coupled to cyclic AMP formation, but the evidence in higher animals is less compelling. The content of cyclic AMP and gastric secretion exhibit the same circadian rhythm. The inhibitors of phosphodiesterase increase secretion. However, dibutyryl cyclic AMP is not always a secretagogue, nor are all the substances that increase cyclic AMP content. The latter may be explainable by different cell types within the mucosa.[44] The issue is no less clear for stimulation of cardiac contraction and adenylate cyclase activity.[45]

Establishing the postulated relationship is easier in studies of a biochemical effect of histamine. In adipocyte homogenates, cyclic AMP and histamine separately activate triglyceride lipase in the presence of kinase, suggesting that adenylate cyclase activation determines the action of histamine. The fact that phosphodiesterase inhibitors produce the effect, and that inhibitors of cyclic AMP synthesis also inhibit lipolysis support the hypothesis.[46]

The histamine linked adenylate cyclase has been most extensively studied in guinea-pig brain homogenates.[47] GTP or 5′-guanylimidodiphosphate, p(NH)ppG is very nearly obligatory for histamine activation of adenylate cyclase. The guanine nucleotides increase $V_{max}$ for histamine (at $10^{-4}$ M histamine), p(NH)ppG being more potent that GTP in this regard. The $ED_{50}$ for histamine is decreased by the guanine nucleotides, p(NH)ppG again being more potent than GTP in this effect. The effects of the guanine nucleotides on the $ED_{50}$ of histamine was greater than on the $ED_{50}$, of dimaprit, which is a partial agonist in this system (S. Maayani, unpublished observations). Further, implicit in the paper by Kanof *et al.*[47] is suggestive evidence that just as the guanine nucleotides influence the action of histamine on the cyclase (and the binding of agonists to the receptor), histamine may influence the action of the guanine nucleotides on the cyclase: a Michaelis–Menten plot of their data suggests that the $ED_{50}$ of p(NH)ppG is decreased three-fold (from $3.5 \times 10^{-6}$ M to $1 \times 10^{-6}$ M) in the presence of histamine, $10^{-4}$ M.

Another important consideration in the action of histamine on adenylate cyclase is magnesium ion. $Mg^{2+}$ complexes with $ATP^{4-}$ to form magnesium-ATP (MgATP), which is recognized as the substrate for the enzyme. Magnesium ions may also stimulate by chelating free ATP, which inhibits the enzyme. Because more $Mg^{2+}$ is required than necessary for stoichiometric formation of MgATP, it was suggested that magnesium ions may act directly on an allosteric site on the enzyme. The stimulatory effect of magnesium ions is seen on histamine sensitive adenylate cyclase e.g. in guinea-pig heart homogenates[48] and in guinea-pig brain homogenates.[47] Plotting the data of Kanof *et al.*[47] shows that increasing magnesium ions increased the affinity of the cyclase for its substrate, MgATP, (the apparent $K_m$ value falls). This plot also suggests that the cyclase is inhibited at high concentrations of MgATP. In guinea-pig heart and brain homogenates, the histamine sensitive adenylate cyclase has a ten-fold sensitivity towards magnesium ions in the presence of histamine, $10^{-3}$ M, than in its absence. Analogously in the presence of histamine, ATP is a more potent inhibitor of the cyclase than in the absence of histamine, $K_i = 0.16$ mM and 0.33 mM.[47]

## VI. MOLECULAR CHARACTERISTICS OF $H_2$ ACTIVATION

One may try to gain insight into the characteristics of receptors by analysing the molecular characteristics of agonists and antagonists. This approach to the $H_1$-receptor has not been very fruitful.[49] More recently hypotheses have been made for the molecular characteristics of $H_2$-receptor activation. $H_2$-activity is related to the tautomerism of the imidazole ring.[22,24]

The N(3)H tautomer is prevalent in the mono-cation and for the free base the preference is shifted towards the N(1)H tautomer. Structural changes,

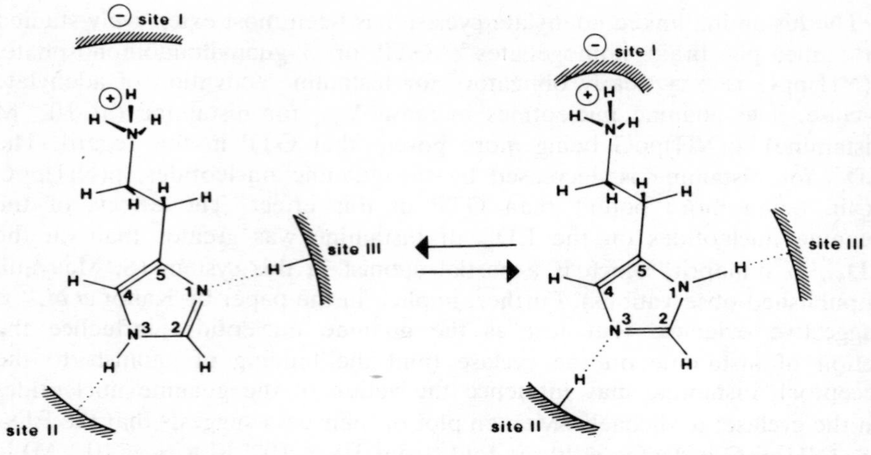

Figure 16.2. Proposed charge–relay mechanism for histamine activation of the $H_2$-receptor. Histamine reacts with the receptor through its cationic head (i.e. the protonated amino group) at site I and through the N(3)H and N(1) groups at sites II and III, respectively. Neutralization of the cationic head triggers transition of a proton from N(3)H to site II and transition of a proton from site III of the receptor to N(1)

both in the imidazole ring and in the side chain, affect the tautomeric ratio. Quantum chemistry showed that certain characteristic changes appear in the electronic structure of the imidazole ring when the proton dissociates from the side chain amine in histamine mono-cation to yield the free base.[50] These changes reveal the electronic mechanism responsible for the shift in the tautomeric ratio: the N(3) nitrogen becomes less basic than the N(1) nitrogen when the electronic charge is redistributed in the imidazole ring and the side chain is neutralized. Identical changes in the electronic structure occur when histamine mono-cation interacts with an anionic site.

These findings suggested[50] a mechanistic model for the role of imidazole tautomerism in the activation of the $H_2$-receptor (Figure 16.2). The interaction of the protonated side chain amine at site I triggers the transition of a proton from the vicinity of the N(3) nitrogen to site II of the receptor model and from site III of the receptor model to the N(1) nitrogen of histamine. Since the interactions are assumed to occur simultaneously at all three sites, the proposed mechanism is fully reversible and the imidazole ring would regain its preferred tautomeric form when the protonated side chain dissociates from the anionic site I.[50] The same charge relay reaction can be written for other $H_2$-agonists (e.g. dimaprit), regardless of their atom-to-atom resemblance to histamine; all contain a proton releasing fragment, a proton accepting fragment and a triggering mechanism.[14] The activities of 4-methylhistamine and 2-methylhistamine are less than histamine on the $H_2$-receptor; the 2-methyl derivative has the least activity. This decrease in potency could be related, at least in part, to changes induced by the methyl groups in the proton affinities of the imidazole nitrogens. The calculations predicted that the N(1) nitrogen of 4-methylhistamine has a slightly increased proton affinity compared to N(1) in histamine, and the 2-methyl derivative has an even more basic N(1) nitrogen. As a result, the di-cation species, in which both ring nitrogens bind hydrogens, will be much more abundant in 2-methylhistamine than in histamine or even in the 4-methyl derivative. Since the charge relay is established by mono-cation molecules, which are diminished in the ring methylated derivatives, they should be less potent than histamine, with 2-methylhistamine being by far the weakest $H_2$-agonist. $H_2$-antagonists cannot undergo these reactions. They lack the capacity to interact at site I; or, because of the many electronegative and polarizable groups attached to the side chain, interaction at site I does not result in a charge redistribution in the imidazole ring.[14]

## ACKNOWLEDGEMENTS

The authors' work referred to was supported by a grant from the National Institute on Drug Abuse (DA-01875). We thank Saul Maayani, Roman Osman, Carl Rosoff, and Harel Weinstein for provocative and helpful discussions.

## VII. REFERENCES

1. Beaven, M. A. *New Engl. J. Med.*, **294**, 30–36, (1976).
2. Kaliner, M., and Austen, F. A. *Ann. Rev. Pharmacol.*, **15**, 177–189, (1975).
3. Goth, A., and Johnson, A. R. *Life Sci.*, **16**, 1201–1214, (1978).
4. Paton, W. D. M. *Pharmacol. Rev.*, **9**, 269–328, (1957).
5. Lorenz, W. *Agents Actions*, **5**, 402–416, (1975).
6. Kahlson, G., and Rosengren, E. in *Biogenesis and Physiology of Histamine*, monographs of the Physiological Society No. 21 (Davson, H., Greenfield, A. D. M. Whitlam, R., and Brindley, G. S., eds.), Edward Arnold, London, pp. 1–318, (1971).
7. Lichtenstein, L. M. in *The Role of Immunological Factors in Infections, Allergic and Autoimmune Processes* (Beers, R. F., Jr., and Bassett, E. G., eds.), Raven Press, New York, pp. 339–354, (1976).
8. Schayer, R. W. in *Handbook of Experimental Pharmacology*, Springer-Verlag, Berlin, Vol. **18/1**, pp. 688–728, (1966).
9. Chand, N., and Eyre, P. *Agents Actions*, **5**, 277–295, (1975).
10. Reite, O. B. *Physiol. Rev.*, **52**, 778–819, (1972).
11. Fordtran, J. S., and Grossman, M. I. *Gastroenterology*, **74**, 338–488, (1978).
12. Editorial, *Br. Med. J.*, **i**, 739–740, (1978).
13. Nicholls, P. E., and Henry, K. *Lancet*; **i** 1095–1096, (1978).

14. Green, J. P., Johnson, C. L., and Weinstein, H. in *Psychopharmacology: A Generation of Progress* (Lipton, M. A., DiMascio, A., and Killam, K. F., eds.), Raven Press, New York, pp. 319–332, (1978).
15. Ryan, M. J., and Brody, M. J. *J. Pharmacol. Exp. Ther.*, **174**, 123–132, (1970).
16. Euler, U.S.v. in *Handbook of Experimental Pharmacology*, (Rocha e Silva, M., and Rothschild, H. A., eds.), Springer-Verlag, Berlin, Vol. **18/1**, pp. 318–333, (1978).
17. Schwartz, J.-C., Baudry, M., Bischoff, S., Martres, M.-P., Pollard, H., Rose, C., and Verdiere, M. in *Drugs and Central Synaptic Transmission* (Bradley, P. B., and Dhawan, B. N., eds.), University Park Press, Baltimore, pp. 371–382, (1976).
18. Weinreich, D. in *Biochemistry of Characterised Neurons* (Osborne, N. N., ed.), Pergamon Press, New York, pp. 153–175, (1978).
19. Ash, A. S. F., and Schild, H. O. *Br. J. Pharmacol.*, **27**, 427–439, (1966).
20. Black, J. W., Duncan, W. A. M., Durant, C. J., Ganellin, C. R., and Parsons, E. M. *Nature (London)*, **236**, 385–390, (1972).
21. Haverback, B. J., Stubrin, M. I., and Dyce, B. J. *Fed. Proc., Fed. Am. Soc. Exp. Biol.*, **24**, 1326, (1965).
22. Durant, G. J., Ganellin, C. R., and Parsons, M. E. *J. Med. Chem.*, **18**, 905–909, (1975).
23. Parsons, M. E., Owen, D. A. A., Ganellin, C. R., and Durant, G. J. *Agents Actions*, **7**, 31–37, (1977).
24. Ganellin, C. R. in *Handbook of Experimental Pharmacology*, (Rocha e Silva, M., and Rothschild, H. A., eds.), Springer-Verlag, Berlin, Vol. **18/2**, pp. 251–294, (1978).
25. Durant, G. J., Emmett, J. C., Ganellin, C. R., Miles, P. D., Parson, M. E., Prain, H. D., and White, G. R. *J. Med. Chem.*, **20**, 901–906, (1977).
26. Schayer, R. W., and Reilly, M. A. in *International Symposium on Histamine $H_2$-Receptor Antagonists* (Wood, C. J., and Simkins, M. A., eds.), Smith, Kline, and French, Ltd., Welwyn Garden City, pp. 87–95, (1973).
27. Fantozzi, R., Mannaioni, P. F., and Moroni, F. in *International Symposium on Histamine $H_2$-Receptor Antagonists* (Wood, C. J. and Simkins, M. A., eds.), pp. 107–113, Smith, Kline and French Ltd., Welwyn Garden City, (1973).
28. Mandsley, D. V., Kobayashi, Y., Williamson, E., and Bovaird, L. *Nature, New Biol.*, **245**, 148–149, (1973).
29. Barth, H., Niemeyer, I., and Lorenz, W. *Agents Actions*, **3**, 138–147, (1973).
30. Black, J. W. in *Proceedings of the 6th International Congress of Pharmacology* (Klinge, G., ed.), Pergamon Press, New York, pp. 3–16, (1976).
31. Green, J. P., Johnson, C. L., Weinstein, H., and Maayani, S. *Proc. Natl. Acad. Sci. U.S.A.*, **74**, 5697–5701, (1977).
32. Green, J. P., and Maayani, S. *Nature (London)*, **269**, 163–165, (1977).
33. Brimblecombe, R. W., Duncan, W. A. M., Owen, D. A. A., and Parsons, M. E. *Fed. Proc., Fed. Am. Soc. Exp. Biol.*, **35**, 1931–1934, (1976).
34. Brimblecombe, R. W., Duncan, W. A. M., Durant, G. J., Emmett, J. C., Ganellin, C. R., and Parsons, M. E. *J. Int. Med. Res.*, **3**, 86–92, (1975).
35. Schild, H. O. *Br. J. Pharmacol.*, **2**, 189–206, (1947).
36. Hill, S. J., Young, J. M., and Marrian, D. H. *Nature (London)*, **270**, 361–363, (1977).
37. Tran, V. T., Chang, R. S. L., and Snyder, S. H. *Proc. Natl. Acad. Sci. U.S.A.*, **75**, 6290–6294, (1978).
38. Hill, S. J., Emson, P. C., and Young, J. M. *J. Neurochem.*, **31**, 997–1004, (1978).

39. Green, J. P., Maayani, S., and Weinstein, H. *Abstracts 7th Int. Congr. Pharmacol.*, Pergamon, New York, p. 339, (1978).
40. Verma, S. C., and McNeill, J. H. *Can. J. Pharm. Sci.*, **13**, 1–3, (1978).
41. Palacios, J. M., Garbarg, M., Barbin, G., and Schwartz, J. C. *Mol. Pharmacol.*, **14**, 971–982, (1978).
42. Dismukes, K., Rogers, M., and Daly, J. W. *J. Neurochem.*, **26**, 785–790, (1976).
43. Richelson, E. *Science*, **201**, 69–71, (1978).
44. Ruoff, H. J., and Sewing, K. E. in *Cyclic 3′, 5′-Nucleotides: Mechanisms of Action* (Cramer, H., and Schultz, J., eds.), John Wiley & Sons, New York, pp. 147–160, (1977).
45. Krause, E. G., and Wollenberger, A. in *Cyclic 3′, 5′-Nucleotides: Mechanisms of Action* (Cramer, H., and Schultz, J., eds.) John Wiley & Sons, New York, pp. 229–250, (1977).
46. Fain, J. N. in *Cyclic 3′, 5′-Nucleotides: Mechanisms of Action* (Cramer, H., and Schultz, J., eds.), John Wiley & Sons, New York, pp. 207–228, (1977).
47. Kanof, P. D., Hegstrand, L. R., and Greengard, P. *Arch. Biochem. Biophys.*, **182**, 321–334, (1977).
48. Alvarez, R., and Bruno, J. J. *Proc. Natl. Acad. Sci. U.S.A.*, **74**, 92–95, (1977).
49. Nauta, W. T., and Rekker, R. F. in *Handbook of Experimental Pharmacology*, (Rocha e Silva, M., and Rothschild, H. A., eds.), Springer-Verlag, Berlin, Vol. **18/2**, pp. 215–249, (1978).
50. Weinstein, H., Chou, D., Johnson, C. L., Kang, S., and Green, J. P. *Mol. Pharmacol.*, **12**, 738–745, (1976).

SECTION IV

# Cell membrane–surface receptors for neurotransmitters

Cellular Receptors for Hormones and Neurotransmitters
Edited by D. Schulster and A. Levitzki


CHAPTER 17

# Opiate receptors and their endogenous ligands

Richard J. Miller
*The University of Chicago, Department of Pharmacological and Physiological Sciences, Chicago, Illinois 60637, U.S.A.*



## I. INTRODUCTION

The story of the discovery of the opiate receptor and its related endogenous ligands, the enkephalins and endorphins, is a fascinating one which is still far from complete. In this case, certain observations on the pharmacology of a group of drugs led to the discovery of a specific class of receptor sites which in turn led to the discovery of an endogenous peptide hormone system. This of course, is a somewhat different sequence of events from many other situations where the endogenous ligand was known prior to the discovery of the receptor.

Narcotic drugs have played a central role in medicine for hundreds of years. The various pharmacological effects of morphine have been extensively documented. Apart from its well known analgesic and euphoria-producing effects, morphine is also an extremely powerful agent for combating diarrhoea. Other effects associated with the use of some or all narcotics include respiratory depression and dysphoria. In addition,

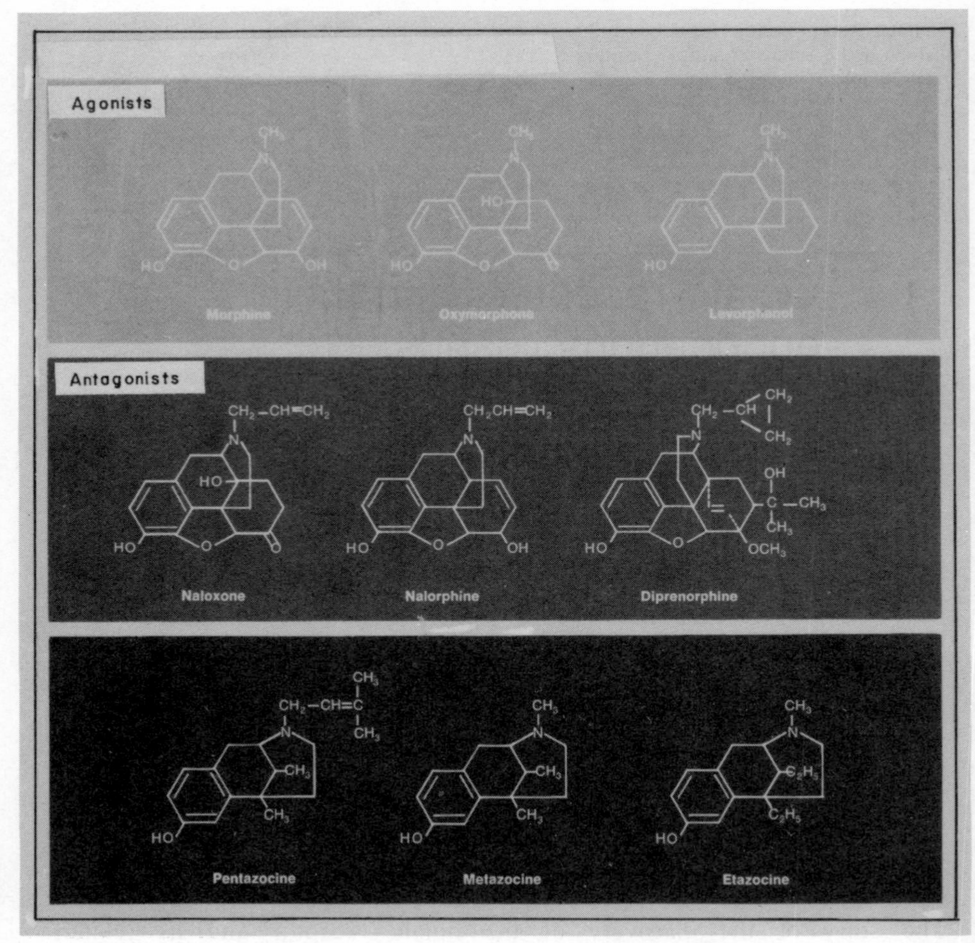


Figure 17.1. Structure of representative opiates

tolerance and physical dependence occur on continued administration of morphine and on discontinuation of the drug a profound withdrawal syndrome manifests itself. Following these basic observations certain features of opiate pharmacology led to the concept of the existence of specific opiate receptors. For example, progress in the chemistry of narcotic drugs led to the realization that structural requirements for opiate agonist activity were extremely stringent. Work in this area was greatly facilitated by the introduction of specific bioassays for opiate agonist activity such as the mouse vas deferens or the guinea-pig ileum.[8] The structural requirements for opiate agonists are illustrated by the activity of pairs of optical enantiomers such as dextrorphan and levorphanol (Figure 17.1). Here it is found that the laevorotatory isomer (levorphanol) has considerably higher biological potency than the dextrorotatory isomer (dextrorphan). This stereospecificity suggests that interaction of narcotics with their locus of action is an extremely precise event. In addition to stereospecificity several synthetic narcotics were found to act as partial agonists and in some cases as pure antagonists. Naloxone is a prime example of a pure opiate antagonist (Figure 17.1). The existence of specific opiate antagonists again suggests the existence of specific opiate receptor sites mediating the action of such drugs.

In the early 1970s several groups of workers utilizing radiolabelled opiate agonists and antagonists of high specific activity were able to demonstrate specific binding of these ligands to some entity in membrane preparations from the brain and gut.[14,18] Intensive investigation of this binding site rapidly revealed that it possessed all the properties expected of the hypothetical opiate receptor. These properties will be discussed in detail below. It now became clear that this opiate receptor was indeed truly specific for narcotic drugs. Thus, known neurotransmitters, steroid, or peptide hormones, and a multitude of other substanced all failed to show appreciable interaction with the binding site. Such observations raised a problem in the minds of several investigators. 'Why should specific receptors exist in mammalian brain for morphine—a plant alkaloid?' One possibility was that, in fact, such receptors were usually acted upon by some as yet unknown endogenous ligand. Other experimental data also favoured such a hypothesis. It had been demonstrated that electrical stimulation of certain parts of the brain, such as the periaqueductal grey area, produced analgesia in animals and man. Moreover, such analgesic effects could be blocked by prior treatment of the animal with naloxone.[1] These data were interpreted to mean that the electrical stimulation was able to release some endogenous analgesic producing substance whose effects could be blocked by naloxone.

Several groups now began to search for some 'morphine-like factor' in extracts of brain. Using bioassays such as the guinea-pig ileum or receptor binding assays it rapidly became clear that crude extracts of brain did indeed contain some opiate agonist activity. Surprisingly, the pituitary gland also proved to contain very high concentrations of opiate-like material. In the central nervous system the opiate-like material was shown to be concentrated in synaptosomal fractions increasing speculation that the compound might be a novel neurotransmitter. Hughes *et al.*[6] were the first to elucidate the chemical structure of the morphine-like material they had isolated from brain and which they had named enkephalin. The results of these investigations were extremely surprising for a number of reasons. To begin with, enkephalin proved to be a mixture of two pentapeptides. These peptides have sequences $H_2N$-Tyr-Gly-Gly-Phe-Met-OH (Methionine enkephalin) and $H_2N$-Tyr-Gly-Gly-Phe-Leu-OH (Leucine enkephalin) (Figure 17.2). On consideration of the structure of morphine it

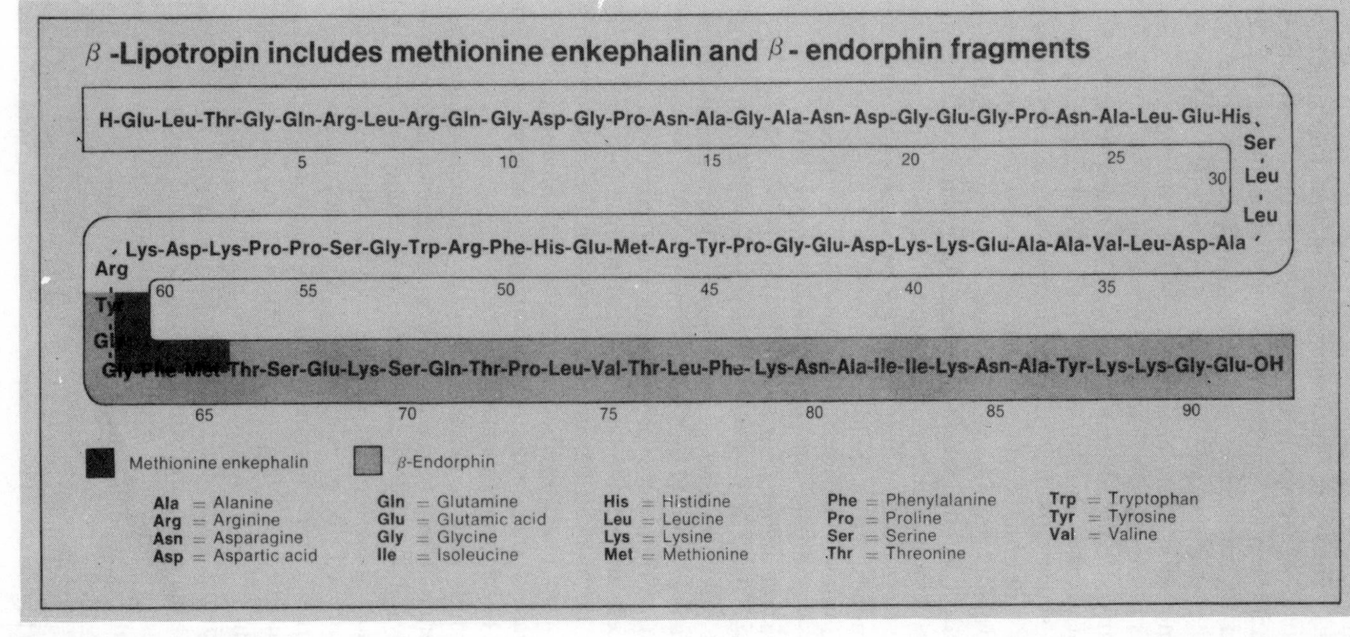

β -Lipotropin includes methionine enkephalin and β - endorphin fragments
H-Glu-Leu-Thr-Gly-Gln-Arg-Leu-Arg-Gln-Gly-Asp-Gly-Pro-Asn-Ala-Gly-Ala-Asn-Asp-Gly-Glu-Gly-Pro-Asn-Ala-Leu-Glu-His
5
10
15
20
25
30
Ser
Leu
Leu
Lys-Asp-Lys-Pro-Pro-Ser-Gly-Trp-Arg-Phe-His-Glu-Met-Arg-Tyr-Pro-Gly-Glu-Asp-Lys-Lys-Glu-Ala-Ala-Val-Leu-Asp-Ala
Arg
60
55
50
45
40
35
Thr-Ser-Glu-Lys-Ser-Gln-Thr-Pro-Leu-Val-Thr-Leu-Phe-Lys-Asn-Ala-Ile-Ile-Lys-Asn-Ala-Tyr-Lys-Lys-Gly-Glu-OH
65
70
75
80
85
90
Methionine enkephalin
β-Endorphin
Ala = Alanine
Arg = Arginine
Asn = Asparagine
Asp = Aspartic acid
Gln = Glutamine
Glu = Glutamic acid
Gly = Glycine
Ile = Isoleucine
His = Histidine
Leu = Leucine
Lys = Lysine
Met = Methionine
Phe = Phenylalanine
Pro = Proline
Ser = Serine
Thr = Threonine
Trp = Tryptophan
Tyr = Tyrosine
Val = Valine


Figure 17.2.

is evident that at first glance one cannot think of anything less likely to have morphine-like properties. Secondly, it was also noticed that the sequence of [Met$^5$]-enkephalin was contained in its entirety within the sequence of $\beta$-lipotropin, a hormone of 91 amino acids previously isolated from the pituitaries of several species. [Met$^5$]-enkephalin represents residues 61–65 of $\beta$-lipotropin. Now it was known at this time that $\beta$-lipotropin was broken down *in vivo* to yield $\gamma$-lipotropin ($\beta$-lipotropin 1–58) and C-fragment ($\beta$-lipotropin 61–91). It was thought that $\gamma$-lipotropin functioned as a precursor to the hormone $\beta$-MSH, however, no function was known for C-fragment. On realizing that the first five amino acids of C-fragment were identical to [Met$^5$]-enkephalin many workers now examined the entire C-fragment 31 amino acids for possible opiate activity. C-fragment proved to be a very potent opiate-like substance, at least as potent as [Met$^5$]-enkephalin.[10] In addition, C-fragment proved to be very stable in contrast to [Met$^5$]- and [Leu$^5$]-enkephalins which are rapidly degraded by enzymes in blood or

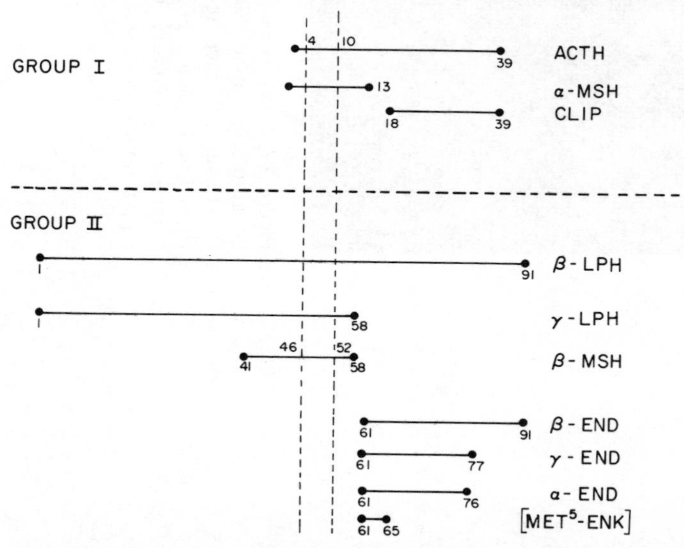


Figure 17.3. Sequence homologies between ACTH, $\beta$-Lipotropin and related peptides.
Group I ACTH and related peptides. ACTH-(Corticotropin), $\alpha$-MSH-($\alpha$-melanotropin); CLIP-(Corticotropin-like intermediate lobe peptide).
Group II $\beta$-LPH ($\beta$-lipotropin); $\gamma$-LPH-($\gamma$-lipotropin); $\beta$-MSH-($\beta$-melanotropin); $\alpha$-END-($\alpha$-endorphin); $\beta$-END-($\beta$-endorphin); $\gamma$-END-($\gamma$-endorphin); [Met$^5$-ENK]-(methionine-enkephalin).
Vertical dashed lines represent seven amino acid sequence (ACTH 4–10, $\beta$-LPH 46–52) which is common to ACTH and $\beta$-LPH

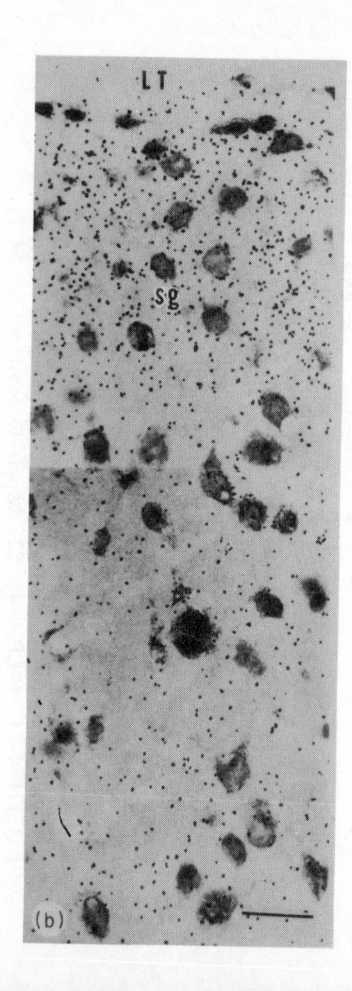


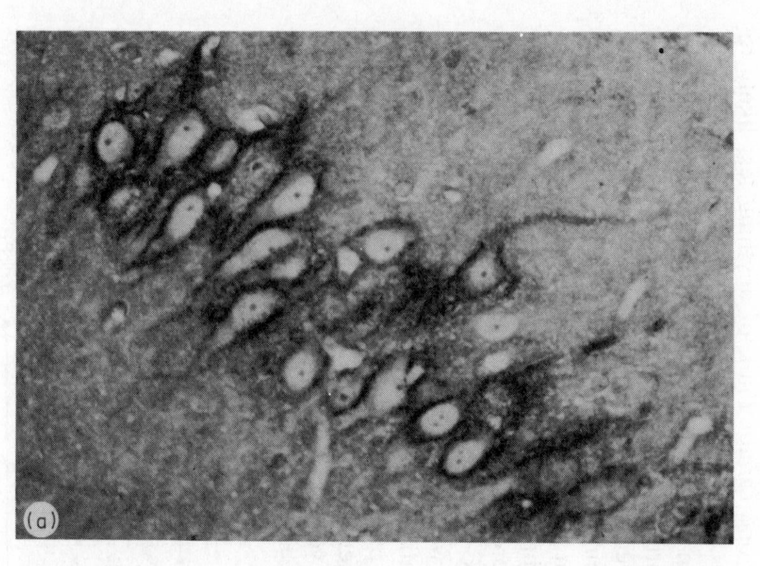


Figure 17.4. (a) Localization of opiate receptors in rat spinal cord by autoradiography. Silver grains represent receptor localization. (LT-Lissauer's Tract, Sg-substantia gelatinosa).
(b) Localization of [Leu$^5$]-enkephalin in rat brain by immunohistochemistry. Enkephalin containing nerve terminals in the hippocampus (dark staining) (Picture courtesy of Dr M. Kuhar, Dr M. Sar, and Dr W. Stumpf)

tissue homogenates. In recognition of its newly found properties C-fragment was renamed β-endorphin. Since that time C-fragment and several related peptides ranging in length from [Met$^5$]-enkephalin (β-LPH 61–65) to β-endorphin (β-LPH 61–91) have been found in brain and pituitary. Such peptides are collectively known as endorphins (Figure 17.3).

The fact that the sequence of [Met$^5$]-enkephalin is found within that of β-LTH has led to considerable speculation that β-LPH/β-endorphin function as biosynthetic precursors to [Met$^5$]-enkephalin. It may be noted in passing that the synthesis of β-LPH itself is linked to that of another peptide hormone, corticotropin (ACTH). A large molecule of approximate molecular weight 31,000 known as 'big ACTH' is processed by the pituitary to give both β-LPH and related peptides on the one hand and ACTH and related peptides on the other hand. Immunohistochemical localization of the two hormones in the pituitary show that they are localized in exactly the same cells. Moreover, physiological manipulations such as adrenalectomy, dexamethasone treatment and stress appear to regulate the release of both hormones in an identical fashion (Figure 17.3).

Recently it has been possible to demonstrate both opiate receptors and their endogenous ligands morphologically. The receptors themselves may be localized by autoradiography following administration of the highly potent opiate antagonist diprenorphine[9] (Figure 17.4). The enkephalins and other endorphins may be localized by using immunohistochemical techniques. Examples of such localizations are illustrated in Figure 17.4. In conjunction with the biochemical data this information gives several clues as to the possible functions of the opiate receptors and their associated ligands.

The above represents a brief introduction to what is a relatively new hormone–receptor system. However, these discoveries have produced such wide interest in the scientific community that our knowledge of them has rapidly increased. In the next part of this chapter we shall be concerned with what is now known about the nature of the opiate receptor. Certain aspects of the above discussion have been reviewed elsewhere.[11]

## II. THE OPIATE RECEPTOR

### (i) Localization

The opiate receptor may be localized either by using receptor binding assays or autoradiography as described above. It can be said that data from the two sets of studies agree very well on the whole. A more precise localization of the receptor can be obtained using autoradiography whereas receptor binding assays are somewhat more accurate in quantitating receptor numbers. High densities of receptors by either method are found in the dorsal horn of the spinal cord, in laminae I and II (substantia gelatinosa). Autoradiographic grains occur in the neuropil suggesting a synaptic association. Elevated levels of receptors are also found in the substantia gelatinosa of the spinal trigeminal nucleus, the ventral area postrema, the vagus nerve, and associated nuclei. In the diencephalon, dorsal medial areas of the thalamus as well as the infundibulum contain high densities of receptors. Extremely high receptor concentrations are also found in the basal ganglia. It is interesting to note that some evidence suggests that opiate receptors may be localized presynaptically on axons in several instances.

Receptors for opiates are also known to exists in the gut. Receptor binding assays have demonstrated the presence of receptors in homogenates of the guinea-pig ileum.[3] However, no receptor binding has yet been found in homogenates of mouse vas deferens. Additional localizations of opiate receptors are the posterior lobe of the pituitary and the kidney.

Apart from occurring *in vivo* opiate receptors have also been shown to occur on certain cell lines in culture.[7] This observation is an important one as such cell lines have gained wide popularity as model systems for investigating the mode of action of opiates at the molecular level. The first line reported to possess opiate receptors was the NG108-15 glioma X neuroblastoma hybrid line. Several other lines have also been reported to possess significant levels of receptors. More recently a neuroblastoma line N4TG1 has also been shown to have high levels of opiate receptors. In non-neuronal lines no cells have been reported to have more than a small number of receptors.

### (ii) Receptor binding assays

Originally opiate receptor binding assays were performed utilizing tritiated narcotic agonists and antagonists. Early attempts to label opiate receptors by incubation of brain membranes with such ligands were unsuccessful. Binding of the ligands to the membranes certainly occurred but such binding did not show any of the characteristics associated with the putative opiate receptor. For example, binding sites should be finite in number and so the binding should be saturable. Secondly, unlabelled drugs should be able to compete for binding to the receptor in a manner consistent with the *in vitro* potencies of the drugs in bioassays. Here the criterion of stereospecificity is particularly important. Thus, among natural and synthetic opiates studied it is always the laevorotatory isomer that is active while its dextrorotatory enantiomer is considerably less effective. In 1971 Goldstein and his collaborators applied this criterion of stereospecificity to the binding of radioactive narcotics to brain membranes.[5] In this study, 2% of the total binding appeared to be stereospecific, i.e. displaced by levorphanol but not by dextrorphan at low concentrations. Within a few years however, three other groups had demonstrated binding of labelled opiates that was almost completely stereospecific, up to 90% being selectively displaced by low concentrations of levorphanol. The opiate antagonist naloxone or the agonists etorphine or dihydromorphine were utilized in these studies.[14,18] Since the original papers, many other tritiated opiates such as naltrexone and diprenorphine have also been successfully employed. It was clear in these early studies of the receptors that apart from stereospecificity drugs interacted with the receptor in a manner that correlated well with their potencies *in vivo*. In many cases, however, drugs were found to be rather more potent in the binding assay than in producing a response in the guinea-pig ileum preparation for example. This, as will be seen, is due to the interaction of sodium ions with the opiate receptor. The effect of sodium, in fact, can be partly deduced by noting a discrepancy in two of the original sets of data to be published. Pert and Snyder[15] observed little effect of sodium ions on the binding of tritiated naloxone to the opiate receptor whereas Simon *et al.*[18] found that NaCl reduced the binding of [$^{3}$H]etorphine. The key thing to notice here is that naloxone is a pure antagonist whereas etorphine is an agonist. Further investigation of the action of sodium revealed that it could indeed alter the binding of ligands to the receptor but that the degree to which this occurred depended on the ligand. In the case of a pure agonist such as morphine, physiological concentrations of sodium ions (i.e. 1–100 mM) decreased binding considerably. On the other hand, the binding of a pure antagonist such as naloxone actually increased. Pert and Snyder consequently devised a 'sodium effect index'. The ability ($IC_{50}$) of a drug to displace [$^{3}$H]naloxone in the absence of sodium is compared to its ability in the presence of 100 mM sodium. For naloxone itself, this ratio is unity. For morphine it is 30. For a partial agonist such as pentazocine it is about three. In general, Pert and Snyder hypothesized that pure opiate agonists would have a sodium ratio $> 10$, partial agonists a ratio 1–10, and pure antagonists a ratio of one. This prediction has held up remarkably well for a large number of narcotic drugs (Table 17.1). However, recent data using opiate peptides have cast some doubt on the validity of such predictions (see below).

Apart from its potential pharmacological utility the sodium effect is certainly one of the most characteristic features of the opiate receptor. The effect is very specific for sodium ions. Potassium ions for example do not produce the same effects. Lithium ions do show some effects but are less potent that sodium ions. Divalent cations do not produce these effects, in fact manganese ions have been observed to potentiate the binding of

Table 17.1. Receptor affinities of opiates as influenced by sodium

| Drug | Relative affinity for opiate receptor binding* (nM) No sodium | 100 mM sodium | Sodium response ratio for opiate receptor binding† |
|---|---|---|---|
| Pure antagonists | | | |
| Naloxone | 1.5 | 1.5 | 1.0 |
| Naltrexone | 0.5 | 0.5 | 1.0 |
| Diprenorphine | 0.5 | 0.5 | 1.0 |
| Antagonists 'contaminated' with agonist activity | | | |
| Cyclazocine | 0.9 | 1.5 | 1.7 |
| Levallorphan | 1.0 | 2.0 | 2.0 |
| Nalorphine | 1.5 | 4.0 | 2.7 |
| Mixed agonist–antagonists | | | |
| Pentazocine | 15 | 50 | 3.3 |
| Ketocyclazocine | 18 | 60 | 3.3 |
| (−)-5-Propyl-5-nor metazocine | 7 | 30 | 4.3 |
| Ethylketocyclazocine | 9 | 59 | 6.4 |
| Agonists | | | |
| Etorphine | 0.5 | 6.0 | 12 |
| Phenazocine | 0.6 | 8.0 | 13 |
| Meperidine | 3000 | 50,000 | 17 |
| Levorphanol | 1.0 | 15 | 15 |
| Methadone | 7.0 | 200 | 28 |
| Oxymorphone | 1.0 | 30 | 30 |
| Morphine | 3.0 | 110 | 37 |
| Dihydromorphine | 3.0 | 140 | 47 |
| Normorphine | 15 | 700 | 47 |
| (+)-Propoxyphene | 200 | 12,000 | 60 |

*Relative affinity is defined by the concentration of drug required to inhibit by 50% the stereospecific binding of [$^3$H]naloxone (1.5 nM) to homogenates of rat brain minus cerebellum in the presence or absence of 100 mM NaCl. Lower affinity values indicate greater potency.
†Sodium response ratio is the ratio of the relative affinity for inhibition by drugs of [$^3$H]naloxone binding in the presence of 100 mM NaCl to the relative affinity value in the absence of added NaCl. (Data courtesy of Dr S. H. Snyder)

opiate agonists to the receptor. Some argument still exists as to the mode of action of sodium ions in this system. Scatchard plots of [$^3$H]naloxone binding or [$^3$H]dihydromorphine binding in the brain in the absence of sodium can be resolved into two linear components suggesting the existence of a high and low affinity binding site for agonists (Figure 17.5). According to Pasternak and Snyder[13] sodium ions decrease the number of high affinity binding sites for agonists such as dihydromorphine and increase the number of sites for antagonists such as naloxone. In contrast to this however, Simon and colleagues[19] using [$^3$H]etorphine obtained

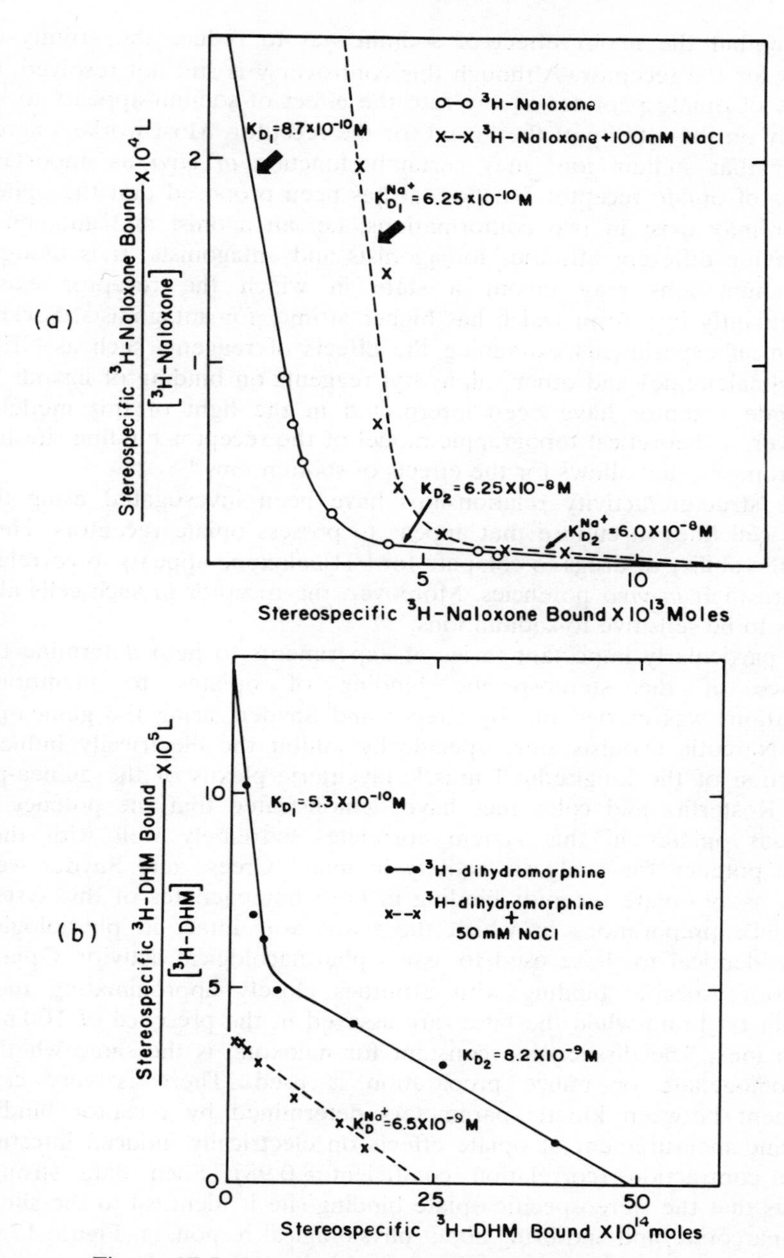


Figure 17.5. Scatchard plots of (a) [$^3$H]naloxone binding in the presence and absence of NaCl (b) [$^3$H]dihydromorphine (DHM) binding in the presence and absence of NaCl (adapted from Reference 13)

evidence that the major effect of sodium was to reduce the affinity of agonists for the receptor. Although this controversy is still not resolved, in the case of opiate peptides at any rate the effect of sodium appears to be primarily on the affinity of the ligand for the receptor. Most workers agree however that sodium ions may certainly function *in vivo* as important effectors of opiate receptor function. It has been proposed that the opiate receptor may exist in two conformations, i.e. an agonist and antagonist state having different affinities for agonists and antagonists. It is thought that sodium ions may favour a state in which the receptor exists predominantly in a form which has higher affinity for antagonists. Several biochemical experiments examining the effects of reagents such as NEM (*N*-ethylmaleimide) and other sulphydryl reagents on binding of ligands to the opiate receptor have been interpreted in the light of this model.[18] Moreover, a theoretical topographic model of the receptor binding site has been proposed that allows for the effects of sodium ions.[4]

Some structure/activity relationships have been investigated using the various cell lines in culture that appear to possess opiate receptors. Here again, the ability of drugs to compete for [$^3$H]naloxone appears to correlate well with their *in vivo* potencies. Moreover, the receptor in such cells also appears to be sensitive to sodium ions.

One particularly important series of experiments to help determine the relevance of the stereospecific binding of opiates to membrane preparations was carried out by Creese and Snyder[3] using the guinea-pig ileum. Narcotic agonists stereospecifically inhibit the electrically induced contraction of the longitudinal muscle/myenteric plexus of the guinea-pig ileum. Kosterlitz and colleagues have demonstrated that the potency of numerous opiates in this system correlates extremely well with their relative potency for analgesic activity in man.[8] Creese and Snyder were able to assay opiate receptor binding in both homogenates of this system and mince preparations in which the tissue was intact in physiological buffers identical to those used to assess pharmacological activity. Opiates show stereospecific binding with affinities closely approximating those found in rat brain when the latter are assayed in the presence of 100 mM sodium ions. The dissociation constant for naloxone is the same whether the homogenate or mince preparation is used. There is very close agreement between kinetic parameters determined by receptor binding assay and measurement of opiate effects on electrically induced intestinal muscle contraction (correlation coefficient = 0.98). Such data strongly suggests that the stereospecific opiate binding site is identical to the site to which narcotics bind and bring about physiological responses (Figure 17.6).

The discovery of the enkephalins has added many new restraints to our criteria for defining opiate receptors and the responses that they mediate. Clearly in order to be sure that a response is opiate receptor mediated the

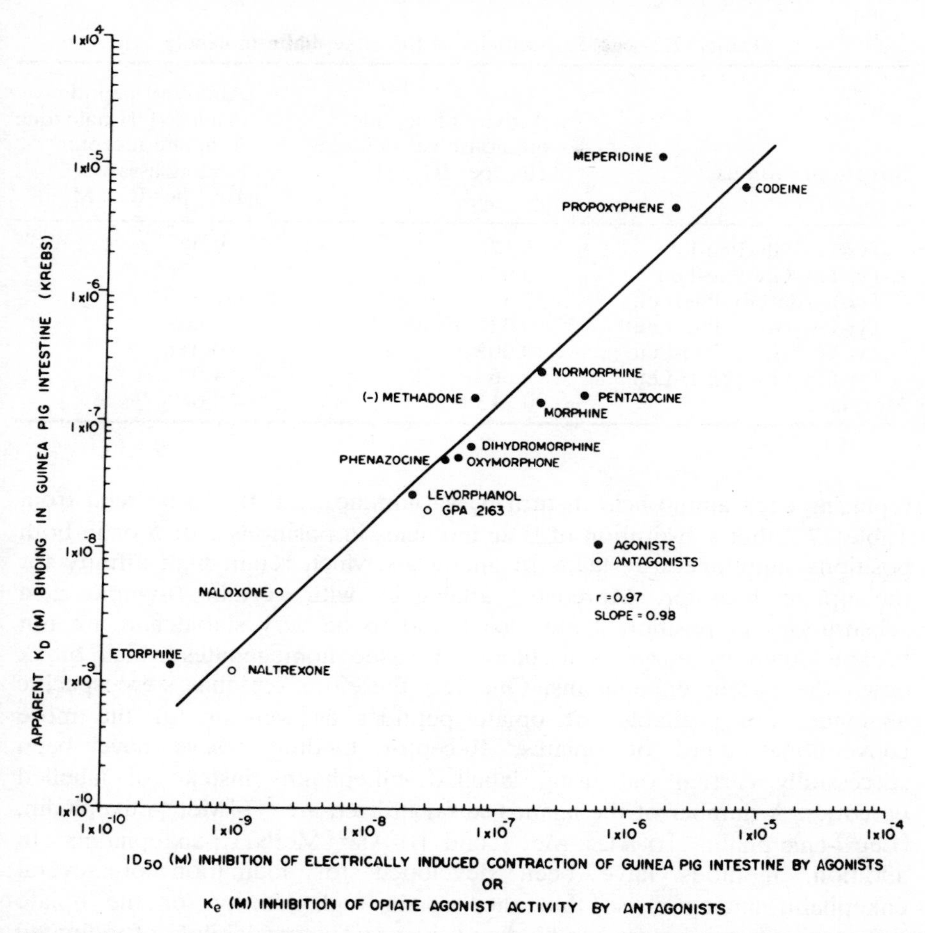


Figure 17.6. Correlation between receptor binding and pharmacological activities of opiates in the guinea-pig intestine (adapted from Reference 3)

response must not only be stereospecific with respect to traditional opiates as mentioned above but must also be brought about by enkephalins or their synthetic analogues. In fact one is probably more correct in speaking of enkephalin or endorphin receptors than of opiate receptors, as the former compounds are the true *in vivo* ligands.

The past few years have seen tremendous activity in investigation of the chemistry of the opiate peptides. Owing to the relative ease of synthesizing a pentapeptide, well over 500 structural analogues of the enkephalins have so far been produced. One of the more interesting aspects of the chemistry of the enkephalins relates to their stereochemical requirements, by

Table 17.2. Stereospecificity of the enkephalin molecule

| Structural formula | Activity of peptide on mouse vas deferens $10^{-7}$ per $IC_{50}$ M | Ability of peptide to displace [$^3$H]naloxone in opiate receptor binding assay $10^{-7}$ per $IC_{50}$ M |
|---|---|---|
| Tyr-Gly-Gly-Phe-Leu | 8.17 | 9.09 |
| D-Tyr-Gly-Gly-Phe-Leu | 0.034 | — |
| Tyr-D-Ala-Gly-Phe-Leu | 57.3 | 31.3 |
| Tyr-Gly-D-Ala-Phe-Leu | 0.021 | 0.002 |
| Tyr-Gly-Gly-D-Phe-Leu | 0.002 | $<0.011$ |
| Tyr-Gly-Gly-Phe-D-Leu | 9.66 | 4.00 |
| Morphine | 0.215 | 28.6 |

replacing each amino acid in turn by a D-amino acid. It can be seen from Table 17.2 that substitution of D-amino acids in positions 2 or 5 or in both positions simultaneously leads to analogues which retain high affinity for the opiate receptor. Moreover, analogues with certain D-amino acid substitutions in position 2 are also found to be very stable and are not broken down by enzymes in blood or tissue homogenates which break down the parent enkephalins. One can therefore see that stereospecific analogues are available for opiate peptides as well as for the more conventional types of opiates. Receptor binding assays have been successfully carried out using labelled enkephalins instead of labelled narcotics. A number of the ligands so far utilized are $^3$H-[Met$^5$]-enkephalin, [Leu$^5$]-enkephalin, [D-Ala$^2$ Met$^5$] and [D-Ala$^2$ MetNH$_2^5$]-enkephalins. In addition, methods have been developed for iodination of several enkephalin analogues so that they retain high affinity for the opiate receptor and may be used in binding assays.[12] The availabiliity of iodinated enkephalins is certainly useful when studying the opiate receptor as such derivatives are of very high specific activity when compared to previously utilized tritiated ligands.

Binding of labelled enkephalins to brain membranes again results in biphasic Scatchard plots (see above). It is interesting to note, however, that when utilizing cultured cells only one binding site for opiates or opiate peptides is manifest from Scatchard analysis (Figure 17.7). The opiate receptor in membranes of NG108-15 or N4TG1 cells as monitored by enkephalin binding shows many of the characteristics described above for the opiate binding site as defined by $^3$H-narcotic agonist or antagonist binding. This indicates that both types of molecule are probably binding to the same receptor. However, there are certainly some differences which are very clear and which may turn out to be of great significance.

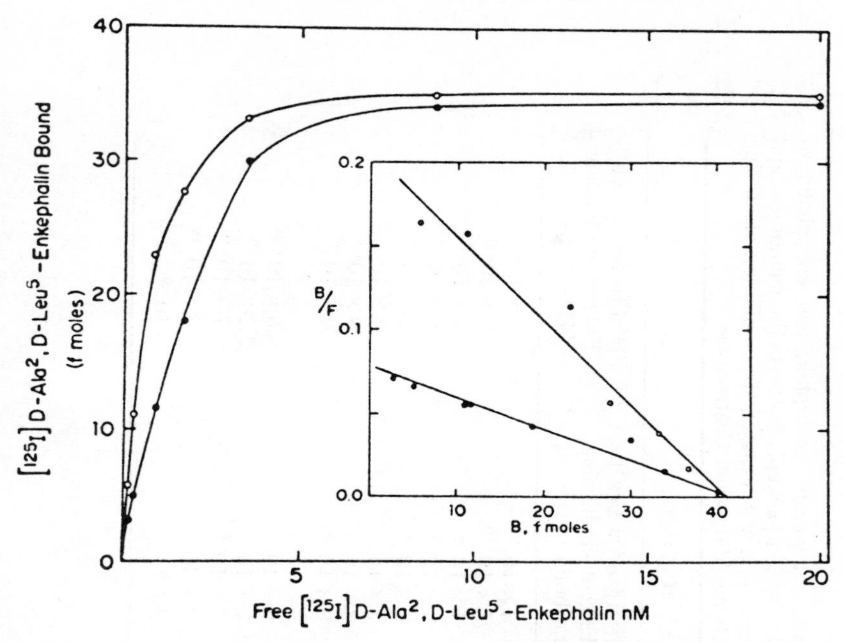


Figure 17.7. Saturation curves and Scatchard plots of [$^{125}$H]D-Ala$^2$, D-Leu$^5$-enkephalin binding to intact N4TGl cells (•) and cell membranes (○). (Chang, K., Miller, R., and Cuatrecasas, P. in press)

First of all there is the effect of sodium ions. Binding of enkephalin is certainly altered by sodium ions just as is binding of narcotic agonists such as morphine. However, it is not clear what this 'means' in the pharmacological sense. As discussed above the magnitude of the sodium effect with traditional narcotics seems highly predictive of certain aspects of their pharmacology. This does not seem to be true with respect to opiate peptides. In general, it has been found that several enkephalin analogues and [Met$^5$]-enkephalin itself have $Na^+$ shifts in the range 1–10 indicating thay they should act in some degree as partial agonists.[11] However, such effects have not been detected although they have been looked for. In bioassays such peptides behave as pure agonists. At this stage therefore, it appears that enkephalins do exhibit sodium shifts but what relationships this index holds to any aspects of their pharmacology, if any, is yet to be ascertained.

A second feature of enkephalin receptor binding assays relates to the effects of various narcotic drugs in displacing bound enkephalins from their receptors. It is found that in general, narcotics are considerably less potent in displacing bound enkephalin from receptors than in displacing naloxone

Table 17.3. Comparison of the potency of narcotic agonists, antagonists, endorphin, and enkephalin in competing with the binding of [$^3$H]naloxone and [$^{125}$I][D-Ala$^2$-D-Leu]-enkephalin to brain membrane and neuroblastoma cells. The potencies are expressed as the concentration which causes 50% inhibition of binding ($IC_{50}$). Reproduced with permission from Chang, Miller, and Cuatrecases

| | $IC_{50}$ (M) | | |
|---|---|---|---|
| Drug | [$^3$H]Naloxone binding rat brain | [$^{125}$I][D-Ala$^2$Leu$^5$]-Enkephalin binding rat brain | [$^{125}$I][D-Ala$^2$D-Leu$^5$]-Enkephalin binding N4TG1 cells |
| Naloxone | $1 \times 10^{-9}$ | $4 \times 10^{-9}$ | $9 \times 10^{-9}$ |
| Etorphine | $8 \times 10^{-10}$ | $5 \times 10^{-10}$ | $8 \times 10^{-10}$ |
| Morphine | $3 \times 10^{-9}$ | $4 \times 10^{-8}$ | $7 \times 10^{-8}$ |
| Levallorphan | $2 \times 10^{-9}$ | $9 \times 10^{-9}$ | $6 \times 10^{-9}$ |
| Pentazocine | $9 \times 10^{-9}$ | $2 \times 10^{-8}$ | $1 \times 10^{-7}$ |
| Phentanyl | $1 \times 10^{-9}$ | $5 \times 10^{-8}$ | $4 \times 10^{-7}$ |
| Pethidine | $1 \times 10^{-6}$ | $2 \times 10^{-5}$ | $9 \times 10^{-5}$ |
| Methadone | $1 \times 10^{-8}$ | $8 \times 10^{-8}$ | $9 \times 10^{-8}$ |
| [Met$^5$]-Enk | $2.0 \times 10^{-8}$ | $6 \times 10^{-9}$ | $9.8 \times 10^{-10}$ |
| [Leu$^5$]-Enk | $1.1 \times 10^{-8}$ | $9.6 \times 10^{-9}$ | $2.2 \times 10^{-9}$ |
| [D-Ala$^2$D-Leu$^5$]-Enk | $2.6 \times 10^{-9}$ | $1 \times 10^{-9}$ | $7 \times 10^{-10}$ |
| $\beta$-Endorphin | $8 \times 10^{-10}$ | $2 \times 10^{-9}$ | $8 \times 10^{-10}$ |
| $\beta$-Lipotropin | $>10^{-6}$ | $>10^{-6}$ | $>10^{-6}$ |

or dihydromorphine. Moreover, the rank order of effectiveness of such drugs also changes when comparing their abilities to displace the two sets of ligands (Table 17.3). On the other hand, opiate peptides are generally better at displacing bound enkephalin than they are at displacing narcotics. There are exceptions to this, however. The potent narcotic agonist etorphine for example is equally effective in displacing both peptide and narcotic ligands. This is also true for β-endorphin. Such data may suggest that narcotics and opiate peptides are interacting with overlapping, but slightly different, portions of the opiate receptor. However, other explanations have also been evoked. In order to understand these alternate explanations one must first mention certain other intriguing features of the pharmacology of opiate peptides. The two main bioassay systems used to evaluate opiate agonist activity are the electrically stimulated mouse vas deferens and guinea-pig ileum preparations.[8] In general, there is an excellent correlation between results in these two systems and with the effectiveness of drugs as narcotic analgesis *in vivo*. However, recently some newer narcotics have been found to give differing quantitative effects in the two systems. Thus, drugs such as ketocyclazocine and related compounds of the 6,7-benzomorphan series are found to be relatively more effective on the guinea-pig ileum than in the mouse vas deferens. More recently it has been found that enkephalins are uniformly more effective on the mouse vas deferens than on the guinea-pig ileum. When looking at the relative effectiveness of a series of enkephalins in other pharmacological systems as well, it is quite clear that different potency series are generated depending on the pharmacological effect examined, i.e. antinociceptive, anti-tussive, or anti-diarrhoeal. These pharmacological data in conjunction with the binding data mentioned previously have led to the suggestion that possibly different subclasses of receptors mediate different pharmacological effects of opiates and opiate peptides. This would be analogous to the situation with the cholinergic system for example where nicotinic and muscarininc receptors are known to exist (Chapters 18 and 19). At the moment, however, all this is rather hypothetical and the possibility of multiple classes of opiate receptors remains one of the currently most interesting research areas in opiate pharmacology.

## III. ISOLATION OF THE OPIATE RECEPTOR

In comparison to the great progress in isolation of many hormone and neurotransmitter receptors, advances in this area have been modest to say the least. This is mainly due to the inability of investigators to devise a method for solubilizing the receptor in a manner that retains its ability to bind ligand. It is also true that even with respect to experiments on receptors in intact membranes, considerable controversy exists. The receptor in intact membranes of both brain and N4TG1 cells is sensitive to sulphydryl reagents such as NEM and to proteolytic enzymes such as trypsin. It is also interesting to note that it is exquisitely sensitive to phospholipase A treatment although not to treatment by phospholipase C or D. Binding is also extremely sensitive to all detergents, both ionic and non-ionic, which is one of the reasons why binding is so hard to follow on solubilization of membranes. The sensitivity of binding to proteolysis, NEM, and temperature denaturation indicates that the receptor is a protein as might be expected. However, one group of workers has made a case for cerebroside sulphate or a similar molecule as a functional opiate receptor. Certainly glycolipids may function as physiological receptors in some cases (e.g. for cholera toxin or interferon). Most of the evidence for the involvement of cerebroside sulphate stems from the observation that labelled narcotics are able to bind in a stereospecific fashion to sulphatide molecules. It is also claimed that in certain mutant mice ('jimpy') which have a substantial depletion of sulphatide in the CNS, morphine is relatively ineffective in producing analgesia. In spite of such evidence it must be said that it seems rather unlikely that cerebroside sulphate per se can act as a functional opiate receptor. For example, recently we have found that neuroblastoma cells N4TG1 which possess functional opiate receptors do not synthesize sulphatide (unpublished observations). It is, of course, possible that the sulphatide molecule may in the same way mimic the steric requirements for the opiate binding portion of some other entity such as a membrane protein. As a further word of caution, it should also be noted that stereospecific binding of labelled opiates to entities such as glass filters has been previously reported. Thus, stereospecific binding by itself cannot be considered a sufficient criterion for opiate receptor identification.

The only other recently reported attempt at receptor isolation was that of Simon and colleagues.[18] These authors took the approach of solubilizing [$^3$H]etorphine prebound to opiate receptors in brain membranes by use of the non-ionic detergent Brij36T. Using this method some radioactivity was solubilized associated with a high molecular weight species. This was not the case when the prebinding of etorphine was carried out in the presence of unlabelled levorphanol. However, the status of this etorphine binding macromolecule is not yet clear.

As can be seen from the above discussion, no data are yet available on interaction of enkephalins with solubilized fractions. It seems possible that the availability of peptide ligands for the opiate receptor may be an aid in isolation of the receptor. It should be possible to produce some type of affinity label based on the enkephalin structure which would enable the receptor to be followed during its solubilization by detergents or other agents.

## IV. MECHANISM OF ACTION OF OPIATES

Most theories concerning the action of narcotics at the molecular level relate the possible interaction of these drugs with the adenylate cyclase system. One of the first observations suggesting that this was the case was reported by Collier and Roy[2] who showed that stimulation of brain adenylate cyclase by prostaglandin could be inhibited by opiates in a stereospecific fashion. However, this particular observation has been rather hard to repeat and little additional information about this system has been reported.

Most information about opiate–cyclase interaction has come from studies of cultured cells that contain opiate receptors and in particular studies using the NG108-15 glioma × neuroblastoma hybrid line. Sharma *et al.*[16] showed that narcotics could inhibit $PGE_1$ stimulated adenylate cyclase in cell homogenates or $PGE_1$ stimulated cyclic AMP concentrations in whole cells. However, these authors also found that the narcotics inhibited basal or adenosine stimulated cyclic AMP production as well. Thus, these observations differ fundamentally from those of Collier and Roy whose results implied that the effects of narcotics were selectively against $PGE_1$ stimulated cyclic AMP production. Another interesting aspect of the cyclase response in NG108-15 cells relates to the possible cooperativity of the interaction. Whereas the slope of the Hill plot for inhibition of [$^3$H]naloxone binding to NG108-15 cells by narcotics is approximately one, it is approximately 2–3 for inhibition of cyclase by narcotics. Further evidence for the linkage of opiate receptors to cyclase in such cells comes from the effects of various guanine nucleotides. Such nucleotides and p(NH)ppG in particular have very characteristic effects on adenylate cyclase and its activation by various hormones such as glucagon or epinephrine. Adenylate cyclase in NG108-15 or N4TG1 cells is considerably activated by p(NH)ppG and in addition this nucleotide reduces the affinity of opiate ligands for the receptor (unpublished observations). The effects of this nucleotide increase the likelihood that opiate receptors are linked to adenylate cyclase. The effects of a wide range of opiate ligands on the adenylate cyclase of NG108-15 cells have been examined. This includes a wide range of enkephalin analogues and other endorphins as well as narcotics. In general, these experiments confirm the original data that the potency of compounds in inhibiting the cyclase correlates closely with their potency in inhibiting [$^3$H]naloxone binding to the receptor. It should be noted that in spite of their differing chemical structures, enkephalins are as effective as many of the most potent narcotics in inhibiting the cyclase. It should also be noted that all the inhibitory effects just mentioned are found to be blocked by the opiate antagonist naloxone again indicating that the effects are opiate receptor mediated. Naloxone alone has no effect. Perhaps the most intriguing aspect of the NG108-15 adenylate cyclase system is its response to chronic treatment by opiates and opiate peptides.[17] It is found that if morphine (or an active endorphin or other active narcotic) is included in the cell culture medium for 12 h or more, then the activity of adenylate cyclase in the cells begins to rise. Morphine still inhibits cyclase activity, however, as the cyclase activity rises one reaches a stage when even with the morphine inhibition, activity is as high as original basal activity in non-treated cells. This phenomenon has been put forward as a model for opiate addiction and tolerance to narcotics. Thus it is possible that the cells become truly addicted to morphine. Even more interesting are observations on cyclic AMP concentrations in intact cells. In this case, if cells are cultured in the presence of morphine and then treated with naloxone, cyclic AMP levels immediately rise to extremely high concentrations. It is possible that this cyclic AMP 'overshoot' is a model for precipitated withdrawal by naloxone in humans. It of course a long conceptual jump to go from cells in culture to animal behaviour, however, the results are certainly suggestive of the behavioural syndromes associated with morphine addiction. Whether changes in brain cyclase activity actually underlie such behaviour however, remains to be elucidated.

## V. FUTURE RESEARCH

The discovery of the opiate receptors and their associated peptide ligands is certainly one of the most fascinating and important discoveries in pharmacology in recent years. Because the whole field is a relatively new one, many questions are still unresolved. In general, however, one may care to speculate as to the possible functions of the opiate receptors found in different parts of the nervous system and elsewhere. The data on the localization of opiate receptors and enkephalin certainly suggest some functions. Certainly opiate receptors in the spinal cord, periaqueductal grey, and some thalamic nuclei may be involved in the processing of sensory information related to pain. However, opiate receptors are certainly not only involved in pain perception. Consider, for example, the localization of enkephalin and opiate receptors in the basal hypothalamus. It is well known that narcotics have potent releasing effects on prolactin and growth hormone from the pituitary. Thus, it is quite possible that the endogenous opiate system is involved in modulating the endocrine system. Other possible functions include roles in locomotion, mood, gastrointestinal motility, and the pathogenesis of psychosis and epilepsy (Table 17.4). A related problem is finding a role for opiate peptides released from the pituitary gland. Although as referred to above, it is known that $\beta$-endorphin is stored in the pituitary and released in response to various stimuli, it is not known what its target tissues may be in the periphery.

Table 17.4. Localization and possible function of opiate receptors. (Table courtesy of Dr. S. H. Snyder)

| Location | Functions influenced by opiates |
|---|---|
| Spinal cord | |
| Laminae I and II | Pain perception in body |
| Brainstem | |
| Substantia gelatinosa of spinal tract of caudal trigeminal | Pain perception, head |
| Nucleus of solitary tract, nucleus commissuralis, nucleus ambiguus | Vagal reflexes, respiratory depression, cough suppression, orthostatic hypotension, inhibition of gastric secretion |
| Area postrema | Nausea and vomiting |
| Locus coeruleus | Euphoria |
| Habenula-interpeduncular nucleus-fasciculus retroflexus | Limbic, emotional effects, euphoria |
| Pretectal area (medial and lateral optic nuclei) | Miosis |
| Superior Colliculus | Miosis |
| Ventral nucleus of lateral geniculate | Miosis |
| Dorsal, lateral, medial terminal nuclei of accessory optic pathway | Endocrine effects through light modulation |
| Dorsal cochlear nucleus | |
| Parabrachial nucleus | Euphoria in a link to locus coeruleus |
| Diencephalon | |
| Infundibulum | ADH secretion |
| Lateral part of medial thalamic nucleus, internal, and external thalamic laminae, intralaminar (centromedian) nuclei periventricular nucleus of thalamus | Pain perception |
| Telencephalon | |
| Amygdala | Emotional effects |
| Caudate, putamen, globus pallidus nucleus accumbens | Motor rigidity |
| Subfornical organ | Hormonal effects |
| Interstitial nucleus of stria terminalis | Emotional effects |

With respect to the receptor itself, some sort of a breakthrough is clearly needed in studies on its isolation and purification. The difficulty in solubilizing the receptor in a state where it can still be assayed by its ability to bind ligand have proved a major stumbling block in its isolation. Although some attempts have been made to synthesize affinity labels for the receptor none has been entirely successful as yet. Possibly the availability of peptide ligands will improve this situation.

There are some interesting indications as to the possible mode of action of enkephalin at the molecular level. In particular studies in cultured cells have indicated some interaction with adenylate cyclase. In addition, such effects also extend to possible explanations of tolerance and dependence. Extension of such observations to the CNS will be of great importance in elucidating the mechanism of action of both exogenous and endogenous opiates.

## VI. REFERENCES

1. Akil, H., Mayer, D. J., and Liebeskind, J. C. *Science,* **141**, 961–962, (1976).
2. Collier, H. O. J., and Roy, A. *Nature (London),* **248**, 24–27, (1974).
3. Creese, I., and Snyder, S. H. *J. Pharmacol. Exp. Ther.,* **194**, 205–219, (1975).
4. Feinberg, A., Creese, I., and Snyder, S. H. *Proc. Natl. Acad. Sci. U.S.A.,* (1976).
5. Goldstein, A., Lowney, L. I., and Pal, B. K. *Proc. Natl. Acad. Sci. U.S.A.,* **68**, 1742–1746, (1971).
6. Hughes, J., Smith, T., Kosterlitz, H. W. Fothergill, L. H., Morgan, B. A., and Morris, H. *Nature (London),* **255**, 577–579, (1975).
7. Klee, W. A., and Nirenberg, M. *Proc. Natl. Acad. Sci. U.S.A.,* **71**, 3474–3478, (1974).
8. Kosterlitz, H. W., and Waterfield, A. A. *Ann. Rev. Pharmacol.,* **15**, 29–47, (1975).
9. Kuhar, M. J. *Fed. Proc.,* **37**, 153–157, (1978).
10. Lazarus, L. H., Ling, N., and Guillemin, R. *Proc. Natl. Acad. Sci. U.S.A.,* **73**, 2156–2159, (1976).
11. Miller, R. J., and Cuatrecasas, P. *Vitamins and Hormones*, **36**, 297–382, (1978).
12. Miller, R. J., Chang, K. -J., Cooper, B., and Cuatrecasas, P. *J. Biol. Chem.,* **253**, 531–541, (1978).
13. Pasternak, G., and Snyder, S. H. *Nature (London),* **253**, 563–565, (1975).
14. Pert, C. B., and Snyder, S. H. *Science,* **179**, 1011–1014, (1973).
15. Pert, C. B., and Snyder, S. H. *Proc. Natl. Acad. Sci. U.S.A.,* **70**, 2243–2247, (1973).
16. Sharma, S. K., Klee, W. A., and Nirenberg, M. *Proc. Natl. Acad. Sci. U.S.A.,* **72**, 3092–3096, (1975).
17. Sharma, S. K., Nirenberg, M., and Klee, W. A. *Proc. Natl. Acad. Sci. U.S.A.,* **72**, 590–594, (1975).
18. Simon, E. J., and Hiller, J. M. *Fed. Proc.,* **37**, 141–146, (1978).
19. Simon, E. J., Hiller, J. M., Groth, J., and Edelman, I. *J. Pharmacol. Exp. Ther.,* **192**, 531–537, (1975).

Cellular Receptors for Hormones and Neurotransmitters
Edited by D. Schulster and A. Levitzki


CHAPTER 18

# Nicotinic acetylcholine receptors

Joav M. Prives
*Department of Anatomical Sciences, Health Services Center, State University of New York at Stony Brook, Stony Brook, Long Island, N.Y. 11794, U.S.A.*

## I. INTRODUCTION

Nicotinic acetylcholine (ACh) receptors mediate transmission of impulses at certain cholinergic synapses. These integral membrane proteins, situated on the surface of the postsynaptic cell, bind the neurotransmitter ACh released by the arrival of an impulse at the presynaptic nerve ending. The binding of ACh by these receptors is translated into an excitatory signal in the postsynaptic cell.

Activation of receptors by ACh triggers a selective rise in permeability of the postsynaptic membrane to small cations (primarily $Na^+$ and $K^+$), which now move across the membrane along their electrochemical gradients producing a decrease in membrane potential. The resulting local depolarization if sufficiently large can give rise to an action potential and produce a general response in the postsynaptic cell.

Nicotinic ACh receptors are present in surface membranes of vertebrate skeletal muscle cells as well as on specialized cells derived from skeletal muscle, namely the electrocytes of certain species of electric fish—*Electrophorus electricus* and *Torpedo*. Similar nicotinic ACh receptors are found on neurons with cell bodies located in autonomic ganglia. Less direct evidence indicates that related receptors may be present on certain neurons in brain and retina.

The distribution of ACh receptors on the cell surface is not uniform. Instead, these receptors are highly localized to the membrane region facing the presynaptic nerve ending which comprises less than 0.1% of the total cell surface in muscle cells.

The ACh receptor has been the focus of an extraordinary amount of fruitful research spanning several decades. The early suggestion by Du Bois Raymond, in 1877, that synaptic transmission might involve a chemical mediator (cited in Reference 1), the discovery of the biological actions of ACh by Dale and coworkers in 1914 (cited in Reference 1), its synthesis by neurons[1,2] and its effects on the postsynaptic membrane[1,2] set the stage for the rapid advances in the last ten years.

The high degree of current interest in ACh receptors reflects the recent progress in its characterization (for recent reviews see References 3–11).

As a result of this progress, the nicotinic ACh receptor is at present the best characterized of any neurotransmitter, hormone, or drug receptor. With the availability of this component in purified form, a stage has been reached at which it is possible to obtain definitive and previously inaccessible answers for basic questions such as the macromolecular basis for ACh receptor function and the regulation of its properties.

The aim of this chapter is to summarize the properties of ACh receptors revealed by recent advances in what is clearly a rich field for future investigation.

## II. PHARMACOLOGICAL PROPERTIES

The pharmacology of synaptic transmission at cholinergic synapses has been studied extensively and ACh receptors are classified as 'nicotinic' or 'muscarinic' according to pharmacologic criteria. At 'muscarinic' synapses, the effects of ACh are reproduced by muscarine. The ACh receptors at nicotinic synapses can be activated by nicotine and inhibited by D-tubocurarine (see Figures 18.1, 18.2, and 18.3).

The pharmacology of ACh receptors has been studied by measurement of the membrane potential or conductance changes induced by ACh and its congeners and by the inhibition of the cholinergic response caused by ACh receptor antagonists which block synaptic transmission. In addition, the binding of ACh receptor activators and antagonists has been measured directly with intact cells, membrane fractions, and detergent-solubilized ACh receptors.

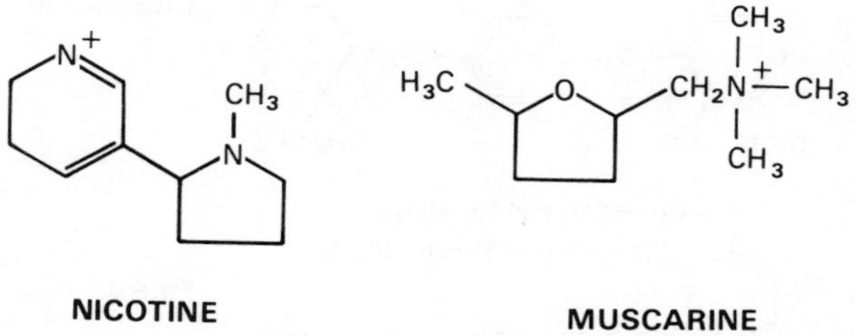

NICOTINE

MUSCARINE

Figure 18.1. Chemical structures of nicotine and muscarine (see Section II)

### (i) Reversible ligands

The ACh receptor response is initiated by the reversible binding of activators (agonists) and blocked by the binding of antagonists. Agonists of the nicotinic ACh receptors include, in addition to the neurotransmitter ACh, nicotine, carbamylcholine, and decamethonium (Figure 18.2). A characteristic antagonist of these receptors is D-tubocurarine which functions as a competitive inhibitor by occupying the ACh binding site and thus

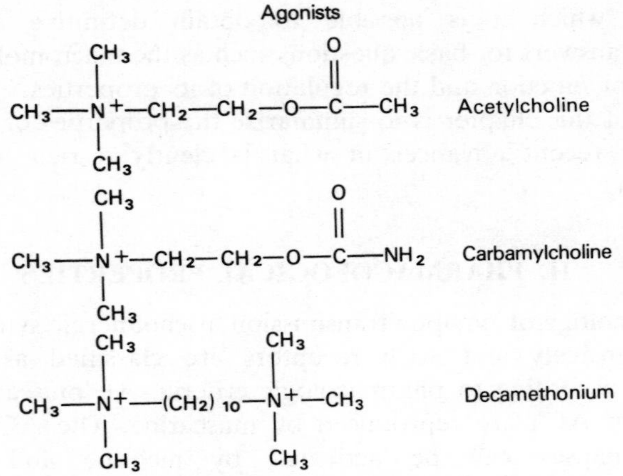

Agonists

$CH_3-N^+(CH_3)_2-CH_2-CH_2-O-C(=O)-CH_3$ Acetylcholine

$CH_3-N^+(CH_3)_2-CH_2-CH_2-O-C(=O)-NH_2$ Carbamylcholine

$CH_3-N^+(CH_3)_2-(CH_2)_{10}-N^+(CH_3)_2-CH_3$ Decamethonium

Figure 18.2. Chemical structures of acetylcholine, carbamylcholine,. and decamethonium. These compounds are activators of nicotine acetylcholine receptors (see Section II)

Antagonists

d-Tubocurarine

Flaxedil

MBTA

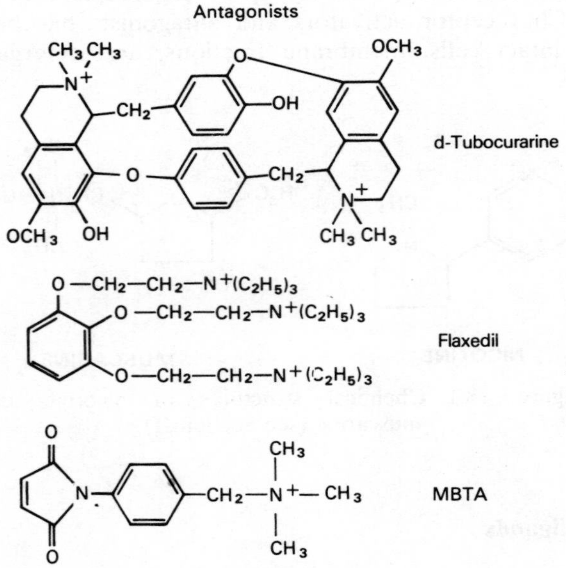

Figure 18.3. Chemical structures of D-tubocurarine, flaxedil, and 4-(*N*-maleimido)benzyl trimethylammonium (MBTA). All three compounds are competitive inhibitors of nicotinic acetylcholine receptors. However, upon reduction of nicotinic acetylcholine receptors, MBTA acts as a site-directed alkylating agent of these receptors (see Section II)

preventing the activation of receptors by ACh. Both agonists and antagonists (with the exception of the $\alpha$-neurotoxins—see below) resemble ACh in having charged onium N-groups. Agonists tend to be slender, flexible molecules with a minimum of steric hindrance. Antagonists are generally more bulky and rigid and may have hindering groups attached to the onium N-atom (see Figures 18.2 and 18.3 and Reference 4). Agonists presumably activate the ACh receptor by inducing a change in its conformation.[12] The magnitude of the change in ACh binding site conformation needed to activate receptors has been inferred in a study using depolarizing affinity labels of ACh receptors.[13]

### (ii) Irreversible ligands

The recent availability of ligands that bind tightly to ACh receptors has led to dramatic advances in the characterization of this component. These consist of affinity labels which attach covalently to ACh receptors[13] and a class of neurotoxic polypeptides from elapid and hydrophidae snake venoms, called $\alpha$-neurotoxins which bind to these receptors with low reversibility and high specificity.[14]

#### *(a) Affinity labels*

Certain maleimide derivatives bearing a quarternary ammonium group were shown to be highly effective site-directed ligands of nicotinic ACh receptors.[13] The development of these covalent labels was based on the finding that ACh receptors contain an easily reducible disulphide bond apparently very near the ACh binding site. Reduction of this bond with dithiothreitol markedly alters the pharmacological specificity of ACh receptors. Its restoration by reoxidation leads to full reversal of the pharmacological effects of reduction.[3,8,13] The sulphhydril groups formed by reduction are susceptible to alkylation by *N*-ethylmaleimide and this alkylation prevents the reversal by oxidizing agents of the pharmacological effects of reduction. Maleimide derivatives of ACh analogues act similarly to *N*-ethylmaleimide in blocking the reversal by oxidizing agents of the effects of dithiothreitol. However, these quarternary ammonium compounds alkylate receptors at apparent rates that are much higher than those of other non-site-directed alkylating agents. For example, 4-(*N*-maleimido)-benzyltrimethylammonium (MBTA) alkylates reduced ACh receptors at approximately, 5,000-fold the rate of alkylation by *N*-ethylmaleimide. This reaction is blocked by other cholinergic ligands, including $\alpha$-neurotoxins, and by affinity reoxidation of the receptor.[3,8,13] Tritium-labelled MBTA has been successfully used to determine the quantity of ACh receptors in intact electrocytes, as well as to identify membrane-bound, detergent solubilized, and purified ACh receptors.[8]

*(b) The α-neurotoxins*

This class of basic polypeptides, made up of 61–74 amino acids (M.W. ~7,000), was discovered to bind ACh receptors non-covalently but extremely tightly ($K_d \sim 10^{-11}$ M) and with high specificity.[14] Though strikingly different in structure from ACh analogues, the toxins bind to the receptors at or very near the ACh binding site and act as competitive antagonists in a similar manner as D-tubocurarine.[9,15] The use of radioactive α-neurotoxins, most commonly labelled with $^{125}I$, is currently the method of choice for the assay of ACh receptors in intact cells, membrane fractions, and solubilized extracts. Radioautography of tissue labelled with radioactive α-neurotoxins is widely used to visualize receptor distribution on the surface of cells. α-Bungarotoxin (α-Btx) from the venom of *Bungarus multicinctus* is usually used for these purposes. Other α-neurotoxins which bind to receptors more reversibly, when immobilized on Sepharose, provide the means for purification of ACh receptors by affinity chromatography.[9,10]

## III. REGULATION OF MEMBRANE PERMEABILITY

The ACh receptor transduces the binding of ACh into the transient opening of ionic channels, allowing the transmembrane movement of $Na^+$ and $K^+$ ions in response to their electrochemical gradients.[2,4]

The resulting decrease in membrane potential is measured by an intracellularly-located glass microelectrode. A more refined measurement of the membrane permeability change triggered by ACh receptor activation consists of maintaining the membrane potential at a set value ('voltage-clamp') and monitoring the current required to prevent depolarization. The amount of depolarization or current measured by these methods is proportional to the net flux of cations and hence to the number of open membrane channels.[4,5]

Two types of responses associated with the presynaptic release of ACh have been recorded in postsynaptic cells. Miniature end-plate potentials (Mepps) of small amplitude are produced by spontaneous liberation of ACh in individual 'quanta' of $\sim 10^4$ molecules. This release is spontaneous in that it is not dependent on the arrival of impulses at the presynaptic nerve ending. The release of ACh that is evoked by an action potential reaching the nerve terminal is 300-fold greater, i.e. involves $\sim 3 \times 10^6$ ACh molecules, and produces a substantial end-plate potential (Epp) associated with a net inward current (Epc).[4] The Epp is usually adequate to initiate an impulse that spreads throughout the muscle cell membrane. The concentration of ACh in the synaptic cleft has been estimated to reach between $10^{-4}$ M and $10^{-3}$ M within 0.3 ms after release of a quantum of ACh. The duration of increase in Mepp amplitude is in the range 0.05–0.3 ms. Stimulation-evoked Epps have a rise time of 0.5–1 ms. In both

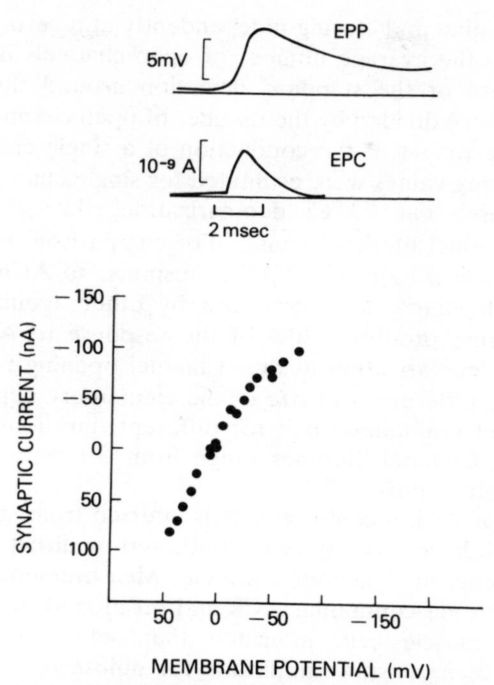


Figure 18.4. Effects of acetylcholine release on membrane potential and current at the muscle end plate.
Epp = Transient change in membrane potential induced by acetylcholine released by the nerve ending; Epc = End plate current induced by acetylcholine with the membrane potential clamped at various values. Negative values for current signify inward currents (see Section III)

cases the decay time is considerably longer with a duration of several milliseconds (see Figure 18.4).

In the last few years, methods have been developed which allow the study of the cholinergic response at the level of single ACh receptors. The measurement of ACh induced membrane noise was used to estimate the electrical conductance contributed by a single open channel in the postsynaptic membrane.[16,4,5] Low concentrations of ACh were found to produce fluctuating or 'noisy' depolarization of muscle membrane, in contrast to the steady resting potential obtained with these cells. The fluctuations in the response to submaximal ACh concentrations are consistent with the notion that the response measured is the summation of many channels opening and closing independently at a certain average rate. In these conditions the average number of open channels is approximately equal to the square of the standard deviation around the average. The measured conductance divided by the number of open channels calculated in this way gives an estimate of the conduction of a single channel. With this method, the following values were estimated for single channels activated by ACh in frog muscle, at 22 °C: depolarization, 0.25 $\mu$V; conductance, 100 pmho; and channel lifetime, 1 ms.[4,5] For comparison, in rat diaphragm an elementary depolarization of 0.7 $\mu$V in response to ACh was estimated. The elementary depolarizations produced by other agonists were lower. Thus decamethonium produced 20% of the response to ACh in frog and 70% of the ACh depolarization in rat.[5] Channel opening is an all or none event; hence these differences in size of the elementary depolarizations are due to the channel remaining open for different durations in response to different agonists. Channel lifetimes range from 0.1 ms to 1.65 ms in the presence of different agonists.[5,10]

The properties of ACh receptor channels inferred from statistical analysis of membrane noise have recently been confirmed by direct measurement of single channel events in denervated muscle. Measurements made with an external microelectrode containing ACh and positioned on the surface of a voltage clamped muscle cell, indicated that activated single channels produce currents of a constant amplitude for millisecond durations.[17] The lifetime of open channels was shown to vary with the agonist, the membrane potential, and the temperature.[5,10]

The findings summarized above suggest the following quantitative relationships:

1. A single channel, opened by the binding of between one and four ACh molecules, allows the passage of $1\text{–}2 \times 10^3$ ions.
2. The spontaneous release of a single quantum ($\sim 10^4$ molecules) of ACh opens approximately $2 \times 10^3$ receptor channels in the muscle cell membrane.
3. Transmission of a single impulse across a synapse involves the release of $\sim 3 \times 10^6$ molecules of ACh, opening $\sim 6 \times 10^5$ channels and allowing the passage of $6 \times 10^8$ ions across the postsynaptic membrane.

## IV. ISOLATION OF ACETYLCHOLINE RECEPTORS

A most significant recent advance in the study of neurotransmitter and hormone receptors has been the successful purification of nicotinic ACh receptors. The availability of purified receptors makes it possible to study the structural and chemical properties responsible for the transduction of ligand binding to a physiological response; in this case an increase in membrane permeability to ions. A variety of uniquely favourable circumstances have made it possible to achieve the purification of ACh receptors at a time when the availability of purified receptors of other types still seems remote. These circumstances include:

1. The availability of a rich source of ACh receptors in the electrogenic tissues of two types of electric fish, *Electrophorus electricus* (electric eel) and *Torpedo*.
2. The ability to use specific ligands to identify and quantify ACh receptors under conditions where the physiological effects of these receptors are lost. Affinity-labelling reagents and the $\alpha$-neurotoxins from elapid snake venoms (see Section II. ii) bind tightly to ACh receptors with extremely high specificity. Moreover, the affinity of these ligands for ACh receptors is retained after fractionation, solubilization and subsequent steps in receptor purification.
3. The effective use of affinity chromatography in purification of soluble ACh receptors.

These methods utilize specific receptor ligands, most commonly $\alpha$-neurotoxins, covalently coupled to solid matrix, to selectively remove receptors from detergent extracts. Consequently, ACh receptors have been purified in several laboratories from electric tissue of several *Torpedo* species and *Electrophorus electricus* as well as from denervated rat diaphragm skeletal muscle and cultured muscle cells to high specific activities ranging from approximately 2–12 mol of toxin bound per g of protein (for recent reviews, see References 8–10). As the average apparent molecular weight of the receptors purified from electric tissue of electrophorus and *Torpedo* is about 100,000 to 125,000 per $\alpha$-neurotoxin binding site (References 8–10 and see below), the best preparations thus may be composed solely of ACh receptor in 'pure' form.

While these purification procedures have been notably effective, the structural and chemical characterization of ACh receptor isolated by these means has encountered difficulties. The problems rise in part from technical obstacles such as the effects on receptor properties of the detergents used in its purification. An appreciable amount of Triton X-100 or cholate remains tightly bound to the isolated ACh receptor molecules (see below, Sections IV.ii and IV.iv) comprising as much as 15–20% of the apparent mass of the purified macromolecules and resulting in spuriously large partial specific volumes for ACh receptors solubilized with these detergents. These as well as the possibility of proteolysis during isolation have produced uncertainty with regard to the molecular weight of receptors, the number and size of subunits, the stoichiometry of subunits, and the location and number of binding sites for ACh, neurotoxins, and affinity ligands. The difficulties in part reflect the complex structure of ACh receptors.

### (i) Fractionation

Purification procedures utilizing affinity chromatography generally do not require extensive subcellular fractionation and can be used with crude extracts of low specific activity.[10] Thus most ACh receptor purifications have involved detergent extraction of a crude membrane fraction. Efficient use of standard fractionation procedure has produced a membrane fraction from *Torpedo* electric tissue in which ACh receptor was enriched to the extent of comprising 50% or more of the total protein content.[10]

### (ii) Solubilization

The role of ACh receptors in regulation of transmembrane ionic permeability suggests that these components are embedded in the cell membrane and penetrate the lipid bilayer. The procedures required to solubilize membrane-bound ACh receptor are consistent with its classification as an integral membrane protein.[18] Namely, the ACh receptors cannot be extracted from membranes by aqueous salt solutions but are readily solubilized by non-ionic detergents.[10] This convenient and widely used solubilization procedure has been something of a mixed blessing. The persistence of detergent tightly associated with the receptor protein has strongly interfered with the elucidation of fundamental molecular properties such as molecular weight and the stoichiometry and features of ligand binding to purified receptors. However, other solubilization procedures, such as those utilizing organic solvents,[19] have been far more problematic (see Reference 10).

### (iii) Purification

ACh receptors have been purified to high specific activity by standard protein purification methods.[9,10] However, affinity chromatography of detergent extracts on columns of receptor ligands covalently coupled to Sepharose has been the standard method for receptor purification. $\alpha$-Neurotoxins, receptor activators such as phenyltrimethylammonium, and a derivative of the competitive inhibitor flaxedil have been used as end groups to selectively remove ACh receptors from detergent extracts of electric tissue or skeletal muscle cell membranes.[9,10] Selective elution of receptors bound to the column is achieved with reversible receptor ligands which remove ACh receptors from the column through competition with the immobilized ligands. $\alpha$-Neurotoxin columns have the advantage of high specificity but the disadvantage of low reversibility of receptor binding so that extensive exposure to relatively high concentrations of cholinergic ligands is necessary for the elution of receptors. As in other applications of affinity chromatography, the yield of purified material is rather low with a range of 20% to 50% recovery of ACh receptor. A more fundamental potential disadvantage of the use of affinity chromatography to purify ACh receptors is that such purification is restricted to the ligand binding site. The possibility that subunits which do not contain ACh-binding sites but are nevertheless essential for ACh receptor function are lost during affinity chromatography will only be ruled out by the successful reconstitution of the purified ACh receptor into membrane with the restoration of receptor activity (see Section IV. iv). To achieve maximal purification, several procedures have combined affinity chromatography with an additional purification step such as sucrose gradients, centrifugation, ion-exchange chromatography or electrophoresis.[9,10] One procedure utilizes two types of affinity chromatography in sequence; using concanavalin A and cobra toxin as immobilized ligands.[20] This procedure takes advantage of the affinity of ACh receptors for concanavalin A. ACh receptors from electrophorus, *Torpedo*, chick, mouse, and rat all bind tightly to concanavalin A since these glycoproteins contain external sugar residues for which this lectin has a high affinity.

### (iv) Properties of purified ACh receptor

#### *(a) Binding properties and stoichiometry*

Binding capability of ACh receptor towards both reversible and irreversible ligands is retained during solubilization and purification of this component. Thus α-neurotoxin-binding has been routinely used to assay receptors during the course of purification and to determine specific activities of purified preparations of ACh receptors. Similarly, the affinity label MBTA has been utilized successfully to monitor the purification of solubilized receptors.[8] The rate of affinity alkylation by MBTA of ACh receptor purified in detergent solution and of receptor in intact membranes is similar. Further evidence that the ligand-binding sites of ACh receptors are relatively unchanged during solubilization and isolation of these components comes from competition-binding studies which indicated that characteristic binding affinity toward neurotoxins and small ligands are not altered.[7] Nevertheless, the interaction of ACh receptors with certain ligands is altered after the extraction of receptors by non-ionic detergents. The affinities of *Torpedo* and *Electrophorus* ACh receptors towards antagonists are unchanged by solubilization in detergent. In contrast, the affinities toward agonists are markedly changed: detergent extraction of *Electrophorus* receptors strongly increases the affinity toward agonists. In the case of *Torpedo*, the identical procedures have the opposite effect; solubilization results in decreased ACh receptor affinities to agonists.[8,10] The reason for these effects if as yet unclear. A partial explanation involves evidence that ACh receptors may exist in low affinity and high affinity states that are interconvertible. The effect of neutral detergent on *Electrophorus* might be to stabilize the ACh receptors in the high affinity state, while in *Torpedo* the low affinity form may be selectively stabilized.[10] Models in which ACh receptor exists in two or more forms have been frequently utilized to describe the actions of agonists and antagonists and to explain the phenomenon of desensitization of ACh receptors.[4,10]

It is fairly well established that agonists, reversible competitive antagonists and $\alpha$-neurotoxins bind to identical or overlapping sites on ACh receptors.[10,15] However, there is uncertainty as to the number of binding sites per receptor molecule, as well as to the extent of possible heterogeneity of binding sites. Several factors contribute to the uncertainty. Firstly, the molecular weight of ACh receptor has not yet been established with sufficient accuracy to allow an unambiguous estimate of the number of $\alpha$-neurotoxin or MBTA binding sites per receptor protomer. Secondly, the binding of both small ligands and $\alpha$-neurotoxins is more complex that would be anticipated if all binding sites were identical. For example, it has been reported that there are apparently two rates of association for both types of ligands. Moreover, the sites to which toxin binds more rapidly (about 50% of the total number of toxin binding sites) are those to which ACh binds more slowly.[8] Thirdly, the interrelationship between binding of neurotoxins and MBTA affinity labelling of purified ACh receptors is difficult to explain on the basis of a single class of binding sites. There is abundant evidence that both ligands attach to common sites, i.e. to the ACh binding sites, based on competition between toxins and MBTA.[15] All the MBTA labelling is blocked by preincubation of receptors with toxin. Surprisingly, however, only half of the toxin binding is eliminated after reaction of (reduced) receptor with MBTA. Moreover, at saturation purified ACh receptors bind twice as much neurotoxin as the number of sites specifically labelled by MBTA.[8] This asymmetry in competitive effects and binding stoichiometry might be due to negative cooperativity of binding resulting in 'half of the sites reactivity', a property common to many multi-subunit proteins. Alternatively, only one of two classes of binding sites might be available to MBTA. These distinct binding sites may be situated on different receptor moieties or in separate quarternary structural domains on the same macromolecules.

### (b) *Molecular weight*

Determination of the size of ACh receptors has been hampered by the presence of detergent necessary to maintain these integral membrane proteins in solution. The significant amount of detergent bound to the receptor (10–20%) results in anomalous hydrodynamic behaviour and inaccurate molecular weight determination by gel filtration.[8,10] Size determination of *Torpedo* ACh receptors by sedimentation in sucrose density gradients showed two major components corresponding to a light form (L) and a heavy form (H).[8,10] There is good evidence that both forms are found in the intact cell membrane and that the H-form is apparently a dimer composed of two L-monomers connected by an intermolecular disulphide bridge.[21] In contrast, sedimentation of *Electrophorus* ACh receptor reveals only a single, monomeric form. The molecular weight of ACh receptor has been determined with a variety of methods, including sedimentation, SDS gel electrophoresis after cross-linking, and osmometry. The estimates of molecular weight obtained with the different methods are in reasonably good agreement. For *Electrophorus* ACh receptor, molecular weights range between 230,000 and 360,000 (for Reference, see 8, 10). For *Torpedo* receptor (L-form) the range is from 200,000 to 400,000. Thus, both *Torpedo* and *Electrophorus* receptors have apparent molecular weights of 300,000–400,000 in detergent. However, the functional molecular weight of receptor when situated in the membrane may be different.[8]

### *(c) Subunit structure*

The subunit composition of purified ACh receptor from *Torpedo* and *Electrophorus* electric tissue, as well as from muscle has been characterized in a number of recent studies.[8–10] In all these cases, SDS polyacrylamide gel electrophoresis—a technique which resolves polypeptides by molecular weight—has been used to show that these receptors are multi-subunit glycoproteins composed of polypeptides of different size. There are differences in the number and size of receptor subunits isolated from the same species in different laboratories. Possible reasons for these discrepancies include the use of different purification procedures leading to different degrees of final homogeneity and proteolysis. However, several general features of the subunit structure of ACh receptors have emerged from these studies.

The ACh receptors from *Electrophorus* and *Torpedo* are different in subunit composition. *Electrophorus* ACh receptors are probably made up of three major subunits with molecular weights ranging between 40,000 and 53,000 (see References 8–10 and references therein). *Torpedo* receptors are most likely composed of four types of subunits with the molecular weight of the smallest being 40,000 and the largest being approximately 65,000.[8] Each of these subunits contains carbohydrate.[11] The individual subunits are immunologically non-crossreactive indicating that they are not closely related in primary structure and that the smaller chains do not arise from proteolysis of the larger subunits.[6]

What is the functional significance of the various subunits? Reaction with

MBTA of reduced ACh receptors purified from *Torpedo* as well as *Electrophorus* results in the labelling of only the 40,000 molecular weight subunit, suggesting that this subunit contains part or all of the ACh binding site.[8] The ionic channel activity was suggested to involve a subunit of 43,000 molecular weight[10] but some uncertainty exists as to the nature and role of this component.[22] ACh receptor occurs in *Torpedo* membrane predominantly as a dimer joined by disulphide bonds between the 65,000 molecular weight chains on separate monomers. Such dimers are not found in *Electrophorus* and their functional significance in *Torpedo* is unknown.

### *(d) Reconstitution*

An important approach is aimed at incorporating purified ACh receptor into lipid membranes of controlled composition.[9] The appearance of ion permeability regulation characteristic of native membrane bound ACh receptor as a result of reconstitution would demonstrate that whole, functional ACh receptor is isolated by the procedures outlined above (Section IV. iii). This methodology may make it possible to establish the roles of the different subunits in the integrated ACh receptor response. To date, reconstitution attempts using highly purified ACh receptors have been largely unsuccessful in achieving restored ACh receptor functions. The reasons may be technical or indicative of loss of an essential component during the purification procedure. Recent procedures, utilizing membrane preparations rich in ACh receptors rather than the purified receptors as starting material, have been more successful.[23] ACh receptors extracted from *Torpedo* membrane and reconstituted into liposomes exhibited rapid agonist-induced $Na^+$ uptake. Furthermore, in the continued presence of agonist a rapid attenuation of the response was observed which was reminiscent of the desensitization characteristic of ACh receptor activity in intact cells.[23]

## V. DISTRIBUTION OF ACh RECEPTOR

### (i) Innervated muscle

The postsynaptic membrane immediately adjacent to the nerve terminal is a highly specialized region of the muscle cell surface, and contains ACh receptors as a major structural component, at extremely high packing density ($10^4$–$10^5$ receptor units $\mu m^{-2}$). Recent studies utilizing negative staining and freeze etch electron microscopy as well as X-ray diffraction indicate that the subsynaptic membrane is constituted primarily of ACh receptors organized in a dense, hexagonal arrangement.[10] In striking contrast, extrajunctional membranes in the same postsynaptic cells are virtually devoid of receptors. The demarcation is abrupt, the ACh receptor density diminishing by orders of magnitude within a short distance of the nerve terminal.[24]

### (ii) Non-innervated muscle

Embryonic muscle and denervated adult muscle have high levels of ACh receptors with markedly more uniform cell surface distribution than on innervated muscle. After denervation of adult muscle there is no marked change in the high density of ACh receptors in the junctional membrane for several days. In contrast, after a lag of 1–2 days, over-all ACh receptor levels on denervated muscle cells increase rapidly, due to synthesis and insertion of new receptors in extrajunctional regions of the cell surface.[24] Conversely, the establishment of new synapses in embryonic muscle and denervated adult muscle results in restriction of receptor distribution to subsynaptic regions and its disappearance from extrajunctional membrane areas.[24]

## VI. COMPARISON OF JUNCTIONAL AND EXTRAJUNCTIONAL ACh RECEPTORS

The ACh receptors that appear in embryonic and denervated adult muscle have generally similar properties to the junctional ACh receptors of innervated muscle. However, the recent introduction of increasingly sensitive methods for studying ACh receptors has allowed the detection of significant differences between junctional and extrajunctional receptors, suggesting that these may be different molecular forms.

### (i) Pharmacological specificity and functional properties

Extrajunctional ACh receptors are less sensitive than junctional receptors to inhibition by D-tubocurarine. This difference is retained after solubilization and purification of receptors from innervated and denervated muscle. Measurements of D-tubocurarine binding to purified receptors by the inhibition of neurotoxin binding revealed a ten-fold higher affinity of D-tubocurarine for junctional receptors than for extrajunctional receptors[25] indicating that intrinsic differences exist between the two receptor types.

Junctional and extrajunctional receptors differ in kinetic channel properties: the mean channel open times of extrajunctional receptors in denervated muscle are greater by three- to five-fold than those of junctional receptors.[5,10]

### (ii) Structural properties

The possibility that junctional and extrajunctional ACh receptors are structurally distinct is supported by the observation that the two receptor types differ slightly (by 0.1–0.2 pH units) in isoelectric points.[20,25] This difference could result from 20–40 charge differences arising from different degrees of phosphorylation[36] or glycosylation in two closely related molecular species of 300,000 molecular weight.[26] In addition, the two receptor subclasses are apparently antigenically different.[27]

### (iii) Rates of degradation

Large difference in the degradation rates of junctional and extrajunctional ACh receptors have been demonstrated.[25] Junctional receptors are degraded relatively slowly, with a catabolic half-life of several days. In contrast, extrajunctional receptors are degraded much more rapidly, with a half-time of less than 24 h. Similar high rates of degradation are displayed by ACh receptors on embryonic muscle *in situ*[28] and in cell culture.[29]

### (iv) Regulation of extrajunctional ACh receptors

There is clear evidence that the elaboration of extrajunctional ACh receptors in denervated muscle and in embryonic muscle is regulated by the level of electrical activity. Direct electrical stimulation of muscle prevents the appearance of extrajunctional ACh receptors after denervation,[30] and electrical activity reduces receptor levels in cultured myotubes.[31,32] Conversely, inhibition of impulse conduction induces the appearance of extrajunctional receptors in innervated muscle and increased accumulation of receptors in cultured muscle.[26,33,34] Rates of degradation of these receptors are unaffected by the procedures that elevate or inhibit electrical activity. Rather, the synthesis of extrajunctional receptors seems to be the parameter regulated by muscle activity.[25]

### (v) Regulation of junctional ACh receptors

The differences between junctional and extrajunctional ACh receptors indicate the existence of different regulatory processes for the two types of receptors. While it is possible that the two forms are products of different genes, it is equally feasible that the differences are epigenetic and involve protein modification associated in some way with innervation. There is some indication that extrajunctional ACh receptors may be precursors of junctional receptors. During formation of nerve–muscle synapses in culture, the high ACh receptor density in subsynaptic membrane involves the recruitment of pre-existing receptors that migrate to the synaptic region.[35] There are other observations indicating that the changes in ACh distribution and degradation rate occur sequentially after synapse formation.[28] The chemical basis for the possible transformation of extrajunctional receptors to junctional receptors is unknown. The capability of ACh receptors to undergo phosphorylation has led to the suggestion that this reaction may be involved.[36] Modification of the carbohydrate constituents of these receptors by lectin-type molecules is a further possibility.[37,38]

## VII. ACh RECEPTOR METABOLISM IN MUSCLE CELL CULTURES

### (i) Synthesis of receptors

Although ACh receptors apparently have no function other than the reception of trans-synaptic signals from motor neurons, differentiating muscle cells do not require neuronal intervention to initiate synthesis of ACh receptors. Muscle cells in culture have proven highly convenient for studying the biosynthesis, incorporation into cell membrane, and catabolism of ACh receptors.[24,39,40]

ACh receptors are synthesized during differentiation of muscle cells in culture. The rapid synthesis underlies the sharp increase in surface ACh receptor levels which occurs at about the time of myoblast fusion, concomitantly with appearance of other muscle-specific proteins. The elaboration of ACh receptors is not dependent on the fusion process.[41–44] The appearance of ACh receptors is unhindered under conditions of fusion arrest in which the synthesis of muscle proteins involved in contractility may be delayed.[41] This suggests that the appearance of ACh receptors is regulated by a different development pathway than that governing the elaboration of muscle cytoplasmic proteins. Newly synthesized ACh receptors require 2–3 h to appear on the external surface of cultured muscle cells.[29,42] As with extrajunctional receptors of intact denervated muscle, the rates of elaboration of ACh receptors in cultured myotubes can be diminished by electrical stimulation and elevated by inhibitors of electrical activity.[31,33] In mature muscle cultures ACh receptor levels diminish due to spontaneous muscle activity.[32] ACh receptors in cultured muscle cells also resemble extrajunctional receptors of adult muscle in having a similarly rapid rate of internalization and degradation by a temperature and energy dependent process.[29]

### (ii) ACh receptor distribution

The surface distribution of ACh receptors on cultured muscle cells has been studied with the hope of identifying regulatory mechanisms which operate in the formation and maintenance of nerve–muscle synapses. Interest has largely focused on the possibility that cellular mechanisms involved in the formation of ACh receptor aggregates in cultured muscle cells are similar to those by which innervation produces localization and stabilization of receptors in the postsynaptic membrane.

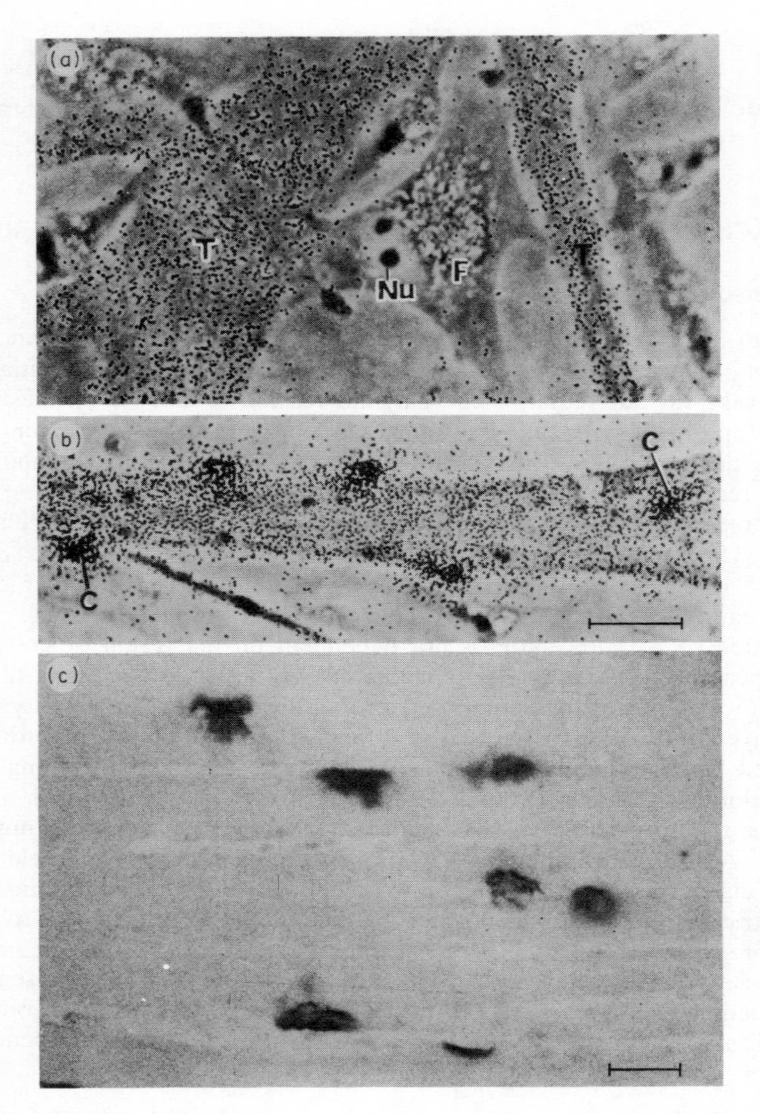


Figure 18.5. Distribution of acetylcholine receptors on the surface of muscle cells differentiated in culture and *in situ*.
(a) Autoradiogram of cultured muscle cells labelled with $[^{125}I]\alpha$-bungarotoxin ($\alpha$-Btx) after four days in culture. Note the homogeneous distribution of silver grains on muscle cells (T) indicating that ACh receptors are uniformly distributed on the cell surface. In contrast, no appreciable labelling of fibroblasts (F) is observed. (Nu): nucleolus. (See Reference 34.)
(b) Autoradiogram of cultured muscle cells labelled with $[^{125}I]\alpha$-Btx after seven days in culture. Acetylcholine receptors are no longer uniformly distributed but are now present as aggregates on the cell surface as evidenced from the large proportion of silver grains found in clusters (C). (Bar = 20 $\mu$m; also applies to (a)); see Reference 34.)

The distribution of ACh receptors on the surface of cultured muscle cells has been studied by electrophysiological measurements of membrane response to localized application of ACh as well as by localization of [$^{125}$I]α-bungarotoxin labelled receptor by radioautography (see Figure 18.5). These methods reveal that even in the absence of neuronal effects the distribution of ACh receptors on cultured myotubes is not uniform. Patches of aggregated ACh receptors of 5–15 μm diameter are interspersed with diffusely distributed receptors on the entire muscle cell surface. The number of these clusters increases with myotube differentiation[34,45] until the disappearance of receptors upon functional maturation of cultured muscle cells in the absence of innervation.[34]

In addition, clustered ACh receptors on myotubes resemble junctional receptors of innervated muscle in their limited mobility as evidenced by diffusion constants of less than $10^{-12}$ $cm^2$ $s^{-1}$ at 35 °C.[46] In contrast non-clustered receptors on these myotubes have a relatively high lateral mobility showing diffusion constants more than 100-fold higher than those of the aggregated ACh receptors.[46] Moreover, as in the specialized regions of subsynaptic membrane, the lifetime of the cluster exceeds the turnover time of the individual ACh receptors which it contains.[46]

In nerve–muscle co-cultures, patches of elevated ACh receptor density have been observed in muscle membrane areas adjacent to neuronal processes. Apparently this juxtaposition does not result from the arrival of nerve endings at pre-existing ACh receptor clusters. Rather, the formation of nerve–muscle contacts can induce migration resulting in formation of high density patches under nerve processes.[35,39,47] A similar sequence of events occurs during nerve–muscle synaptogenesis *in vivo*.[28] How do nerve endings induce the appearance of high ACh receptor density regions at underlying regions of muscle membrane? The mobilization of receptors is not induced by the appearance of synaptic transmission: it occurs in co-cultures incapable of forming functional synapses and under conditions in which receptor function is blocked by specific inhibitors such as α-neurotoxin.[9,40] Recent evidence indicates that a soluble factor released by neurons may be the agent that induces the aggregation of ACh receptors in postsynaptic membrane of muscle cells. A factor in medium conditioned by cultured neurons caused an increase in the number of ACh receptor clusters in muscle cell cultures.[38,48] These findings indicate that junctional ACh receptors may arise as a result of a localized interaction of a neuronal factor with the muscle cell surface. The distinctive properties of junctional as compared to extrajunc-

---

(c) Neuromuscular junctions in mouse diaphragm visualized by staining acetylcholine receptors with horseradish peroxidase coupled to α-Btx (see Reference 50). (Bar = 20 μm) (Figure 18.5(c) was kindly supplied by M. Daniel and Z. Vogel)

tional receptors include a high packing density and a low degradation rate. What type of mechanism might induce these properties? Several recent observations suggest that the aggregation and stability of ACh receptors are subject to external regulation by means of specific ligands. Antibodies to ACh receptors accelerate their degradation in cultured myotubes by a process similar to the modulation of surface levels of specific membrane components by antibodies in a variety of cell types.[6,40] This mechanism has recently been implicated in the expression of myasthenia gravis.[49] The increased degradation of ACh receptors is associated with the aggregation of these components due to cross-linking by the divalent anti-ACh receptor antibodies.[37,49] Monovalent antibody derivatives bind ACh receptor on the muscle cell surface but induce neither aggregation nor accelerated degradation.[37] In contrast to the effects of antibodies, the plant lectin concanavalin A which binds to ACh receptors, exerts an effect similar to innervation, namely, a decreased rate of degradation of these receptors in cultured muscle cells.[37] These findings support the plausibility of a mechanism in which specific ligands, such as neuronal factors or structural components of the synaptic cleft, can regulate the distribution and turnover of ACh receptors in postsynaptic membranes.

## VIII. REFERENCES


1. Waser, P. G. (ed.) in *Cholinergic Mechanisms*, Raven Press, New York, (1975).
2. Aidley, D. J. *The Physiology of Excitable Cells.* Cambridge University Press, (1971).
3. Karlin, A. *Life Sci.,* **14**, 1385–1415, (1974).
4. Rang, H. P. *Quart. Rev. Biophys.,* **7**, 283–399, (1975).
5. Colquhoun, D. *Ann. Rev. Pharmacol.,* **15**, 307–325, (1975).
6. Lindstrom, J. in *Receptors and Recognition* (Cuatrecasas, P., and Greaves, M. F., eds.), Chapman and Hall, London, pp. 1–45, (1976).
7. Maelicke, A., Fulpius, B. W., and Reich, E. in *Handbook of Physiology, Section I, The Nervous System* (Kandel, E. R. ed.), American Physiological Society, Bethesda, Maryland, pp. 493–520, (1977).
8. Karlin, A. in *Pathogenesis of Human Muscular Dystrophies* (Rowland, L. P., ed.), Excerpta Medica, Amsterdam and Oxford, pp. 73–84, (1977).
9. Briley, M. S., and Changeux, J. P. *Int. Rev. Neurobiol.,* **20**, 31–63, (1977).
10. Heidmann, T., and Changeux, J. P. *Ann. Rev. Biochem.,* **47**, 317–357, (1978).
11. Fambrough, D. M. *Physiol. Rev.,* **59**, 165–227, (1979).
12. Nachmansohn, D. *Harvey Lect.,* **49**, 57–99, (1955).
13. Karlin, A. *J. Gen. Physiol.,* **54**, 245s–270s, (1969).
14. Lee, C. Y. *Ann. Rev. Pharmacol.,* **12**, 265–286, (1972).
15. Prives, J., Reiter, M., Cowburn, D., and Karlin, A. *Mol. Pharmacol.,* **8**, 786–789, (1972).
16. Katz, B., and Miledi, R. *Nature (London),* **226**, 926–963, (1970).
17. Neher, E., and Sakmann, B. *Nature (London),* **260**, 799–802, (1976).
18. Singer, S. J., and Nicolson, G. I. *Science,* **175**, 720–731, (1972).
19. De Robertis, E. *Science,* **171**, 963–971, (1971).
20. Brockes, J. P., and Hall, Z. W. *Biochemistry,* **14**, 2100–2106, (1975).

21. Hamilton, S. L., McLaughlin, M., and Karlin, A. *Biochem. Biophys. Res. Commun.,* **79**, 692, (1977).
22. Neubig, R., Krodel, E., Boyd, N., and Cohen, J. B. *Proc. Natl. Acad. Sci. U.S.A.,* **76**, 690–694, (1979).
23. Epstein, M., and Racker, E. *J. Biol. Chem.,* **253**, 6660–6662, (1978).
24. Fambrough, D. M., Hartzell, H. C., Powell, J. A., Rash, J. E., and Joseph, N. in *Synaptic Transmission and Neuronal Interaction* (Bennett, M. V. L. ed.), Raven Press, New York, p. 285, (1974).
25. Brockes, J. P., Berg, D. K., and Hall, Z. W. *Cold Spring Harbor Symp. Quant. Biol.,* **40**, 253–262, (1976).
26. Fambrough, D. M. in *Biology of Cholinergic Function* (Goldberg, A. N., and Hanin, I., eds.), Raven Press, New York, pp. 101–160, (1976).
27. Weinberg, C. B., and Hall, Z. W. *Proc. Natl. Acad. Sci. U.S.A.,* **76**, 504–508, (1979).
28. Burden, S. *Dev. Biol.,* **57**, 317–329, (1977).
29. Devreotes, P. M., and Fambrough, D. M. *Cold Spring Harbor Symp. Quant. Biol.,* **40**, 237–252, (1976).
30. Lømø, T., and Rosenthal, J. *J. Physiol.,* **252**, 493–515, (1972).
31. Cohen, S. A., and Fischbach, G. D. *Science,* **181**, 76–78, (1973).
32. Spector, I., and Prives, J. *Proc. Natl. Acad. Sci. U.S.A.,* **74**, 5166–5170, (1977).
33. Shainberg, A., Cohen, S. A., and Nelson, P. G. *Pfluegers Arch.,* **361**, 255–261, (1976).
34. Prives, J., Silman, I., and Amsterdam, A. *Cell,* **7**, 543–550, (1976).
35. Anderson, M. J., and Cohen, M. W. *J. Physiol.,* **268**, 757–773, (1977).
36. Teichberg, V., and Changeux, J. P. *FEBS Letters,* **67**, 265–268, (1976).
37. Prives, J., Hoffman, L., Tarrab-Hazdai, R., Fuchs, S., and Amsterdam, A. *Life Sci.,* **24**, 1713–1718, (1979).
38. Christian, C., Daniels, M., Sugiyama, H., Vogel, Z., Jacques, L., and Nelson, P. *Proc. Natl. Acad. Sci. U.S.A.,* **75**, 4011–4015, (1978).
39. Fischbach, G. D., and Nelson, P. G. in *Handbook of Physiology, Section I, The Nervous System,* (Kandel, E. R., ed.), American Physiological Society, Bethesda, Maryland, pp. 719–774, (1977).
40. Patrick, J., Heinemann, S., and Schubert, D. *Ann. Rev. Neurosci.,* **1**, 417–443, (1978).
41. Paterson, B., and Prives, J. *J. Cell Biol.,* **59**, 241–245, (1973).
42. Prives, J. in *Surface Membrane Receptors*, (Bradshaw, R., Frazier, W., Merrell, R., Gottlieb, D., and Hogue-Angeletti, R., eds.), Plenum Press, New York, pp. 363–375, (1976).
43. Prives, J., and Paterson, B. *Proc. Natl. Acad. Sci. U.S.A.,* **71**, 3208–3211, (1974).
44. Patrick, J., McMillan, J., Wolfson, H., and O'Brien, J. C. *J. Biol. Chem.,* **252**, 2143–2153, (1977).
45. Sytkowski, A. J., Vogel, Z., and Nirenberg, M. W. *Proc. Natl. Acad. Sci. U.S.A.,* **70**, 270–274, (1973).
46. Axelrod, D., Ravdin, P., Koppel, D. E., Schlessinger, J., Webb, W. W., Elson, E. L., and Podleski, T. R. *Proc. Natl. Acad. Sci. U.S.A.,* **73**, 4594–4598, (1976).
47. Cohen, S. A., and Fischbach, G. D. *Dev. Biol.,* **59**, 24–38, (1977).
48. Podleski, T., Axelrod, D., Ravdin, P., Greenberg, I., Johnson, M., and Salpeter, M. *Proc. Natl. Acad. Sci. U.S.A.,* **75**, 2035–2039, (1978).
49. Drachman, D. B. *New Engl. J. Med.,* **298**, 136–141, (1978).
50. Vogel, Z., Maloney, G., Ling, A., and Daniels, M. *Proc. Natl. Acad. Sci. U.S.A.,* **74**, 3268–3272, (1977).

Cellular Receptors for Hormones and Neurotransmitters
Edited by D. Schulster and A. Levitzki


CHAPTER 19

# Muscarinic acetylcholine receptors

Robert H. Michell
*Department of Biochemistry, University of Birmingham, PO Box 363, Birmingham B15 2TT, U.K.*



## I. INTRODUCTION

Loewi's original evidence for the release of a chemical transmitter substance from the stimulated vagus nerve was based on the ability of the released material ('Vagusstoff', later identified as acetylcholine) to mimic the heart-slowing effect of stimulating this parasympathetic nerve: a historical sketch of this key discovery is given by Bacq.[1]

This is one of the many muscarinic actions of acetylcholine: such effects are mimicked by muscarine, pilocarpine, and acetyl-$\beta$-methylcholine, cholinergic agonists which do not appreciably stimulate nicotinic receptors, and are prevented by atropine and related compounds (Figure 19.1). These pharmacological characteristics are shared by the responses to parasympathetic nerve stimulation (or to an added muscarinic agonist) of a variety of organs, e.g. exocrine secretory glands (parotid, pancreas, sweat glands, gastric parietal cells, etc.) and many smooth muscles (including

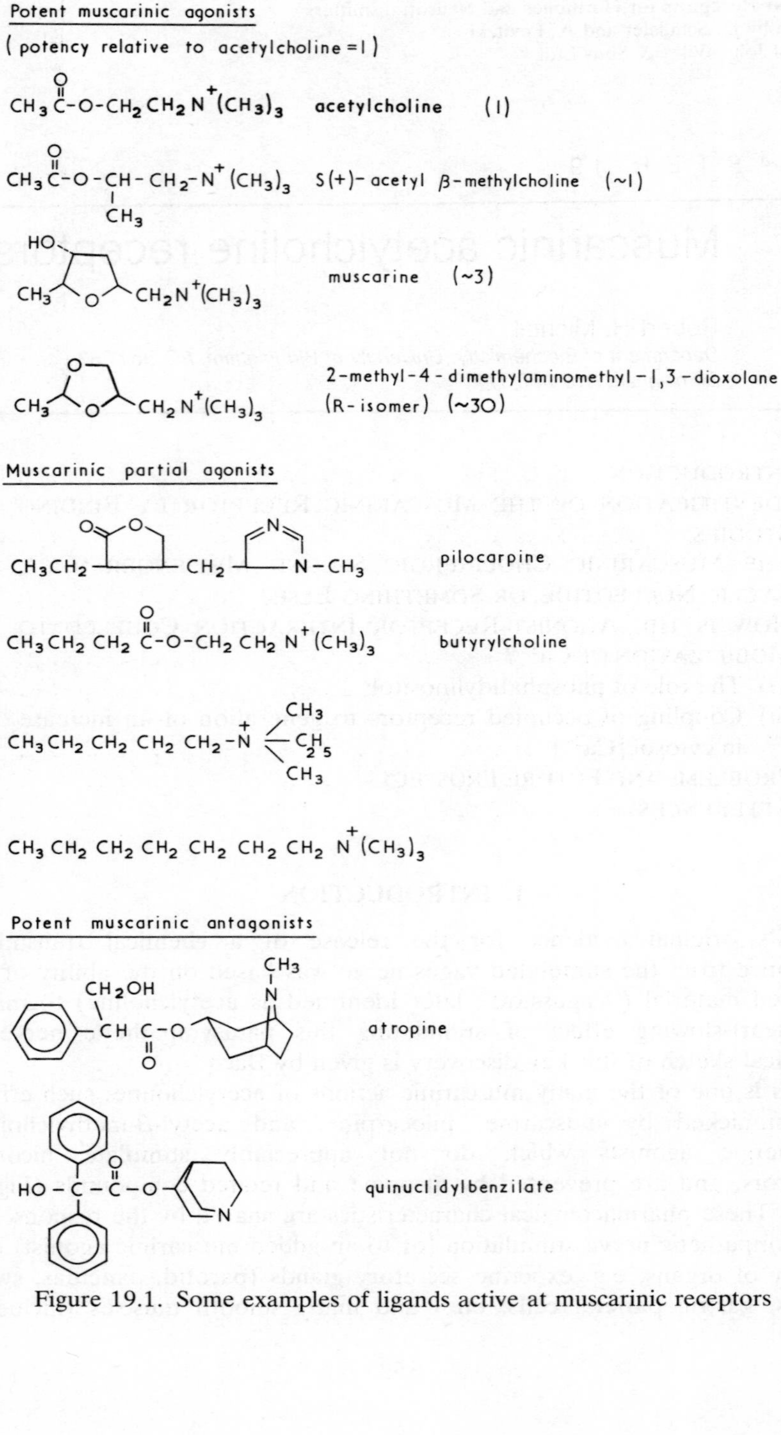

Potent muscarinic agonists

( potency relative to acetylcholine = 1 )

$CH_3C(=O)-O-CH_2CH_2N^+(CH_3)_3$ acetylcholine (1)

$CH_3C(=O)-O-CH(CH_3)-CH_2-N^+(CH_3)_3$ S(+)-acetyl β-methylcholine (~1)

muscarine (~3)

2-methyl-4-dimethylaminomethyl-1,3-dioxolane (R-isomer) (~30)

Muscarinic partial agonists

pilocarpine

$CH_3CH_2CH_2C(=O)-O-CH_2CH_2N^+(CH_3)_3$ butyrylcholine

$CH_3CH_2CH_2CH_2CH_2-N^+(CH_3)(C_2H_5)(CH_3)$

$CH_3CH_2CH_2CH_2CH_2CH_2CH_2N^+(CH_3)_3$

Potent muscarinic antagonists

atropine

quinuclidylbenzilate

Figure 19.1. Some examples of ligands active at muscarinic receptors

those of the stomach, ileum, iris, and gall bladder) (see, for example References 2–4).

In addition, there are now known to be a variety of effects of acetylcholine (or of specific muscarinic agonists and antagonists) which indicate the existence of muscarinic acetylcholine receptors in cells which are not innervated by the parasympathetic nervous system. In the brain there are neurons which are excited (depolarized) by acetylcholine and others which acetylcholine inhibits (hyperpolarizes).[5] Throughout the nervous system there also appears to be presynaptic muscarinic receptors on nerve terminals and these are responsible for feedback control of neurotransmitter output from the nerve terminals.[6] Even in the synapses between afferent and efferent neurons in sympathetic ganglia, a classical situation in which cell–cell communication is by acetylcholine acting on the nicotinic receptors of the postsynaptic cell, there is clear evidence of an additional and slower excitatory action of acetylcholine which is mediated through muscarinic receptors (e.g. Reference 7).

For many years, these diverse responses to muscarinic receptor stimulation have been analysed in two fundamentally different ways. First, the effects of muscarinic stimulation on ion fluxes through the plasma membranes of target cells have been investigated electrophysiologically.[8,9] It is often argued that analyses of these membrane events gives information on events closely coupled to the activation of the receptor; for example, the earliest responses that have been detected in muscarinically stimulated cells are changes in membrane electrical characteristics whose onset typically occurs 0.1–0.3 s after application of the agonist.[9] However, there is a substantial variety of muscarinically provoked electrical events and their explanation has required the proposal of diverse mechanisms. For example, there is an increase in $Na^+$ conductance during depolarization of ileum smooth muscle[9] or uterine smooth muscle,[10] a decrease in $K^+$ conductance during depolarization of some cortical neurons,[5] and an increase in $K^+$ conductance during hyperpolarization in heart[11] and in superficial cerebral cortical neurons.[5] It seems that either the same receptor can be coupled to several different mechanisms which control ion conductances of the plasma membrane in different ways or there must be some even more primary mechanism, as yet unrecognized, to which all of these ion gates can be linked.

The second experimental approach, and one that is often considered to assay events that are more remote from the acetylcholine–receptor interaction, is to study physiological or biochemical events which are either stimulated or inhibited by muscarinic ligands.[12] Examples include contractility, secretion of proteins or ions, release of neurotransmitters from nerve endings or changes in the metabolism or molecules such as phospholipids or cyclic nucleotides. Many of these events are clearly the end results of complex trains of events that are initiated by receptor activation, but there are observations (such as the relatively long period of latency between receptor activation and the first observed electrophysiological changes in the membrane of the stimulated cell) which might mean that these biochemical changes should be carefully examined to see if they include reactions which might participate in the coupling between receptor stimulation and cellular response.

In recent years these two classical approaches to functional analysis of the effects of stimulating muscarinic receptors have been joined by a third approach, the direct examination of muscarinic receptors by ligand binding techniques.[8,12,13] Sites with characteristics expected of muscarinic receptors have proved readily demonstrable by these techniques, but once again an unexpected degree of complexity has been revealed.

The remainder of this chapter will deal mainly with a body of biochemical and physiological information from which a working model has

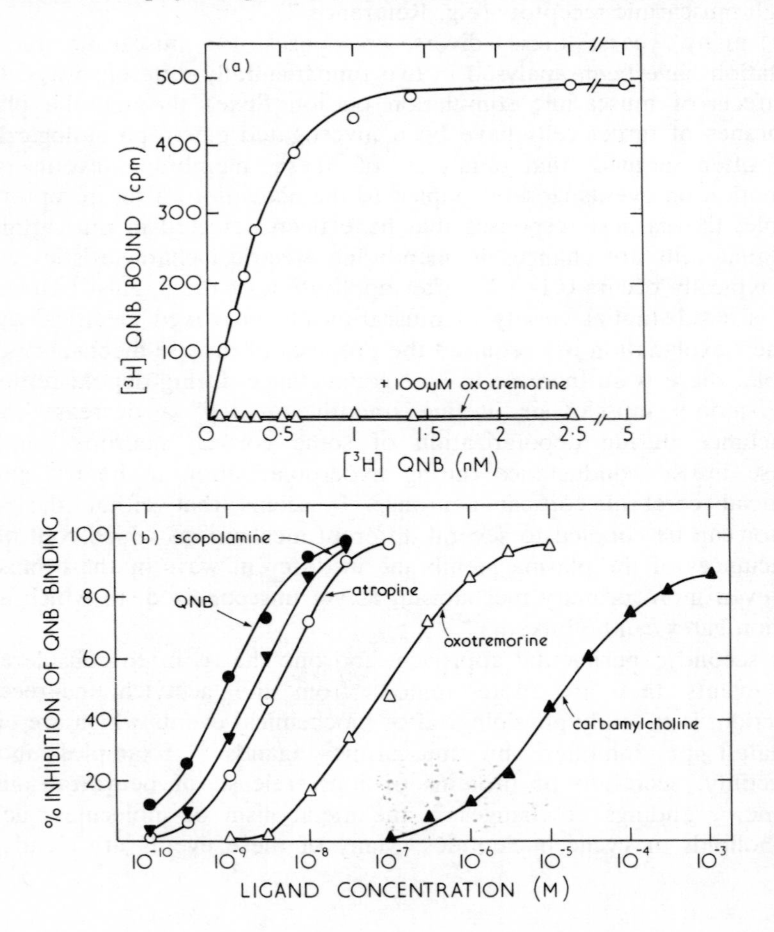

emerged which appears to account for many, though not all, effects of muscarinic cholinergic stimuli on target cells. In this model it is envisaged that the final cellular responses to such stimuli are usually controlled by a rise in the cytosal $Ca^{2+}$ ion concentration ($[Ca^{2+}]$) within the cell and that the key events provoked by receptor activation are those that play some role in bringing about this rise in cytosol $[Ca^{2+}]$.[12,14,15]

## II. IDENTIFICATION OF THE MUSCARINIC RECEPTOR BY BINDING STUDIES

Fortunately there are a substantial number of high affinity antagonists for muscarinic receptors, and several of these have been used as radioligands in binding studies. The earliest successful studies were with atropine,[8] and

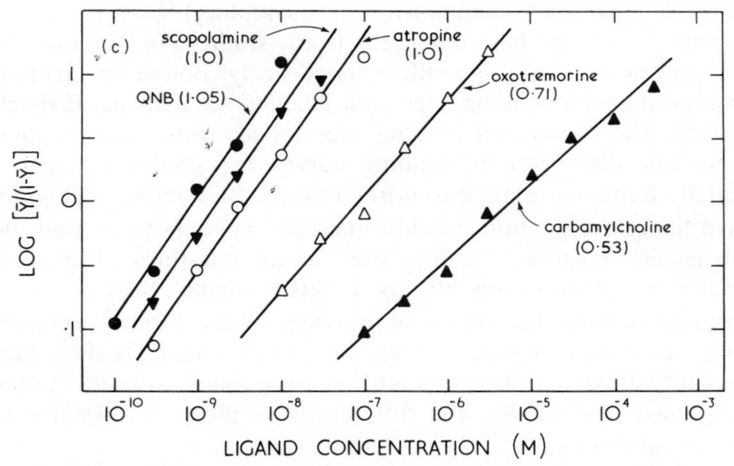


Figure 19.2. Binding of agonists and antagonists at the muscarinic receptors of nine-day old embryo chick hearts (redrawn from information in Reference 37).
(a) The saturable binding of QNB (quinuclidylbenzilate) to muscarinic receptors and the effective competition for these binding sites that is achieved by a high concentration of oxotremorine: non-specific binding of QNB is very low in this tissue.
(b) Ability of agonists and antagonists to bind to muscarinic receptors and thus to inhibit QNB binding. Note that the binding curves of the two agonists (oxotremorine and carbamylcholine) are markedly flattened by contrast with the curves for the three antagonists.
(c) Hill plots of the information in (b), with the slopes of the curves yielding the Hill coefficients which are given in brackets: a value of 1.0 is characteristic of a normal mass action curve and values below 1.0 describe the flattening of the agonist binding curves

more recently labelled quinuclidyl benzilate and propylbenzilylcholine mustard (an irreversible antagonist) have been widely used.[12,13] Binding of all antagonists shows a saturable component which occurs at the same ligand concentrations as are required for antagonism of physiological responses to muscarinic stimuli. The antagonist binding curves conform to the pattern expected from application of the law of mass action to the interaction of a ligand with a homogenous population of binding sites: this is true both for direct studies of labelled antagonist binding and for antagonists whose binding has been assessed indirectly by their ability to compete with labelled ligands for the available sites (Figure 19.2). Moreover, the binding characteristics of the muscarinic receptors in the caudate nucleus and cerebral cortex (two brain regions rich in muscarinic receptors) are essentially identical to those of the receptors in the longitudinal smooth muscle of the ileum (a tissue which is very sensitive to acetylcholine and which is the pharmacologists' favourite system in which to investigate muscarinic stimulus–response coupling).[12]

The picture which has emerged from studies of agonist binding, measured either directly (e.g. with tritiated acetylcholine or oxotremorine) or by competition for binding sites with labelled antagonists, is much more complicated. The number of binding sites for agonists and antagonists are the same, but the observed binding curves for potent agonists are all substantially flattened relative to normal mass action curves[12] (Figure 19.2). The most likely explanation for this 'anomaly' appears to be that there are two subclasses of agonist binding sites, about one-third of them of high affinity and two-thirds of low affinity. If this is correct, then potent agonists must bind more than 100 times more avidly to the former population but antagonists bind equally well to both. Less potent agonists show behaviour intermediate between antagonists and potent agonists: the less potent they are as agonists, the smaller the differences in their affinities for the two agonist site subclasses.

Physiological responses of cells to stimulation by muscarinic agonists show a variety of different dose–response curves for muscarinic stimulation,[12,16,17] so that any attempt to interpret these observations on agonist binding poses a problem: are the high or the low affinity sites (or both) involved in cellular responses to agonists? We shall return to this point later.

## III. THE MUSCARINIC CHOLINERGIC SECOND MESSENGER: $Ca^{2+}$, CYCLIC NUCLEOTIDE, OR SOMETHING ELSE?

Although electrophysiologically detectable changes in cell surface ion permeabilities might (or might not) be mediated by mechanisms that are directly coupled to receptors at the cell surface, the triggering of responses involving intracellular events, such as exocytotic secretion or contractility, clearly calls for some type of intracellular second messenger. Muscarinic stimuli rarely have any effect on cellular cyclic AMP levels or on adenylate cyclase activity: when they do, it is usually to depress the cyclic AMP concentration. By contrast, there is substantial evidence that $Ca^{2+}$ is implicated in the muscarinic responses of a variety of tissues and that indicates that intracellular cyclic GMP levels often rise as a result of muscarinic stimulation; $Ca^{2+}$ or cyclic GMP might therefore be a muscarinic second messenger.

The major evidence pointing to a central role of $Ca^{2+}$ ions in muscarinic stimulus–response coupling comes from studies of responses that are substantially reduced when cells are deprived of extracellular calcium. Examples include contraction of various smooth muscles;[12,17] (also see Reference 52, Chapter 4) and the efflux of $K^+$ from muscarinically stimulated parotid gland.[18,19] Such studies simply demonstrate the sensitivity of cell responsiveness to $Ca^{2+}$ deprivation; they do not show that $Ca^{2+}$ is the intracellular messenger for such responses. However, evidence in support of the idea that intracellular $Ca^{2+}$ is the major ligand responsible for control of each of these processes has accumulated from a variety of sources. For example, muscle contraction seems always to be caused by a rise in the cytosol $Ca^{2+}$ concentration in the vicinity of the contractile apparatus, and in many electrically excitable smooth muscles depolarization of the cells by elevation of the extracellular $K^+$ concentration causes a contraction which is entirely dependent on the presence of $Ca^{2+}$ in the extracellular medium (see Reference 52, Chapter 4).

Such arguments gain considerable extra cogency from several types of experiment. First, ionophores (e.g. A23187) have become available which allow $Ca^{2+}$ to pass across membranes via the lipid phase and therefore bypass any channels that are controlled by receptors. In general, it has been found that the application of these compounds to tissues or cells evokes the same responses as are elicited by stimulation of muscarinic receptors.[4,18] This is an important observation since it demonstrates that a rise in the cytosol $Ca^{2+}$ concentration is sufficient on its own to mimic the effects of muscarinic receptor stimulation in such cells. The simplest interpretation of the $Ca^{2+}$-dependence of tissue responsiveness to muscarinic stimuli is therefore that $Ca^{2+}$ is the muscarinic second messenger, and that the effect of muscarinic stimulation is to open cell surface $Ca^{2+}$ channels which allow $Ca^{2+}$ to move into cells down its electrochemical gradient and thus raise the cytosol concentration of this ion. This general view of events receives support from two directions. The first is that certain ions (e.g. $La^{3+}$ and other lanthanides, $Mn^{2+}$, $Co^{2+}$) which inhibit movement of $Ca^{2+}$ ions through potential-sensitive $Ca^{2+}$ channels in excitable membranes such as that of the squid axon also inhibit cellular responses to stimulation of muscarinic receptors.[20] Second, in some systems direct evidence has been obtained using $^{45}Ca^{2+}$ which shows that $Ca^{2+}$ from the extracellular medium moves into muscarinically stimulated cells.[21]

There seems to be little disagreement that responses to muscarinic stimuli that are as diverse as exocytotic secretion, $K^+$ efflux, contractility, and glycogen breakdown are all ultimately controlled by changes in cytosol [$Ca^{2+}$], but the preceding discussion probably over-simplifies the situation. This is because in most tissues the responses are not entirely dependent on the presence of extracellular $Ca^{2+}$. Often there is a rapid and transient response (known pharmacologically as a 'phasic' response) which persists even when $Ca^{2+}$ has been absent from the extracellular medium for a brief period, but under such circumstances the sustained response of the same tissue when it is continuously stimulated (the 'tonic' response) is usually lost.[14,19] The most common interpretation of this pattern of behaviour suggests that a single family of muscarinic receptors can bring about a mobilization of $Ca^{2+}$ ions into the cytosol in two ways; it is not clear whether both effects are consequences of activation of the same effector mechanism by the stimulated receptors. It is suggested that immediately after receptor activation there is release of a limited pool of bound $Ca^{2+}$, probably held at the inner surface of the plasma membrane, and that this is followed by an influx of $Ca^{2+}$ through cell surface $Ca^{2+}$ channels which are opened as a result of receptor activation (Figure 19.3).[14,19]

In a few tissues, typified by the exocrine pancreas of some species, the situation is more extreme, and even quite prolonged $Ca^{2+}$ deprivation makes little mark on the magnitude or time-course of the response. In such cases, workers tend to have polarized around two alternative interpretations.[21–25] One school maintains that the $Ca^{2+}$ required to trigger the secretory response must be released from some intracellular $Ca^{2+}$ source such as either mitochondria or the endoplasmic reticulum, whereas the other argues that these tissues represent an extreme version of the situation described above, with sufficient membrane-bound $Ca^{2+}$ available to sustain a substantial response for quite a long period. Although many of the experimental data seem to favour the former alternative, there is also experimental support for the second view. This latter view seems more likely on grounds of biological unity; in most tissues muscarinic control of cell responses seems to be exerted by mobilization of $Ca^{2+}$ from and through the plasma membrane, but some additional (and so far unknown) messenger would be needed if receptor activation was in some tissues to release $Ca^{2+}$ from an intracellular reservoir remote from the plasma membrane.

What then of cyclic GMP? Numerous experiments have shown that a rise in cyclic GMP level is a very frequent, if not universal, response to

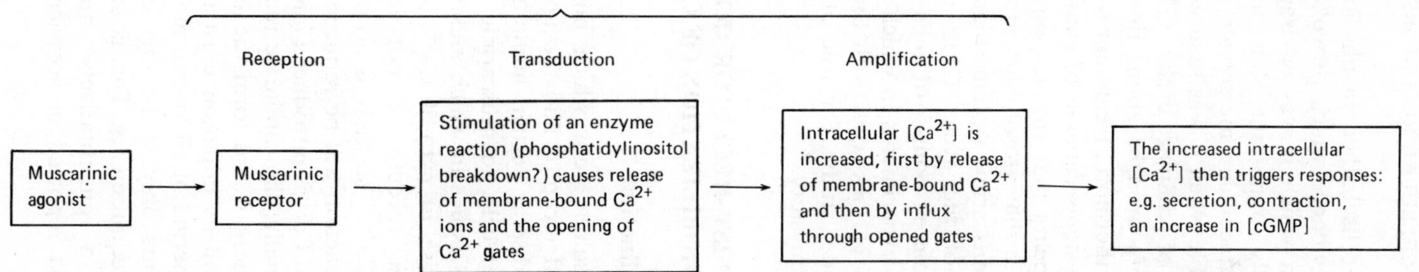


Figure 19.3. Proposed scheme of events involved in stimulus–response coupling at the muscarinic cholinergic receptor (reproduced with permission from Reference 15)

muscarinic stimulation of cells bathed in physiological media.[26] This was among the key items of evidence which provoked the formulation of the 'Yin–Yang' hypothesis[27] in which it was envisaged that cyclic AMP and cyclic GMP were equally important, but independently (and often oppositely) controlled, intracellular messenger molecules. However, this hypothesis became untenable when it was realized that many of the stimuli which caused elevation of cellular cyclic GMP levels, among them muscarinic cholinergic, only brought about this response if extracellular $Ca^{2+}$ was available.[26,28] Furthermore, treatment with the ionophore A23187 and $Ca^{2+}$ was in several systems capable of causing a similar rise in cyclic GMP levels.[26] Thus it appears that this response, like others considered above, is caused by a rise in the cytosol concentration of $Ca^{2+}$, the primary second messenger of the muscarinic system (Figure 19.3). Confirmation of the non-essential nature of the cyclic GMP changes, at least for the immediate and most obvious responses of cells to muscarinic stimuli, has recently been provided by experiments in which cellular cyclic GMP levels have been manipulated in various ways with little apparent effect on the functions of the tissues under study. The role of cyclic GMP thus remains an intriguing but unsolved mystery.[26]

## IV. HOW IS THE AGONIST–RECEPTOR INTERACTION COUPLED TO MOBILIZATION OF $Ca^{2+}$?

### (i) The role of phosphatidylinositol

If one is to search for some event which might be implicated in the coupling between activated receptor and the unidentified $Ca^{2+}$-mobilizing mechanism, then it must display certain minimum characteristics. It must be a universal response of cells to muscarinic stimulation (or at least a response that occurs in all tissues in which muscarinic receptors are linked to responses through a rise in cytosol $Ca^{2+}$ concentration). Binding of ligands to the muscarinic receptor is not influenced appreciably by depletion of extracellular $Ca^{2+}$,[12] so there is no reason to expect either that the rate of any coupling reaction will be reduced under such circumstances or that it will be provoked if $Ca^{2+}$ is introduced into cells by an artificial route (e.g. by an ionophore). Finally, the candidate reaction must be localized at the plasma membrane, where it can form the link between receptor and $Ca^{2+}$-mobilizing system, and its initiation must take 0.1 s or less (i.e., be within the known latency period of all known muscarinic responses).

At present, there appears only to be one event which fulfils a substantial number of these requirements. This is an increase in the turnover of phosphatidylinositol, a quantitatively minor anionic membrane glycerophospholipid.[29] First reported in acetylcholine-stimulated pancreas

Table 19.1. Tissues in which cholinergic agonists stimulate phosphatidylinositol metabolism

| | |
|---|---|
| Cerebral cortex | Electroplaque |
| Cerebral cortical synaptosomes | Parotid gland* |
| Corpus striatum | Submaxillary gland |
| Hypothalamus | Lacrimal gland* |
| Thalamus | Peptic mucosa |
| Stellate ganglion | Sweat glands |
| Superior cervical ganglion | Pineal gland |
| Anterior pituitary* | Thyroid gland |
| Adrenal medulla* | Ileum smooth muscle* |
| Salt gland | Iris smooth muscle* |

In most, but not all, of these tissues it has been shown that the receptor responsible for stimulation of phosphatidylinositol metabolism is muscarinic in character. (In no system has it been shown that the phosphatidylinositol response can be nicotinically evoked).
*In these systems there is evidence that the stimulation of phosphatidylinositol metabolism occurs even in situations (e.g. absence of extracellular $Ca^{2+}$) where receptor activation would not be able to bring about the normal rise in cytosol [$Ca^{2+}$]

by M. R. and L. E. Hokin about 25 years ago, this muscarinic response, which is usually detected first as an increase in $^{32}P_i$ incorporation into phosphatidylinositol in the stimulated tissue, has now been seen in many tissues of diverse types (Table 19.1). The initial reaction appears to be receptor-stimulated removal of the phosphorylinositol headgroup of phosphatidylinositol by a specific phospholipase C, followed by a compensatory increase in phosphatidylinositol biosynthesis (Figure 19.4).[29,30] It has recently become clear that in several tissues this response is not prevented by removal of extracellular $Ca^{2+}$ and that it cannot be provoked by an ionophore-mediated influx of $Ca^{2+}$ (see Table 19.1). In relation to the other requirements of any coupling reaction, it is not yet clear whether receptor-stimulated phosphatidylinositol breakdown is localized at the plasma membrane and no-one has yet devised experiments sensitive enough to demonstrate phosphatidylinositol breakdown in less than a few minutes.

As a result of these unusual characteristics it has been suggested that phosphatidylinositol breakdown might be involved in some essential way in the coupling between the activated muscarinic receptor and the mobilization of $Ca^{2+}$ from and through the plasma membrane (but it should be noted that a contrary view has also been expressed[31]). An additional indication which may be in favour of its involvement in some such $Ca^{2+}$-mobilizing mechanism comes from examination of the patterns of responses provoked in target cells by activation of a variety of different

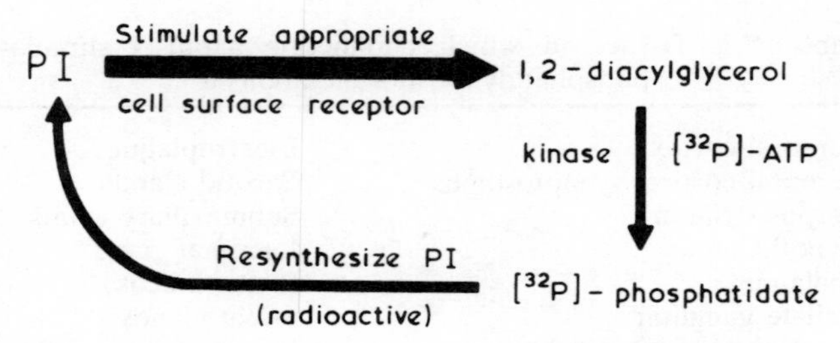


Figure 19.4. A summary of the probable events involved in receptor-stimulated breakdown of phosphatidylinositol (PI) and resynthesis of labelled PI

receptors. Some receptors control adenylate cyclase; these usually do not stimulate phosphatidylinositol metabolism.[29,32] Several other types of receptor seem to share a second mode of action which brings about an increase in the cytosol $Ca^{2+}$ concentration in target cells,[29,30,32] i.e. they utilize the same second messenger as does the muscarinic cholinergic receptor. Examples of such receptors include those for $H_1$-histaminergic stimuli (in smooth muscle), $\alpha$-adrenergic stimuli (in many tissues), pancreozymin (in pancreas), substance P (in parotid gland), vasopressin (at least in liver), glucose (in islets of Langerhans), 5-hydroxytryptamine (in smooth muscle and insect salivary gland), and mitogenic lectins (in lymphocytes). It is remarkable that stimulation of phosphatidylinositol turnover is a response that is rapidly evoked by stimulation of any of these receptors. Presumably it is either (1) a response to some cellular messenger whose level is controlled by all of these receptors (the only current candidate for such a role would be $Ca^{2+}$); or (2) it is involved in the $Ca^{2+}$-mobilizing effector mechanisms that are linked to each of these receptors. In view of the $Ca^{2+}$-independent nature of the phosphatidylinositol response, the latter seems a real possibility.[29,30,32]

### (ii) Coupling of occupied receptors to generation of an increase in cytosol [$Ca^{2+}$]

Many years ago it was realized that a complex series of events might be involved in this coupling, and therefore that a cell's normal physiological response was unlikely to be an exact reflection of the number of productive interactions between agonist molecules and receptors. Indeed, it was found that treatment with low concentrations of several alkylating receptor antagonists (e.g. dibenamine), whose action appeared to irreversibly reduce the number of receptors, did not always reduce the maximum response of tissues to agonists. Instead, they shifted the

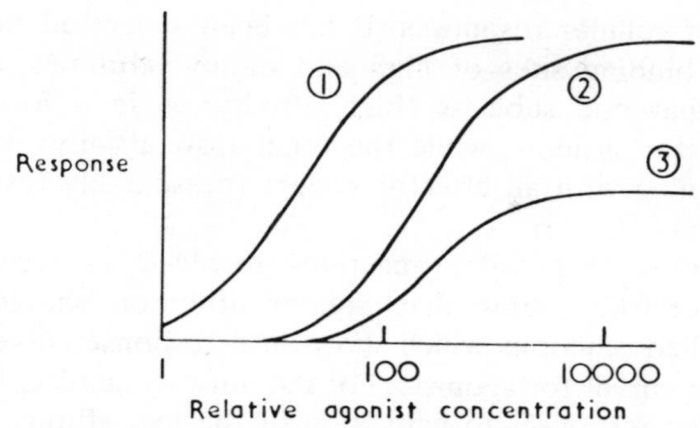


Figure 19.5. A typical pattern of changes in cell responsiveness after treatment of a tissue possessing a substantial 'receptor reserve' with an irreversible antagonist (e.g. dibenamine or phenoxybenzamine). Curve 1 represents the responsiveness of the native tissue. When many of the receptors are put out of action by the antagonist, the dose–response curve shifts so that full response is only achieved at much higher agonist concentrations (curve 2). Further treatment with the antagonist then reduces the residual receptor population to such a small number that activation of all of the residual receptors is insufficient to produce a change in intracellular effector concentration sufficient for a full physiological response (curve 3)

dose–response curve to higher agonist concentrations in a way reminiscent of the effects of reversible competitive antagonists (e.g. Reference 33). Only after extensive receptor blockade (sometimes of 99% or more of the receptor population) did the dose–response curve remain in the same concentration range and show a fall in the maximum response (Figure 19.5). Thus it appeared that the true affinity for the functional agonist–receptor interaction was much lower than the apparent affinity which had been suggested by the dose–response curves of the more sensitive physiological responses (e.g. contraction of ileum smooth muscle, secretion from pancreas), but the presence of large number of receptors in excess of the minimum number needed to elicit a full cellular response allows the cell to respond effectively at agonist concentrations which achieve only a low extent of receptor occupation.

Examination of the heterogenous agonist–receptor binding curves discussed earlier and of the affinity constants calculated indirectly from studies of irreversible antagonism suggest that the subpopulation with the lower binding affinity is more likely to be the population responsible for the triggering of cellular responses. It has been suggested that maybe the populations of binding sites of high and of low affinities are essentially identical, but that one subclass (high affinity) is in a form that is not coupled to effector systems, while the other (low affinity) are involved in productive coupling with an effector system (presumably responsibly for a rise in intracellular [$Ca^{2+}$]).[34]

In consideration of possible reactions involved in receptor–response coupling, it therefore seems that greater attention should be paid to receptor-controlled reactions which show dose–response curves close to the receptor binding curves for agonists. For the muscarinic receptor, this would mean looking for responses to activation of the low affinity population of receptors, an approach quite alien to the usual philosophy of only giving serious consideration to events which occur at 'physiologically meaningful' drug concentrations. So far, two events evoked by the muscarinic receptor have been discussed in this light. One is phosphatidylinositol breakdown, which, on the basis of the limited available information, appears likely to be activated by both the high and low affinity receptor subpopulations.[17] The second is production of cyclic GMP, but here there is disagreement: one group's observations suggest it may be coupled only to the low affinity subpopulation,[35] while another study reports its activation at much lower agonist concentrations.[36]

## V. PROBLEMS AND FUTURE PROSPECTS

The conceptual picture of muscarinic receptor action mediated by $Ca^{2+}$ as a second messenger provides a ready explanation for many muscarinic responses. However, as pointed out at the beginning of this chapter, it certainly does not explain everything. A particularly clear illustration of one of the remaining uncertainties is given by the muscarinic receptors on the exterior of cholinergic nerve terminals, where they constitute a mechanism by which released acetylcholine exerts a feedback inhibitory effect upon the continued output of the neurotransmitters.[6] There is general agreement that the secretion of neurotransmitters from nerve-endings is evoked by an increase in intrasynaptic [$Ca^{2+}$], this being a result of influx of $Ca^{2+}$ through potential-sensitive $Ca^{2+}$ 'gates' which open in response to the arrival of an action potential. In this situation, therefore, the most obvious interpretation of the action of muscarinic receptors is that they reduce the influx of $Ca^{2+}$ into the activated nerve-terminal and thus the rate of acetylcholine release. Despite this apparent difference in mode of action, these presynaptic receptors have a pattern of pharmacological specificity for agonists and antagonists that is essentially indistinguishable from that of muscarinic receptors on other cells.

Even within the broad group of phenomena which can be accounted for on the basis of the 'Ca$^+$ as second messenger' scheme, with phosphatidylinositol breakdown as the most promising potential coupling reaction, little progress has yet been made beyond the study of muscarinic responses in intact cells. There is detailed information from binding studies on the precise specificity of the muscarinic binding sites, but few steps have yet been made towards the isolation and characterization of the membrane component which carries these binding sites. Muscarinic control of phosphatidylinositol breakdown has not been reproducibly achieved in broken-cell systems, and there is no information on the molecular mechanism by which the plasma membrane of a muscarinically stimulated cell changes either its $Ca^{2+}$-binding characteristics or its $Ca^{2+}$ permeability. These are all problems for the future.

## VI. REFERENCES


1. Bacq, Z. M. *Chemical Transmission of Nerve Impulses: a Historical Sketch*, Pergamon Press, Oxford, (1975).
2. Bevan, J. A. (ed.) *Essentials of Pharmacology*, Harper & Row, New York, pp. 89–100; 103–106; 113–121, (1969).
3. Triggle, D. J. *Neurotransmitter–Receptor Interactions*, Academic Press, London, (1971).
4. Case, R. M., and Goebell, H. (eds.) *Stimulus–Secretion Coupling in the Gastrointestinal Tract*, MTP Press, Lancaster, (1976).
5. Krnjevic, K. in *Handbook of Psychopharmacology* (Iversen, L. L., Iversen, S. D., and Snyder, S. H., eds.) Vol. 6, pp. 97–126, Plenum Press, New York, (1975).
6. Muscholl, E. *Pharmacology and the Future of Man*, Karger, Basel, Vol. 4, pp. 440–457, (1973).
7. Greengard, P., and Kebabian, J. W. *Fed. Proc.*, **33**, 1059–1067, (1974).
8. Rang, H. P. *Quant. Rev. Biophys.*, **7**, 283–399, (1975).
9. Bolton, T. B. *Proc. Roy. Soc. Lond. B.*, **194**, 99–119, (1976).
10. Worcel, M., and Hamon, G. in *Physiology of Smooth Muscle*, (Bülbring, E., and Shuba, M. F., eds), Raven Press, New York, pp. 339–345, (1976).
11. TenEick, R. E., Nawrath, H., McDonald, T. F., and Trautwein, W. *Pflügers Arch. ges. Physiol.*, **361**, 207–213, (1976).
12. Birdsall, N. J. M., and Hulme, E. C. *J. Neurochem.*, **27**, 7–16, (1976).
13. Snyder, S. H., Chang, K. J., Kuhar, M. J., and Yamamura, H. I. *Fed. Proc.*, **34**, 1915–1921, (1975).
14. Chang, K.-J., and Triggle, D. J. *J. Theor. Biol.*, **40**, 125–154; 155–172, (1973).
15. Michell, R. H., Jones, L. M., and Jafferji, S. S. *Biochem. Soc. Trans.*, **5**, 77–81, (1977).
16. Bolton, T. B. in *Drug Receptors*, (Rang, H. P., ed.), Macmillan, London, pp. 87–104, (1973).
17. Michell, R. H., Jafferji, S. S., and Jones, L. M. *FEBS Letters*, **69**, 1–5, (1976).
18. Schramm, M., and Selinger, Z. in *Stimulus–Secretion Coupling in the Gastrointestinal Tract* (Case, R. M., and Goebell, H., eds.), MTP Press, Lancaster, pp. 49–64, (1976).

19. Leslie, B. A., Putney, J. W., and Sherman, J. M. *J. Physiol. (London),* **260**, 351–370, (1976).
20. Triggle, C. R., and Triggle, D. J. *J. Physiol. (London),* **254**, 39–54, (1976).
21. Kondo, S., and Schultz, I. *J. Membrane Biol.,* **29**, 185–204, (1976).
22. Williams, J. A., and Chandler, D. *Am. J. Physiol.,* **288**, 1729–1732, (1975).
23. Lucas, M., Schmid, G., Kromas, R., and Löffler, G. *Eur. J. Biochem.,* **85**, 609–619, (1978).
24. Kanno, T., and Nishimura, O. *J. Physiol. (London),* **257**, 309–324, (1976).
25. Iwatsuki, N., and Petersen, O. H. *J. Physiol. (London),* **274**, 81–96, (1978).
26. Goldberg, N. D., and Haddox, M. C. *Ann. Rev. Biochem.,* **46**, 823–896, (1977).
27. Goldberg, N. D., Haddox, M. C., Hartle, D. K., and Hadden, J. W. in *Pharmacology and the Future of Man*, Karger, Basel, Vol. 5, pp. 149–169, (1973).
28. Schultz, K.-D., Schultz, K., and Schultz, G. *Nature* (*London*), **265**, 750–751, (1977).
29. Michell, R. H. *Biochim. Biophys. Acta,* **415**, 81–147, (1975).
30. Michell, R. H. in *Function and Biosynthesis of Lipids* (Bazan, N. G., Brenner, R. R., and Giusto, N. M., eds.), Plenum Press, New York, pp. 447–464, (1977).
31. Hokin-Neaverson, M. R. in *Function and Biosynthesis of Lipids* (Bazan, N. G., Brenner, R. R., and Giusto, N. M. eds.), Plenum Press, New York, pp. 429–446, (1977).
32. Michell, R. H. in *Companion to Biochemistry*, (Bull, A. T., Lugnado, J., Tipton, K., and Thomas, J. O., eds.), Longmans, London, Vol. 2, pp. 205–228, (1979).
33. Furchgott, R. F., and Bursztyn, R. F. *Ann. N.Y. Acad. Sci.,* **144**, 882–899, (1967).
34. Birdsall, E. J. M., Burgen, A. S. V., and Hulme, E. C. in *Cholinergic Mechanisms and Psychopharmacology* (Jenden, D. J., ed.), Plenum Press, New York, pp. 25–33, (1978).
35. Strange, P. G., Birdsall, N. J. M., and Burgen, A. J. V. *Trans. Biochem. Soc.,* **5**, 189–191, (1977).
36. Hanley, M. R., and Iversen, L. L. *Mol. Pharmacol.,* **14**, 246–255, (1978).
37. Galper, J. B., Klein, W., and Catterall, W. A. *J. Biol. Chem.,* **252**, 8592–8599, (1977).

Cellular Receptors for Hormones and Neurotransmitters
Edited by D. Schulster and A. Levitzki


CHAPTER 20

# Amino acid receptors

Vivian I. Teichberg
*Department of Neurobiology, The Weizmann Institute of Science, Rehovot, Israel*

## I. INTRODUCTION

Much of the neurochemical literature is concerned with the analysis of the chemical signals involved in neurotransmission. When one looks at the considerable progress recently made in this field it is clear that advances were often linked with the identification of new neurotransmitters.

Since the early 1950s, several brain metabolites have been thought to be natural neurotransmitters. Among the strong contenders are the amino acids, glutamic acid, γ-aminobutyric acid (GABA), and glycine which are supposed to mediate most of the synaptic transmission within the mammalian CNS circuitry. Other amino acids such as aspartic acid and taurine are also considered as putative excitatory and inhibitory neurotransmitters respectively, but the evidence for such assignment is not yet compelling. In this chapter, we will deal only with the most probably and well-documented neurotransmitters; we will describe their receptors and analyse their suspected role in the physiology and pathology of the mammalian central nervous system.

## II. GLUTAMATE

### (i) Glutamate as a neurotransmitter

The potent excitable properties of glutamate on cortical neurons were observed in 1954 by Hayashi who was the first to propose that this amino acid may have, in addition to its role in cerebral metabolism, a modulating

Table 20.1. Criteria for identification of neurotransmitters

1. The putative neurotransmitter and its biosynthetic enzymes must be present in the neurons from which the neurotransmitter is released.
2. It must be released after stimulation.
3. It should have a postsynaptic action either excitatory or inhibitory that mimics the postsynaptic effects observed upon presynaptic stimulation. It should bind to the same receptor as the natural neurotransmitter and trigger a change in permeability to the same ion(s).
4. It must be removed from the synaptic cleft either by an enzymic inactivation process or by a specific uptake mechanism into synaptic structures.

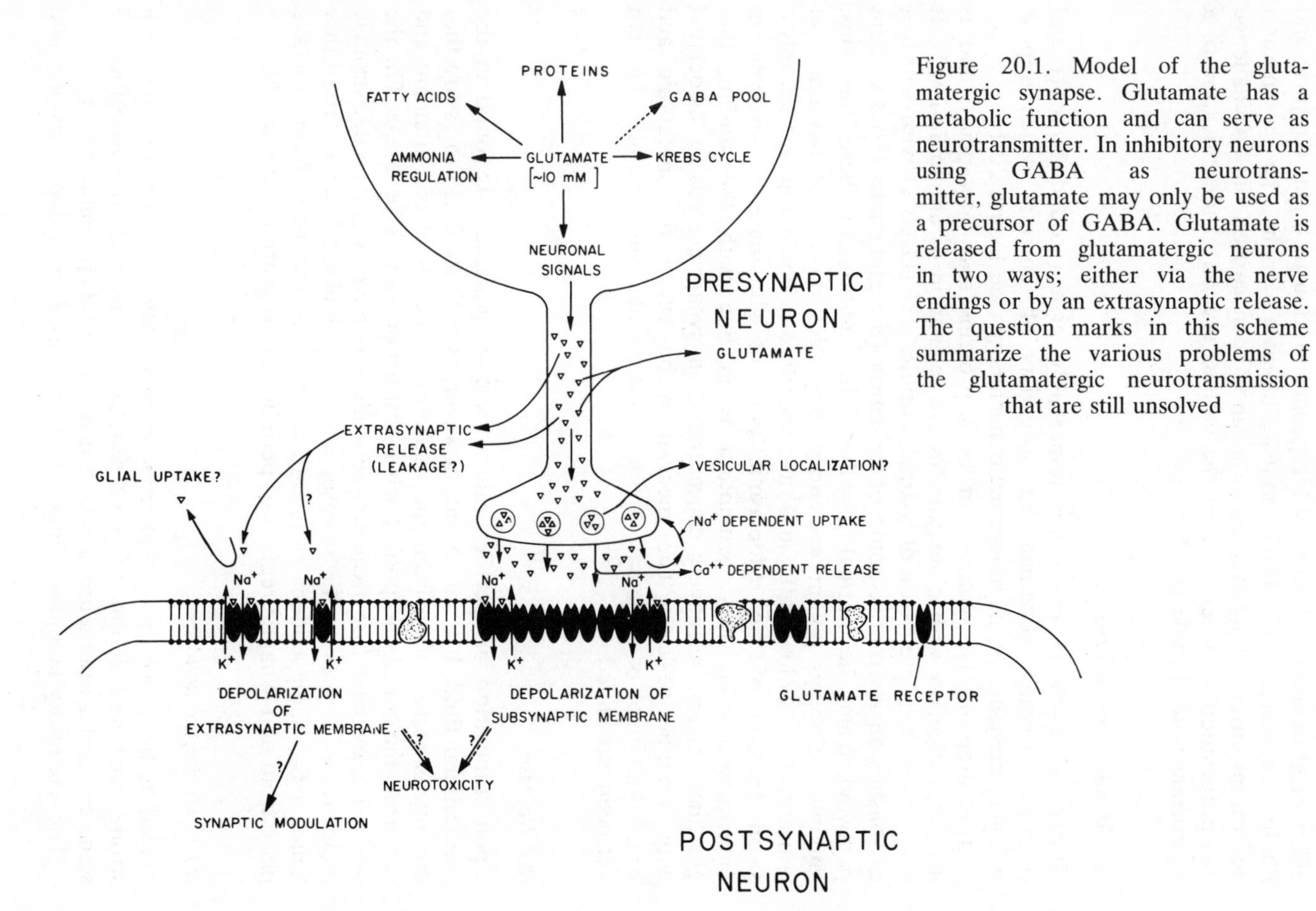


Figure 20.1. Model of the glutamatergic synapse. Glutamate has a metabolic function and can serve as neurotransmitter. In inhibitory neurons using GABA as neurotransmitter, glutamate may only be used as a precursor of GABA. Glutamate is released from glutamatergic neurons in two ways; either via the nerve endings or by an extrasynaptic release. The question marks in this scheme summarize the various problems of the glutamatergic neurotransmission that are still unsolved

activity on brain excitability. Considering both the strong depolarizing effect of glutamate and its very high concentration in brain, Krnjevic and Phillis suggested in 1963 that glutamate might be a natural neurotransmitter. Many studies have since been carried out to challenge this proposition in order to meet the various criteria of identification of a neurotransmitter (Table 20.1).

### *(a) Metabolism and storage*

Glutamate levels in brain are remarkably high (around 10 mM) and constant, although glutamate has a variety of functions which impose stringent demands on its biosynthetic machinery (see Figure 20.1).

It is used in the synthesis of both proteins and fatty acids, and is involved, together with glutamine, in the regulation of ammonia levels. It also serves as a precursor of $\gamma$-aminobutyric acid (GABA) with which it establishes an alternative route of oxidation (the glutamate–GABA shunt pathway) from the normal glycolysis. In addition to these functions, glutamate displays potent excitatory properties on central neurons[1,2] at concentrations as low as 0.1 mM[3] and therefore it would appear that only a small fraction of the cytoplasmic pool of glutamate is involved in neurotransmission. This assumption is in line with the finding that glutamate is not selectively concentrated in synaptic areas or associated with synaptic vesicles. The mechanism by which the metabolic and transmitter pool of glutamate are each maintained and regulated has not yet been elucidated.

### *(b) Release*

Upon stimulation of specific brain structures, glutamate is found in the extracellular fluid. It is important, however, (see Table 20.1) to assure that this release takes place from the neurotransmitter pool of glutamate and not from the metabolic pool. Such a distinction cannot be made. On the one hand, evidence for a calcium-dependent release of glutamate, strongly indicative of a presynaptic release, has been obtained,[4] but on the other hand, a $Ca^{2+}$-independent glutamate release has also been found to take place from non-synaptic regions of peripheral and central nerve trunks.[5]

### *(c) Postsynaptic action*

Tested by the powerful iontophoretic method, glutamate is found to have a strong excitatory action on most central neurons with the exception of some cerebellar interneurons, where it seems to be an inhibitor.[6]

The excitatory action of glutamate is mediated by a depolarization and decreased resistance of the postsynaptic membrane. This depolarization is accompanied by an increase in membrane permeability to sodium ions and to a lesser extent to potassium ions. It has not yet been possible to demonstrate that the reversal potential for the natural neurotransmitter is identical to that produced by iontophoretically applied glutamate, and it is around this infringement to the third criterion of identification of neurotransmitters, that the debate on the role of glutamate is centred.

*(d) Inactivation*

The inhibition of the enzymes involved in glutamate catabolism does not seem to affect the time-course of inactivation of the depolarizing response

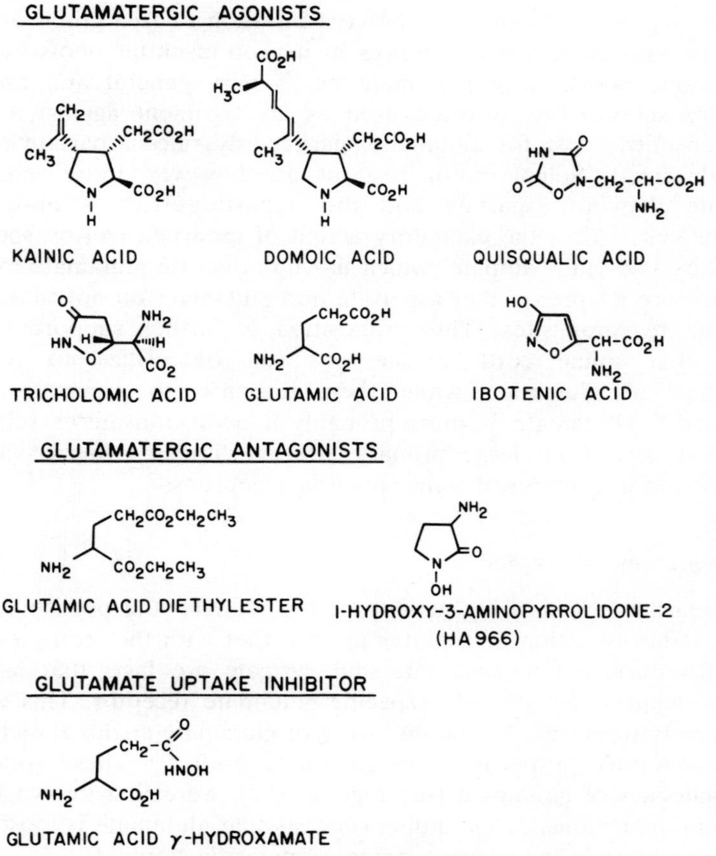

GLUTAMATERGIC AGONISTS

KAINIC ACID

DOMOIC ACID

QUISQUALIC ACID

TRICHOLOMIC ACID

GLUTAMIC ACID

IBOTENIC ACID

GLUTAMATERGIC ANTAGONISTS

GLUTAMIC ACID DIETHYLESTER

1-HYDROXY-3-AMINOPYRROLIDONE-2 (HA 966)

GLUTAMATE UPTAKE INHIBITOR

GLUTAMIC ACID $\gamma$-HYDROXAMATE

Figure 20.2. The glutamatergic ligands

to applied glutamate.[7] Inactivation takes place via specific $Na^+$-dependent high affinity uptake systems present both in presynaptic terminals and in glial cells. The latter cells are probably the main contributors to the removal of glutamate from the extracellular fluid. The presynaptic glutamate uptake sites have a specificity for glutamate analogues different from that of the postsynaptic glutamate receptor sites.[8] Indeed, kainic acid, an excitatory glutamate analogue, inhibits neither the neuronal nor the glial uptake of glutamate, whereas L-glutamate-γ-hydroxamate (Figure 20.2), a substance with only weak neuronal excitatory activity, is a very potent uptake inhibitor.

### *(e) Distribution*

Glutamate is widely distributed in the central nervous system. Although there are regional variations in the concentration of free glutamate, these do not necessarily reflect differences in neurotransmitter pools but rather in metabolic pools, since glutamate has a very general and ubiquitous excitatory action. This is often used as an argument against a specific neurotransmitter role for glutamate, particularly since aspartate produces identical effects. Spinal motor neurons are however, more sensitive to glutamate than to aspartate and the opposite effect is observed on Renshaw cells. Also, the excitatory action of aspartate can be specifically blocked by D-α-aminoadipate[9] which has no effect on glutamate excitation and therefore it appears that aspartate and glutamate do not interact with the same receptor sites. This proposition is further supported by the finding that spinal cord interneurons can desensitize to prolonged application of glutamate while their response to aspartate remains unaffected.[44] Glutamate is most probably a neurotransmitter released in the spinal cord from large primary afferent fibres,[10] whereas aspartate would be the transmitter of some spinal interneurons.

## (ii) The glutamate receptor

The ubiquitous depolarizing activity of L-glutamate, the persisting reports of the excitatory action of D-glutamate together with the strong excitatory properties displayed by aspartate and cysteate are facts that have long militated against the idea of a specific glutamate receptor. This view has been largely weakened by the discovery of glutamate analogues with more potent excitatory properties than glutamate itself.[11,12] These compounds, rigid analogues of glutamate (see Figure 20.2), were first known for their anthelminthic actions. (This finding suggests that glutamate is most likely a neurotransmitter in the nervous system of parasitic worms.)

A comparative study of the potency of glutamate and of its analogues to depolarize frog motor neurons and rat cortical neurons indicated that domoate > quisqualate > kainate > tricholomate > L-glutamate.

Since all these compounds are structurally related to L-glutamate, it is likely that they act on the glutamate receptor sites. Whether these sites are identical to the receptor sites of the natural neurotransmitter is a question that could be answered through the use of specific antagonists of synaptic and glutamate-induced excitation. However, besides compounds such as piperazine (an anthelminthic agent) glutamate diethylester, or 1-hydroxy-3-aminopyrrolidone-2(HA-966) (see Figure 20.2) which display some antagonistic activities at very high concentrations, no high affinity glutamate antagonist has yet been found. Nevertheless, the excitatory effects of microiontophoretically administered kainate on CNS neurons were found to be less affected by glutamate diethylester than those produced by glutamate.[45] This observation could indicate an heterogeneity in the population of glutamate receptors.

In any case, the finding of a specific glutamate antagonist is critical in assigning a status of natural neurotransmitter to glutamate.

Direct biochemical evidence has been sought for the existence of glutamate receptors in the central nervous system and tritiated glutamic acid has been used by several authors to detect and assay its receptor sites in brain.[13,14] These studies demonstrate clearly the existence of specific glutamate binding sites in brain, but do not afford distinction between presynaptic neuronal uptake sites, glial uptake sites, or postsynaptic binding sites related to the glutamate receptor *sensu stricto*, all sites which do contribute to the observed glutamate binding. This problem was partly overcome with the use of kainic acid.[15,46] This glutamate analogue does not inhibit the uptake of glutamate into neurons or glial cells and does not bind to the active site of glutamate decarboxylase and therefore one may safely assume that it interacts specifically with glutamate binding sites representing either the entire population or a subpopulation of glutamate receptors. A saturable binding of tritiated kainic acid to rat brain membranes was observed with dissociation constants at 2 °C of 4–10 nM and 24–50 nM for high and low affinity binding respectively. The maximum specific kainic acid binding capacity of brain membranes is 1.5 pmol $mg^{-1}$ of protein, and approximately one-third of the binding sites display a high affinity. The association of kainic acid to the high and low affinity sites is non-cooperative (Hill coefficient close to 1.0). Tritiated kainic acid can be displaced from its high affinity binding sites by 23 nM quisqualic acid and 63 nM L-glutamate whereas ten-fold higher concentrations are needed to displace it from its low affinity binding sites.

The binding of L-glutamic acid is stereospecific since D-glutamate is only effective in displacing kainic acid at concentrations higher than 100 $\mu$m. Anticonvulsants agents including phenobarbital, diphenylhydantoin, diazepan, and trimethadione suspected to interact with the glutamate receptor display a negligible potency in displacing bound kainic acid. The same applies to the neuroexcitatory amino acids L- and D-aspartic acid and *N*-methyl-D,L-aspartic acid. The specific binding of kainic acid is low in the thalamus, midbrain, and medulla pons; intermediate in the hippocampus, cerebral cortex, hypothalamus, and cerebellum; and high in the corpus striatum.[15,46]

Since glutamate and its rigid analogues are strong depolarizing agents and cause a change in the postsynaptic membrane permeability to $Na^+$ ions and, to a lesser extent, to $K^+$ ions, it is obvious that the glutamate receptor must be tightly linked to an ionic permeability modulator, i.e. to an ionophore. No biochemical or pharmacological data are available on the glutamate ionophore. However, the possibility of an interaction between glutamate and a purine nucleotide cyclase (adenylate or guanylate) has been raised.[16] Addition of L-glutamate to cerebellum slices produces an increase in both cyclic AMP and cyclic GMP. The level of these cyclic nucleotides in the cerebellum is also increased when putative glutamatergic pathways such as the parallel fibres are stimulated. The effects of kainic acid are essentially similar to those of glutamate but are more pronounced. The stimulation of cyclic nucleotide synthesis by glutamate or kainate is strictly dependent on the presence of $Ca^{2+}$ and of adenosine, and is observed only with the addition of relatively high concentrations of glutamate or of kainate (mM range) to cerebellum slices, but not to homogenates.

The effects of glutamate exclusively observed on slices and in the presence of $Ca^{2+}$—an essential ion in all neurosecretory processes—suggest that the relation between glutamate and cyclic nucleotide levels, most likely results from the activation of a polysynaptic pathway, in which another neurotransmitter is released and upon binding to the postsynaptic membrane activates a nucleotide cyclase.

### (iii) Glutamate in neuropathology

Glutamate is neurotoxic. It is surprising and paradoxical that a metabolite ubiquitously present in the brain at very high concentrations, should constitute by its neurotoxicity, a permanent hazard with serious neuropathological consequences to brain function.

Monosodium glutamate administered subcutaneously to infant animals in which the blood–brain barrier is not fully developed, causes the destruction of neurons in the retina and in the hypothalamus.[17] Large amounts of monosodium glutamate in a meal have been suggested to be the cause of a ‘Chinese restaurant syndrome’. When brain slices are incubated with glutamate, a considerable cell swelling is observed. This effect is specific to glutamate and is accompanied by a measurable increase of $Na^+$ in the slice.

Microinjections of kainic acid in brain produce a rapid degeneration of dendritic and somal structures in the injected area, but spare axons terminating in, or passing through the area.[18] The neurotoxic effects of glutamate and kainate are probably due to their ability to depolarize neurons (excitotoxicity) since the neurotoxicity of other dicarboxylic acids and sulphur-containing amino acids structurally related to glutamate correlates with the potency of their excitatory properties.[19] Necrosis may possibly result from a sustained increase in membrane permeability and concomitant swelling of the depolarized cells.

### *(a) Epilepsy*

A significant loss of glutamate is observed in focal epileptogenic tissue in comparison with its level in the surrounding cortex[20] and histological studies with Golgi staining reveal the presence of dendritic deafferentations in brain from both human and experimental epilepsies. Whether these deafferentations are the result of neuronal death from glutamate excitotoxicity is still an unanswered question. One can observe, however, a release of glutamate from cobalt-induced epileptic foci at the onset of seizures and therefore one cannot rule out the possibility that this massive glutamate release might not have neurotoxic consequences causing permanent damages. In such an event, deafferentations caused by neuronal death could induce a phenomenon of supersensitivity of denervation by which the dendritic surface of the deafferented neurons would be enriched in glutamate receptors. This would then increase the susceptibility of the denervated neuron to glutamate and even low levels of extracellular glutamate could become convulsive.[21]

### *(b) Huntington's chorea*

The neurotoxic properties of kainic acid or glutamic acid have been used to create an experimental model of Huntington's chorea.[22] The stereotaxic intrastriatal injection of kainic acid or glutamic acid duplicates faithfully the neurochemical, histological, and behavioural changes that take place in Huntington's chorea (see Figure 20.7). This disease is an autosomal dominant disorder, characterized by dyskinesia and mental deterioration. Histologically, one observes a widespread loss of neurons in particular in the cerebral cortex and corpus striatum. Neuronal death in the cerebral cortex is believed to cause dementia, whereas the striatal cell loss is held responsible for the chorea. Because of the high binding of tritiated kainic acid to the striatum[15] and the possibility that the massive corticostriatal pathway might be glutamatergic, it has been suggested that Huntington's chorea might result from a chronic overstimulation of striatal glutamate receptors which may mediate and trigger the degenerative process.[22] In this respect it is of interest to mention that the kainate induced degeneration of striatal neurons was found to depend on the integrity of the corticostriatal tract.[47] It is therefore conceivable that the neurotoxicity of kainate results from the synergistic actions of kainate and glutamate (or another presynaptic releasable factor) on the glutamate receptor of the target neuron.

The neurotoxicity of glutamate might be at the origin, or be the propagating vector of other degenerative conditions such as amyotrophic lateral sclerosis, for instance, where a progressive loss of glutamate sensitive motor neurons takes place in the spinal cord and in the motor cortex.

### (iv) The glutamate questions

The question of whether glutamate is a true natural neurotransmitter has not yet been completely and convincingly established. It is quite possible that ultimately glutamate will not fill all the criteria required to identify it as a neurotransmitter and therefore one is left with either one of the two following options: to consider that, after all, glutamate is not a natural neurotransmitter, or to question the validity of the criteria of identification of neurotransmitters when applied to glutamate. Historically, these criteria have been outlined to fit what was then the only known neurotransmitter, acetylcholine, and in the haste to establish the universal rules of neurotransmission, one may have overlooked the fact that uniformity is not necessarily a law of nature. Essentially, two main obstacles block the way to a demonstration of the neurotransmitter role of glutamate. The main one concerns the difference between the reversal potential of iontophoretically applied glutamate and that of the natural neurotransmitter. The second problem is a methodological one resulting from the absence of high affinity glutamate antagonists. It is indeed essential to be able to demonstrate an antagonism common to the natural neurotransmitter as well as to iontophoretically applied glutamate. There are also other questions that deserve consideration, such as the mechanism of regulation of the putative neurotransmitter pool of glutamate in presynaptic endings or the role of the non-synaptic release of glutamate. The existence of a non-synaptic release of glutamate indeed creates a paradoxical situation, in which the stimulation of a given neuron could induce the synaptic release of an inhibitory neurotransmitter (GABA for instance) together with the non-synaptic leakage of an excitatory agent (glutamate). This observation seems to be an obvious departure from Dale's law which states that each neuron can use only one neurotransmitter in its intercellular transactions. However, one should realize that the message propagated through synaptic transmission is not necessarily lost because of the existence of competing extrasynaptic events. The latter events could modulate synaptic transmission and therefore enrich the information potential of a given neuronal network. One might therefore, envisage a dual role for glutamate, one in synaptic transmission and another in synaptic modulation. If so, the question arises as to what the biochemical and pharmacological properties of the glutamate receptors involved in these two processes are. Are the glutamate receptors localized exclusively under the nerve terminals or are they distributed in a diffuse fashion over the entire neuronal surface? Whether the glutamate excitotoxicity is related to a release of glutamate from non-synaptic regions and/or to its interactions with extrasynaptic rather than subsynaptic receptors is another important question. Is glutamate responsible for the pathological process observed in Huntington's chorea and in other degenerative conditions?

The answers to these questions are a most serious and difficult challenge, but there is no doubt that they hold the key to fundamental aspects of brain function and dysfunction.

## III. GLYCINE

### (i) Glycine as a neurotransmitter

It may have been daring to speculate several years ago that glycine, the simplest amino acid, might have, in addition to its many metabolic functions, a role in neurotransmission,[23] but today the proposition is largely accepted and there are no more doubts that glycine is indeed an inhibitory neurotransmitter in the spinal cord. Several factors have contributed to this unambiguous assignment.

#### *(a) Electrophysiological data*

Iontophoretic applications of glycine elicit hyperpolarizing responses of neurons in the spinal cord and in the brainstem, but not in higher regions of the neuraxis. The glycine-induced hyperpolarization of spinal motor neurons is accompanied by an increase in the membrane conductance to chloride ions, which is identical to that produced by the natural neurotransmitter released upon stimulation of inhibitory inputs to the motor neurons. If the anionic environment of the motor neuron is altered by intracellular injections of $Br^-$ or $I^-$ anions, then an identical depolarizing response is observed with both glycine and the natural neurotransmitter. Considering that glycine can also saturate the receptor sites normally available to the natural inhibitory neurotransmitter, one is led to conclude that glycine and the natural neurotransmitter act in an identical manner on the postsynaptic membrane. This series of observations constitutes the evidence required to fulfil the third criterion of identification of neurotransmitters as outlined in Table 20.1.

### *(b) Glycine distribution*

Glycine is widely distributed throughout the central nervous system, but its concentration varies from region to region, increasing along a rostrocaudal axis. In the cat cerebrum its concentration is only 1.28 $\mu$mol $g^{-1}$ whereas in the cat spinal cord it reaches 4.42 $\mu$mol $g^{-1}$. Also within the spinal cord, one observes marked variations of glycine concentrations along the dorsoventral axis, the ventral grey matter containing 7.08 $\mu$mol $g^{-1}$ of glycine compared with 5.65 $\mu$mol $g^{-1}$ in the dorsal grey matter. This peculiar glycine distribution in the spinal cord can be unambiguously associated with the presence of interneurons in the grey matter that mediate inhibitory inputs from the dorsal roots to neurons in the ventral horn.[23]

### *(c) Uptake studies*

The proposition that a pool of free glycine can be directly related to the presence of inhibitory interneurons is corroborated by the existence of high affinity $Na^+$-dependent uptake systems for glycine which display the same rostrocaudal and dorsoventral distribution patterns characteristic of free glycine and of the neuronal sensitivity to iontophoretically applied glycine. Nerve terminals represent the major sites of uptake in the spinal cord.

### *(d) Release studies*

Electrical stimulation or high $K^+$ ion concentrations can induce a specific release of glycine from slices of spinal cord, but not, as expected, from slices of cerebral cortex. There is still some debate as to whether the release of glycine is strictly dependent on $Ca^{2+}$ ions.

### *(e) Strychnine blocks the physiological effects of glycine*

Strychnine (Figure 20.3) has been utilized as a poison for rodents since its introduction from India to Europe in the sixteenth century. This vegetable alkaloid, derived from the seeds of *Strychnos nux vomica*, has been part of the pharmacopoeia for a long time, but it is only more recently that its blocking effect on central inhibition could be related to its specific inhibitory action on the hyperpolarization response triggered by glycine.

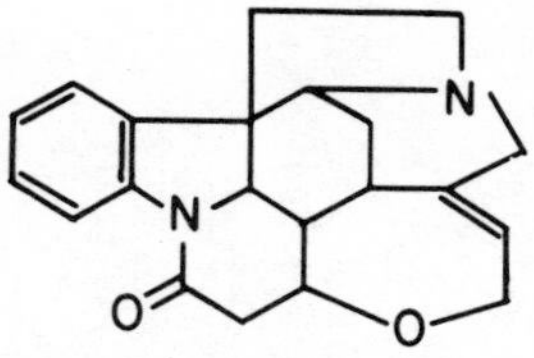

STRYCHNINE

$NH_2-CH_2-COOH$

GLYCINE

Figure 20.3. The glycinergic ligands

Strychnine is a potent convulsant, since it reduces or abolishes the central control on muscle activity including, for instance, the reciprocal inhibition existing between antagonist muscles.

### (ii) The glycine receptor

The confrontation of the data available on glycine with the requirements for its identification as a neurotransmitter reveals that glycine has many, if not all the properties of the natural inhibitory neurotransmitter and therefore one might expect that glycine possesses a specific receptor in the spinal cord. The specific blocking actions of strychnine on the hyperpolarizing response of spinal neurons to iontophoretically applied glycine has served as a basis for a biochemical characterization of the glycine receptors in the central nervous system. Radioactive strychnine has been found to bind selectively to synaptic membrane fractions of the spinal cord and this binding can be prevented by glycine.[25,26]

The regional distribution of specific strychnine binding in the central nervous system correlated closely with the high affinity $Na^+$-dependent uptake for glycine, the endogenous free pool of glycine and the ability of glycine to mimic natural inhibitory transmission. The binding is high in the spinal cord and the medulla oblongata-pons, intermediate in the midbrain, hypothalamus, and thalamus, and low in the cerebellum, hippocampus, corpus striatum, and cerebral cortex. The binding of strychnine to membrane fractions of the spinal cord or of the brainstem is saturable with an affinity of 2.6–4 nM. Since strychnine has very little affinity for the glycine uptake site, the observed binding of strychnine can be unambiguously associated with postsynaptic glycine receptors. Analysis of

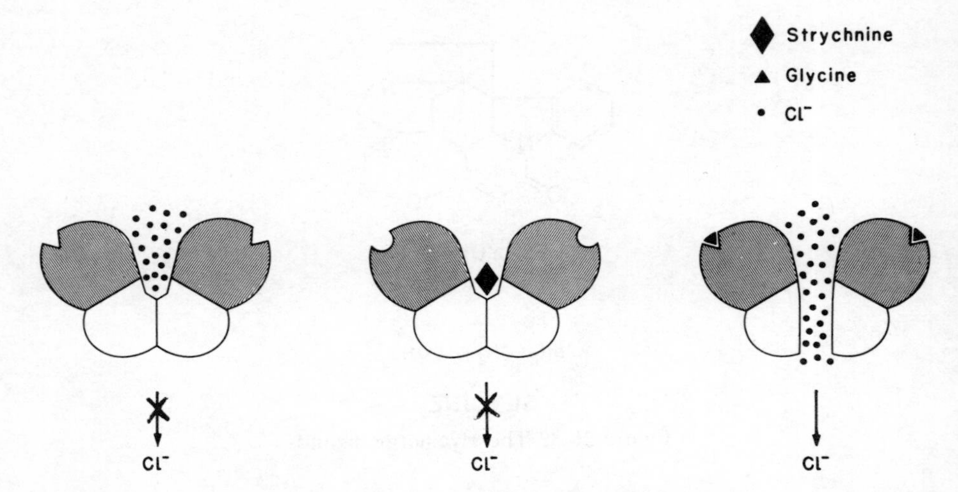


Figure 20.4. A model of the glycine receptor. Strychnine binds at the top of the chloride channel and prevents the binding of glycine

the binding data by double reciprocal and Scatchard plots indicate that strychnine binds with a single high affinity to a population of independent binding sites. The maximum specific strychnine binding capacity is about 1.8 pmol $mg^{-1}$ of crude synaptic membrane protein.

Glycine displaces strychnine binding with half maximal displacement at 25 $\mu$m. The binding affinity of strychnine is therefore four orders of magnitude greater than that of glycine. These two compounds appear, however, to bind to distinct, but mutually interacting portions of the receptor. Indeed, some protein reagents differentially alter the ability of glycine and non-radioactive strychnine to compete with radioactive strychnine binding. Treatment with 2.5 mM diazonium tetrazole for instance, prevents glycine from displacing radioactive strychnine, but does not affect the ability of non-radioactive strychnine from displacing the labelled strychnine. It is likely that diazonium tetrazole modifies sites closely related or identical to the glycine binding sites and does not affect strychnine binding sites, since only glycine, and not strychnine, can protect against the effects of this modifying reagent.

Glycine probably interacts with the glycine recognition site, whereas strychnine should bind to the glycine-related ionophore. This hypothesis is based essentially on the cooperative interactions of glycine with the strychnine binding sites.[26] Glycine displaces strychnine from its binding sites in a cooperative fashion (Hill coefficient 1.7) whereas the binding of strychnine appears to be strictly non-cooperative. The cooperative interactions of the glycine and strychnine binding sites can be altered by protein-modifying reagents, pH changes and denaturants which decrease the Hill coefficient from 1.7 to 0.7–1.2 without affecting the inhibitory effect of glycine on strychnine binding. Chloride ions and other anions which can permeate the subsynaptic membrane upon application of glycine, have also been found to inhibit strychnine binding, but with a Hill coefficient of 2.3–2.7. It appears, therefore that the glycine receptor is composed of at least two functional entities. A glycine recognition component (the glycine receptor site *sensu stricto*) and a glycine related ionophore component. The glycine recognition component is made of at least two subunits (probably four)[38] mutually cooperating in the binding of glycine and in the activation of the glycine-related ionophore. The latter entity is also composed of several subunits cooperating in the binding of permeable anions. Strychnine binds to a site distinct from the glycine binding site as well as from the sites occupied by chloride ions. Strychnine could interact for instance with a site at the exterior of a chloride channel from whence it would be displaced only when several chloride anions occupy the interior of the channel.[27] (Figure 20.4.)

### (iii) The glycine questions

The large body of electrophysiological, pharmacological, and biochemical data that was collected over the years to characterize the glycine receptor has been entirely based on a pair of compounds: glycine and strychnine. No other glycinergic ligand, agonist or antagonist has been found, and as a result of the absence of a glycinergic pharmacopoeia, the biochemical and pharmacological investigations of the glycine receptor have been brought to a premature stand.

Nothing is known, for example, about the molecular structure of the glycine binding sites. These could only be investigated through the use of various glycine analogues. The same could be said about the strychnine binding sites. Evidently the glycine receptor could be best studied from a biochemical viewpoint in a soluble form. However, detergents at concentrations that solubilize membrane proteins markedly affect strychnine binding and in the absence of a specific binding assay of the glycine receptor in its soluble form, only very limited progress can be made.

Although glycinergic nerve endings can be revealed by electron microscopic autoradiography, and postsynaptic glycine receptors be detected by iontophoretic methods, a direct visualization of glycine receptor containing neurons by the use of fluorescent glycinergic ligands would greatly facilitate a mapping of glycine sensitive neurons in the central nervous system. Such ligands could also be of great use in screening within the animal kingdom for sources of glycine receptors. Unless a reasonably rich source of glycine receptor is discovered, very little biochemical progress can be expected.

## IV. γ-AMINOBUTYRIC ACID

### (i) γ-Aminobutyric acid as a neurotransmitter

The wealth of information that has been accumulated over the last 30 years since its discovery establishes γ-aminobutyric acid (GABA) as one of the best understood neurotransmitters of the mammalian central nervous system. The contention today is not longer centred around the question of whether GABA indeed fulfils all the criteria required to identify it as a

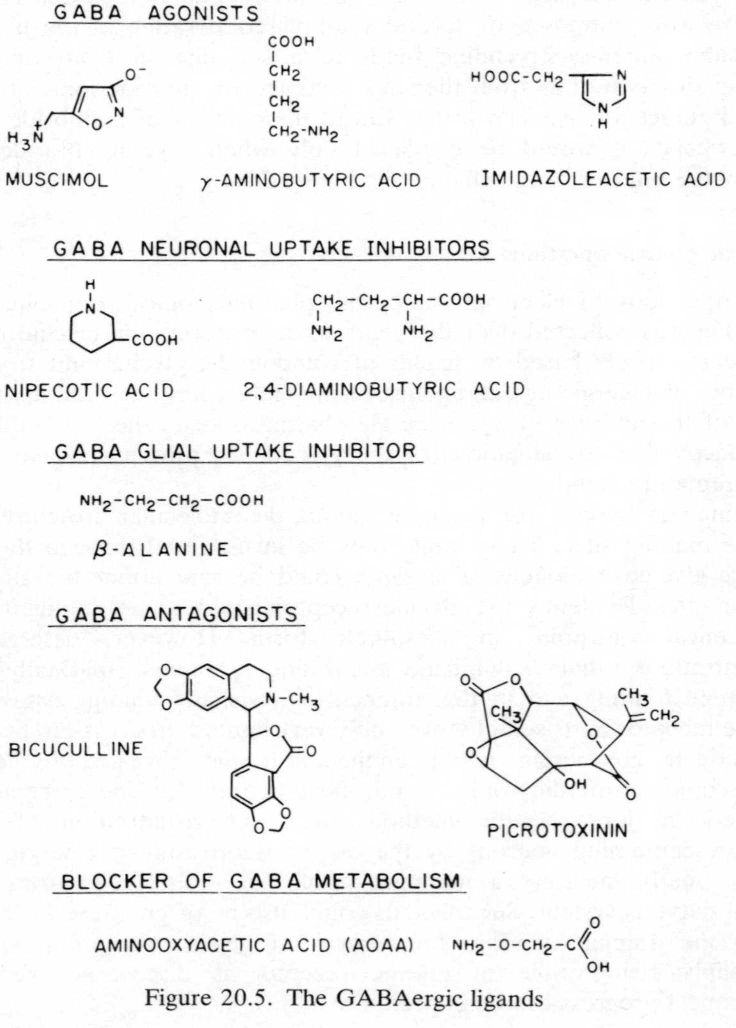

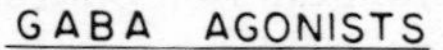


Figure 20.5. The GABAergic ligands

neurotransmitter, but rather addresses itself to the precise role of GABA-releasing neurons in the structure and function of specific neuronal networks, or to the involvement of GABA in the aetiology of neurological and mental disorders.

GABA is probably the major inhibitory neurotransmitter in the central nervous system. In contradistinction to glycine, which exerts its inhibitory action almost exclusively in the spinal cord, GABA mediates neuronal inhibition throughout the various brain regions.

The best way to describe the GABA function is probably through a biochemical and pharmacological dissection of the GABAergic synapse. The availability of a variety of drugs interacting with this synapse (Figure 20.5) has greatly contributed to the understanding of its mode of action, although the specificity of some of these drugs has been and still is a matter of controversy.

The GABAergic synapse is amenable to pharmacological manipulations at several levels. It can be regulated through the synthesis or the degradation of GABA and can be affected by interfering either with the mechanism of synaptic release or with the mechanisms of termination of the GABA action, i.e. the GABA uptake systems into presynaptic nerve terminals and into glial cells. Finally, it is possible to modulate the postsynaptic response to GABA.

### *(a) Synthesis and degradation of GABA*

$\gamma$-Aminobutyric acid is synthesized almost exclusively by decarboxylation of L-glutamic acid, a reaction catalyzed by the enzyme glutamic acid decarboxylase (GAD) (see Figure 20.6). Pyridoxal phosphate is an essential cofactor of GAD. Hydrazides can react with the aldehyde group of pyridoxal phosphate and inhibit GAD activity by decreasing the available levels of pyridoxal phosphate. The convulsant effects of hydrazides can possibly be related to the decrease in GABA levels but may also be due to interference with GABA catabolism. Indeed, GABA, can be degraded by a transamination reaction with $\alpha$-ketoglutarate catalysed by GABA-$\alpha$-ketoglutarate transaminase (GABA-T). Pyridoxal phosphate is an obligatory cofactor of GABA-T and its affinity to this enzyme is much higher than to GAD. This explains the preferential effect of hydrazides on GAD rather than on GABA-T. Both enzymes can be inhibited *in vitro* by amino-oxyacetic acid but only GABA-T seems to be inhibited *in vivo*.

Intraperitoneal administration of amino-oxyacetic acid produces a large increase in cortical endogenous levels of GABA as well as a reduction in the firing rate of cortical neurons establishing a plausible correlation between GABA levels and cortical inhibition.[28]

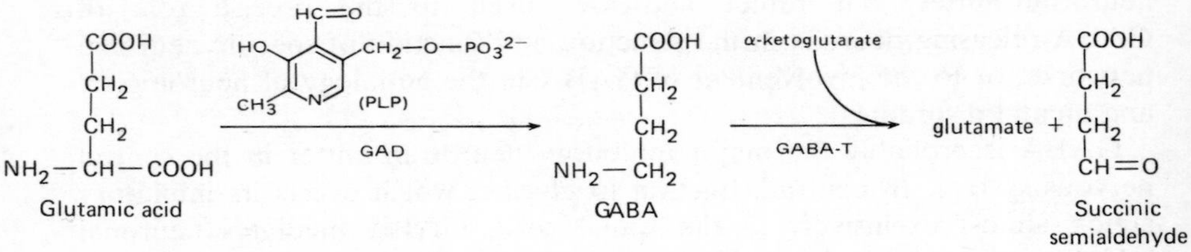

$$HOOC-CH_2-CH_2-CH(NH_2)-COOH \xrightarrow[\text{GAD}]{\text{PLP}} HOOC-CH_2-CH_2-CH_2-NH_2 \xrightarrow[\text{GABA-T}]{\alpha\text{-ketoglutarate}} \text{glutamate} + HOOC-CH_2-CH_2-CH{=}O$$

Glutamic acid → GABA → Succinic semialdehyde

(PLP: HC=O, HO, $CH_2-O-PO_3^{2-}$, $CH_3$, N)

Figure 20.6.

### *(b) Synaptic release*

GABA is released from central nervous system synaptosomes by a $Ca^{2+}$-dependent process. Although a variety of psychoactive drugs such as chlorpromazine, imipramine, and diazepam have been found to inhibit $Ca^{2+}$-stimulated GABA release, it appears, however, that this effect is not entirely specific to the GABAergic synapse. Since these drugs equally affect the catecholamine release, it is probable that they interfere either with $Ca^{2+}$ permeability or with the exocytotic process.[29]

### *(c) GABA uptake into nerve terminals and glial cells*

The termination of GABA action does not take place via a diffusion of GABA out of the synaptic gap or by an enzymic inactivation reaction, but by very efficient uptake mechanisms into nerve terminals and glial cells. The existence of two separate uptake mechanisms has been inferred by uptake studies of radioactive GABA into brain tissue and into dorsal root ganglia. The latter represent a useful model tissue since ganglia contain only large sensory neurons surrounded by satellite glial cells with no synaptic terminals. GABA is taken up in sensory ganglia only by the satellite glial cells and this uptake can be blocked by $\beta$-alanine and to some extent by 2,4-diaminobutyric acid (DABA). When brain slices are used, the GABA neuronal uptake is fast and is more sensitive to the inhibitory effect of DABA than of $\beta$-alanine.[30] (Figure 20.5.)

In brain synaptosomes, the GABA uptake system is $Na^+$-dependent, and has an apparent $K_m$ of 20 $\mu$m. The glial GABA transport systems has a very similar affinity. Although DABA and $\beta$-alanine can be considered as specific inhibitors of GABA uptake they are effective only at relatively high concentration (50–120 $\mu$M).[30] Investigation of GABA analogues has led to the finding that nipecotic acid (piperidine-3-carboxylic acid) is a potent, non-competitive inhibitor of GABA uptake ($K_m$ = 11 $\mu$M) (Figure 20.5). The uptake of nipecotic acid by slices of cerebral cortex is $Na^+$-dependent and is inhibited by GABA and by substances that can also inhibit the uptake of GABA. Nipecotic acid can release preloaded GABA from brain slices and GABA can release preloaded nipecotic acid. These results indicate that GABA and nipecotic acid can be counter-transported using the same mobile carrier.[31]

*(d) GABA inhibitory actions*

Applied to postsynaptic regions of neurons, GABA produces an increase of the membrane permeability to $Cl^-$ ions. If the equilibrium potential of $Cl^-$ is very close to the resting potential of the neuron, the increase in membrane conductance to $Cl^-$ does not modify the membrane potential that stays near its resting level. If the equilibrium potential of $Cl^-$ ions is more negative than the resting potential of the neuron, the application of GABA will lead to a hyperpolarization of the membrane. In both cases, the sensitivity of the neuron to depolarizing stimuli decreases.

The GABA-induced hyperpolarization is reversed if the anionic environment of the cell is altered by intracellular injections of chloride ions, which indicates that GABA normally produces an inward $Cl^-$ flux. There is also evidence, however, that GABA depolarizes primary afferent terminals in the spinal cord and is responsible for the resulting reduction in excitatory transmission. The GABA mediated synaptic inhibition involves an increase in $Cl^-$ permeability, but since the $Cl^-$ equilibrium potential in the primary afferent neuron is probably less negative than the resting potential, the increase in $Cl^-$ conductance and the resulting depolarization are accompanied by an outward $Cl^-$ flux. Picrotoxin and bicuculline, two potent convulsant alkaloids, block the presynaptic inhibition. Both compounds are GABA antagonists.[32]

**(ii) The GABA receptor**

*(a) Binding studies*

The existence of a postsynaptic GABA receptor has been investigated by measuring the binding of [$^3$H]GABA to crude synaptic membrane of the rat central nervous system.[33,34] When binding of [$^3$H]GABA is measured in an incubation medium containing $Na^+$, one observes a saturable class of binding sites for [$^3$H]GABA with $K_D$ of 1.2 $\mu$M. The apparent number of binding sites is 30 ± 2 pmol $mg^{-1}$ of protein. These sites are sensitive to various drugs which are known to interfere with GABA uptake mechanism. In particular, $\beta$-alanine 2,4-diaminobutyric acid and chlorpromazine markedly reduce the amount of $Na^+$-dependent GABA binding. Other experimental observations are consistent with the idea that the $Na^+$-dependent binding is not related to the interaction of GABA with postsynaptic receptors but associated with glial and neuronal uptake sites. The $Na^+$-dependent binding is a time-dependent phenomenon proceeding with a rate constant of association of $6 \times 10^{-7}$ $M^{-1}$ $min^{-1}$. This binding is also extremely sensitive to treatments such as freezing and thawing, or to slight pH changes around 7.0. Since the early studies of GABA binding were carried out in a $Na^+$-containing medium, caution should be exerted in relating these results to interactions of GABA with postsynaptic receptors. Specific GABA binding to the GABA receptor can only be studied after the exhaustive removal of $Na^+$ or after destruction of the GABA uptake compartments by osmotic shocks, cycles of freezing and thawing, and also by treatment of the synaptic membranes with the detergent Triton X-100.[35] Unless such treatments are made, the contribution of the GABA uptake compartments to GABA binding masks that of the GABA receptor. When GABA binding is studied on synaptic membranes in the absence of $Na^+$ one observes a saturable class of GABA binding sites with a $K_D$ of ~0.4 μM. In the retina, however, two distinct components of GABA binding are found: the binding site of high affinity displays a $K_D$ of ~18 nM, whereas the binding site of low affinity has a $K_D$ of ~220 nM.[33] This $Na^+$-independent GABA binding undoubtedly involves an interaction with the GABA receptor since there exists a close correlation between the ability of amino acids such as imidazoleacetic acid and 3-aminopropane sulphonic acid to compete for $Na^+$-independent GABA binding and their neurophysiologic ability to mimic the synaptic effects of GABA.

When GABA binding is studied on synaptic membranes treated with low concentrations of Triton X-100, a heterogenous population of binding sites is also observed. A lower affinity component with a $K_D$ for GABA of approximately 130 nM (which value is close to that found in the absence of Triton X-100) and a high affinity component with a $K_D$ of 16 nM. The high and low affinity sodium-independent GABA binding components are sensitive to various GABAergic ligands but not to GABA uptake inhibitors, and therefore it is suggested that those two affinity states correspond to separate conformations of the same GABA receptor. It appears that the Triton X-100 treatment of synaptic membranes markedly increases the affinity of GABAergic agonists but does not affect the affinity of GABA antagonists such as bicuculline.[35] This situation is very reminiscent of that observed upon treatment of acetylcholine receptor-rich membrane fragments with detergents and might therefore be the expression of a detergent-induced transition of the receptor to a desensitized state. The nature of the interaction of bicuculline with the GABA receptor is still a matter of controversy. At the crayfish neuromuscular junction, bicuculline behaves as a non-competitive inhibitor of GABA, whereas in the vertebrate central nervous system, convincing electrophysiological data indicate a competitive antagonism between bicuculline and GABA; it is conceivable that the GABA receptor in the vertebrates differs from that of invertebrates. From the data of displacement of radioactive GABA by bicuculline, it appears that the two agents interact at the postsynaptic GABA receptor in a competitive fashion and that their interactions are not cooperative. However, the GABA and bicuculline binding sites might not completely overlap; anions such as thiocyanate, iodide, and nitrate are found to increase by ten-fold the potency of bicuculline to displace GABA from its synaptic membrane binding sites, whereas they do not affect the potency of GABAergic agonists to compete with GABA binding.

The effects of picrotoxin are also a matter of debate. In invertebrates picrotoxin behaves as a non-competitive inhibitor of GABA but in vertebrates conflicting reports on the effectiveness of picrotoxin in blocking the depressant action of GABA prevent any definite conclusion. Picrotoxin does not displace GABA from its binding sites on brain synaptic membranes, nor does it inhibit GABA uptake, biosynthesis or release, but it does block the GABA-induced chloride flux in crayfish muscle. It is therefore conceivable that picrotoxin interacts with a site on the GABA ionophore.[37]

Tritiated dihydropicrotoxinin, which has a pharmacological activity similar to picrotoxin has recently been found to bind to particulate fractions of rat brain homogenates with an apparent $K_D$ of 1 $\mu$M. The density of dihydropicrotoxinin binding sites is about 3–4 pmol $mg^{-1}$ of protein which approximately equals the density of GABA receptor sites. The binding of dihydropicrotoxinin is not affected by GABA or by muscimol, a potent GABAergic agonist, but is strongly inhibited in a competitive fashion by barbiturates.[48] This interaction of barbiturates with dihydropicrotoxinin binding sites is consistent with neurophysiological data suggesting that the depressant or convulsant action of barbiturates involves the modulation of inhibitory synaptic transmission at the level of GABA receptor-ionophores.

From the data on GABA binding to synaptic membranes it appears that the binding of GABA to its receptor does not involve cooperative interactions. However, measurements of the membrane conductance increases produced in cat motor neurons by GABA indicate that the activation of the GABA receptor has a high positive cooperativity, and requires four molecules of GABA.[38] It is probable therefore, that the GABA receptor is composed of at least four GABA receptor sites that need to be fully occupied in order to trigger the conformation change of the GABA ionophore that leads to the opening of the chloride channel.

### *(b) Regional distribution of GABA receptors in brain*

Several histological methods have been used to trace GABAergic neuronal networks in the mammalian central nervous system. They are all based on the visualization of specific markers of the GABA neuron and therefore provide information on the localization of presynaptic GABAergic elements and not of postsynaptic specializations involved in GABA neurotransmission. Uptake of [$^3$H]GABA into brain tissues followed by light and electron microscopic autoradiographic studies has been an extensively used technique. Another autoradiographic method consists of injecting [$^{14}$C]thiosemicarbazide, a substance capable of trapping pyridoxal phosphate and displacing it from its binding site on glutamate decarboxylase (GAD). An immunocytochemical technique to visualize GAD and GABA-transaminase (GABA-T) is also available. In this case, use is made of specific rabbit anti-GAD (or anti GABA-T) antibodies that are incubated with the neural tissue before being revealed by goat antibodies coupled to horseradish peroxidase. So far, the regional distribution of GABA receptors can be studied be electrophysiological methods or biochemically, by using for assay purposes the $Na^+$-independent binding of GABA to synaptic membranes isolated from various central nervous system regions. This GABA binding to postsynaptic receptors can be correlated with endogenous GABA levels, synaptosomal GABA uptake and GAD activity. In monkey brain a satisfactory correlation is obtained, whereas discrepancies are observed with rat brain.[39] The poor correlation is explained by the fact that the surface of the postsynaptic membrane and its receptor density may not always match the volume or transmitter content of presynaptic boutons. In the monkey brain, the highest density of GABA receptors is found in the extrapyramidal areas (caudate, putamen, globus pallidus) whereas the cerebellum is the richest area in the rat brain. In both cases, the lowest density of GABA receptors is found in the spinal cord.

Because of the poor resolution of the biochemical localization method, it is not possible in most cases to identify the neurons bearing the GABA receptors. However, evidence for the presence of GABA receptors on cerebellar granule cells has been obtained by measuring the cerebellar level of receptors before and after a viral-induced selective depletion of granule cells in hamsters.[40]

### *(c) GABA receptors in neuropathology*

The normal function of neuronal networks involves an interplay of excitatory and inhibitory processes. Therefore it is expected that any metabolic or structural defect weakening the harmony of these interrelations will lead to dysfunction and often to disease.

GABA has been suspected as playing a role in several neuropathological conditions particularly since central inhibition is almost exclusively carried out by this neurotransmitter. However, in no instance was it possible to establish a causal relationship between the deficiency of a specific

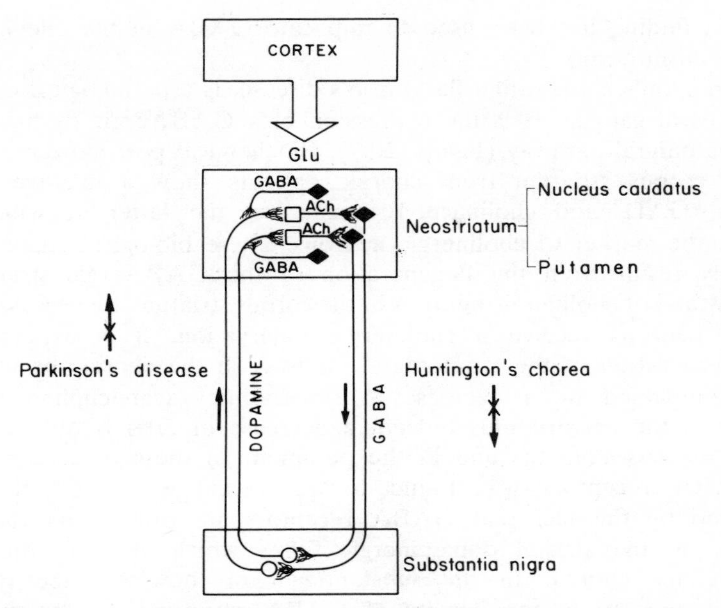


Figure 20.7. Scheme of the corticostriatal pathway and of the striatonigral relationship. The dopaminergic nigostriatal pathway is affected in Parkinsonism. In Huntington's chorea, a degeneration of cholinergic and GABAergic neurons is observed in the corpus striatum. This neuronal degeneration could be related to glutamate neurotoxicity

GABAergic component and pathogenesis. A great number of GABA neurons seem to be present in the extrapyramidal areas (basal ganglia: nucleus caudatus, putamen, globus pallidus) and therefore the role of GABA in extrapyramidal disorders such as Parkinson's disease and Huntington's chorea has been the subject of intensive investigations. Parkinson's disease is the pathological expression of a dopaminergic dysfunction of the nigrostriatal pathway (Figure 20.7). Post-mortem analyses of the corpus striatum and substantia nigra of Parkinsonian patients indicate a massive loss of dopamine which results from a degeneration of dopaminergic cell bodies in the substantia nigra. Reduced levels of GAD has also been reported in both the striatum and the substantia nigra, and it is possible that this loss follows a transynaptic degeneration of GABAergic striatal neurons secondary to the removal of the inhibitory action of dopamine on cholinergic neurons. The L-dopa theory ameliorates Parkinsonism by re-establishing the dopaminergic inhibition of the striatum. In long term therapies, a normal and even elevated GAD striatal activity is observed and this finding has been used to implicate GABA in the alleviation of Parkinsonism tremor.[41]

Huntington's chorea, like Parkinson's disease, is a pathological condition of the basal ganglia. It is the expression of a GABAergic dysfunction of the striatonigral pathway (Figure 20.7). Biochemical post-mortem analyses of the corpus striatum from choreic patients show a massive loss of GABA, GAD, and cholinacetyl transferase, the latter of which is a presynaptic marker of cholinergic neurons. These biological changes most probably result from the degeneration of the GABAergic striatonigral pathway and of cholinergic neurons in the corpus striatum. Since GABAergic striatal neurons receive a cholinergic innervation, it is expected that the degeneration of the striatonigral pathway in the choreic brain should be accompanied by a decrease in (muscarinic) acetylcholine receptor density in the neostriatum. Indeed, a decrease of 50% is observed both in the nucleus caudatus and in the putamen. In these areas the density of GABA receptors was found to be unchanged.[42] This finding is explained by the fact that GABA receptors are present on the nerve endings of nigrostriatal dopaminergic fibres which are not affected in Huntington's chorea. In the substantia nigra, however, the observed two-fold increase in the density of GABA receptors can be explained either by a shrinkage of the volume of the substantia nigra or by a phenomenon of denervation supersensitivity to GABA, secondary to the degeneration of the striatonigral pathway.

As stated above (Section II.iii.b), it is suspected that the massive corticostriatal glutamatergic pathway is involved in the aetiology of Huntington's chorea. Intrastriatal injections of kainic acid cause the destruction of the cholinergic as well as GABAergic neurons and this cellular death finds its expression in the massive decrease in levels of both cholinergic markers (cholinacetyl transferase, acetylcholine, choline uptake, and muscarinic receptor) and GABAergic neuronal markers (GAD, GABA, and GABA uptake). In contrast, the GABA receptor levels are unaffected, although the affinity of these receptors for GABA is altered.[43] Instead of displaying high and low affinity components, the kainate-lesioned striata reveal a single GABA binding component with an intermediate affinity. Since the low affinity GABA binding sites seem to be more abundant in the normal striatum than the high affinity ones, it is considered that the low affinity sites generate the GABA binding sites of intermediate affinity and that the kainate lesions lead to an effective increase in affinity which could represent a form of denervation supersensitivity. This phenomenon is by no means similar to the denervation supersensitivity of the neuromuscular junction, as it is not characterized by an increase in the total number of receptors.

### (iii) The GABA questions

Advances in the study of neurotransmitter receptors are linked with the development of specific pharmacopoeias. A great variety of drugs are available for interaction with the components of the GABAergic synapse, but very few with high binding affinities are known to affect the GABA receptor. Progress in the understanding of the structure and function of the GABA receptor will depend on highly specific drugs with high binding affinities amenable to radioactive labelling with high specific radioactivity. In this respect, binding studies with antagonists will be essential to establish the nature of their regulatory role.

Affinity labels could be of great practical interest for the isolation and characterization of the GABA receptor. So far, one puzzling and unexpected aspect of the GABA receptor studies is the observation that the receptor affinity for GABA increases upon denervation. Although preliminary, this finding deserves further consideration as it may underlie a regulatory mechanism of denervation supersensitivity different from that affecting the acetylcholine receptor at the neuromuscular junction. Another point of interest concerns the physiological role filled by the high and low affinity states of the GABA receptor. Do they represent conformations of the receptor in resting and active states? What is the affinity for GABA, displayed by desensitized GABA receptors?

Among other aspects of the GABA problem we note the question of the localization of GABA receptors not at a tissue level but at a neuronal level. There can be no doubt that the unambiguous association of GABA receptors with specific neurons will enlighten the intimate function of various neuronal circuits and will have crucial implications in physiopathology. Obviously the final aim beyond the understanding of the nature and role of GABAergic synapse should be to fully control their function. GABA is involved in various degenerative diseases of the central nervous system such as Parkinsonism and Huntington's chorea, but it might also play a role in the aetiology or the evolution of psychosomatic disorders or of pathological conditions such as schizophrenia, epilepsy, and depression. Future advances in therapeutics will depend on our ability to manipulate separately and exclusively each component involved in GABA transmission. At this stage, the synthesis of new GABAergic ligands may be a most urgent task.

## ACKNOWLEDGEMENTS

The author acknowledges the support of the French Délégation Générale à la Recherche Scientifique et Technique.

## V. REFERENCES


1. Johnson, J. L. *Brain Res.,* **37**, 1–19, (1972).
2. Curtis, D. R., Duggan, A. W., Felix, D., Johnston, G. A. R., Tebeus, A. K., and Watkins, J. C. *Brain Res.,* **41**, 283–301, (1972).
3. Curtis, D. R., and Watkins, J. C. *Pharmacol. Rev.,* **17**, 347–391, (1965).
4. Roberts, P. J. *Brain Res.,* **67**, 419–428, (1974).
5. Weinreich, D., and Hammerschlag, R. *Brain Res.,* **84**, 137–142, (1975).
6. Yamamoto, C., Yamashita, H., and Chujo, T. *Nature (London),* **262**, 786–787, (1976).
7. Curtis, D. R., Phillis, J. W., and Watkins, J. C. *J. Physiol.,* **150**, 656–682, (1960).
8. Roberts, P. J., and Watkins, J. C. *Brain Res.,* **85**, 120–125, (1975).
9. Biscoe, T. J., Evans, R. H., Francis, A. A., Martin, M. R., Watkins, J. C., Davis, J., and Dray, A. *Nature (London),* **270**, 743–745, (1977).
10. Hammerschlag, R., and Weinreich, D. *Adv. Biochem. Psychopharmacol.* **6**, 165–180, (1972).
11. Shinozaki, M., and Konishi, S. *Brain Res.,* **24**, 363–371, (1970).
12. Biscoe, T. J., Evans, R. H., Headley, P. M., Martin, M. R., and Watkins, J. C. *Br. J. Pharmacol.,* **58**, 373–382, (1976).
13. Roberts, P. J. *Nature (London),* **252**, 399–401, (1974).
14. Michaelis, E. K., Michaelis, M. L., and Boyarsky, L. L. *Biochim. Biophys. Acta,* **367**, 338–348, (1974).
15. Simon, J.-R., Contrera, J. F., and Kuhar, M. J. *J. Neurochem.,* **26**, 141–147, (1976).
16. Schmidt, M. G., Thornberry, J. F., and Molloy, B. B. *Brain Res.,* **121**, 182–189, (1977) (and references therein).
17. Olney, J. W. *J. Neuropath. Expl. Neurol.,* **28**, 455–474, (1969).
18. Olney, J. W., Sharpe, L. G., and deGubareff, T. *Neurosci. Abstr.,* **5**, 371 (1975).
19. Olney, J. W., Rhee, V., and Ho, O. L. *Expl. Brain Res.,* **14**, 61–76, (1972).
20. Tower, D. B. *The Neurochemistry of Epilepsy,* Charles C. Thomas, Springfield, Illinois, (1960).
21. Bradford, H. F. *Biochemistry and Neurology*, (Bradford, H. F., and Marsden, C. D., eds.), Academic Press, New York, Ch. 16, pp. 195–212, (1976).
22. Coyle, J. T., Schwarcz, R., Bennett, J. P., and Campochiaro, P. *Prog. Neuro-Psychopharmac.,* **1**, 13–30, (1977).
23. Aprison, M. H., Davidoff, R. A., and Werman, R. *Handbook of Neurochemistry*, Plenum Press, New York, Vol. 3, pp. 381–397, (1970).
24. Aprison, M. H., Daly, E. C., Shank, R. P., and McBride, W. J. in *Metabolic Compartmentation and Neurotransmission. Relation to Brain Structure and Functions*. (S. Berl, Clarke, D. D., and Schneider, D., eds.), Plenum Press, New York, pp. 37–63, (1974).
25. Young, A. B., and Snyder, S. H. *Proc. Natl. Acad. Sci. U.S.A.,* **70**, 2832–2836, (1973).
26. Young, A. B., and Snyder, S. H. *Mol. Pharmacol.,* **10**, 790–809, (1974).
27. Young, A. B., and Snyder, S. H. *Proc. Natl. Acad. Sci. U.S.A.,* **71**, 4002–4005, (1974).
28. Gottesfeld, A., Kelly, J. S., and Renaud, L. P. *Brain Res.,* **42**, 319–335, (1972).
29. Olsen, R. W., Lamar, E. E., and Bayless, J. D. *J. Neurochem.,* **28**, 299–305, (1977).

30. Iversen, L. L., Dick, F., Kelly, J. S., and Schon, F. in *Metabolic Compartmentation and Neurotransmission. Relation to Brain Structure and Function*, (Berl, S., Clarke, D. D., and Schneider, D., eds.), Plenum Press, New York, pp. 65–89, (1974).
31. Johnston, G. A. R., Stephanson, A. L., and Twitchin, B. *J. Neurochem.*, **26**, 83–87, (1976).
32. Curtis, D. R., Duggan, A. W., Felix, D., and Johnston, G. A. R: *Brain Res.*, **32**, 69–96, (1971).
33. Zukin, S. R., Young, A. B., and Snyder, S. H. *Proc. Natl. Acad. Sci. U.S.A.*, **71**, 4802–4807, (1974).
34. Enna, S. J., and Snyder, S. H. *Brain Res.*, **100**, 81–97, (1975).
35. Enna, S. J., and Snyder, S. H. *Mol. Pharmacol.*, **13**, 442–453, (1977).
36. Enna, S. J., and Snyder, S. H. *Brain Res.*, **115**, 174–179, (1976).
37. Ticku, M. K., and Olsen, R. W. *Biochim. Biophys. Acta*, **464**, 519–529, (1977).
38. Werman, B. *Comp. Biochem. Physiol.*, **30**, 997–1017, (1969).
39. Enna, S. J., Kuhar, M. J., and Snyder, S. H. *Brain Res.*, **93**, 168–174, (1975).
40. Simantov, R., Oster-Granite, M. H., Herndon, R. M., and Snyder, S. H. *Brain Res.*, **105**, 365–371, (1971).
41. Lloyd, K. G., and Hornykiewicz, O. *Nature (London)*, **243**, 521–523, (1973).
42. Enna, S. J., Bennett, J. P., Jr., Bylund, D. B., Snyder, S. H. Bird, E. D., and Iversen, L. L. *Brain Res.*, **116**, 531–537, (1976).
43. Campochiaro, P., Schwarcz, R., and Coyle, J. T. *Brain Res.*, **136**, 501–511, (1977).
44. Dostrovsky, J. O., and Pomeranz, B. *Neuroscience Letters*, **4**, 315–319, (1977).
45. Hall, J. G., Hicks, T. P., and McLennan, H. *Neuroscience Letters*, **8**, 171–175, (1978).
46. London, E. D., and Coyle, J. T. *Mol. Pharmacol.*, **15**, 492–505, (1978).
47. McGeer, E. G., McGeer, P. L., and Singh, K. *Brain Res.*, **139**, 381–383, (1978).
48. Ticku, M. K., and Olsen, R. W. *Life Sciences*, **22**, 1643–1652, (1978).

Cellular Receptors for Hormones and Neurotransmitters
Edited by D. Schulster and A. Levitzki


CHAPTER 21

# Substance P receptors

Vivian I. Teichberg
*Department of Neurobiology, The Weizmann Institute of Science, Rehovot, Israel*
Shmaryahu Blumberg
*Department of Biophysics, The Weizmann Institute of Science, Rehovot, Israel*



## I. INTRODUCTION

In 1931, von Euler and Gaddum[1] reported the presence of a hypotensive, vasodilatatory, and smooth muscle contracting agent in extracts of equine intestine and brain. To this active principle was given the name substance P (P for preparation). Forty years after its discovery, the structure of substance P (SP) was finally unravelled.[2] It is an undecapeptide with the following amino acid sequence: Arg-Pro-Lys-Pro-Gln-Gln-Phe-Phe-Gly-Leu-Met·$NH_2$. Although the elucidation of the structure of SP in 1971 gave a renewed impetus to the investigations of SP, previously formulated ideas concerning the physiological role of SP, continued to lead and orient the research.

In the 1950s, SP was reported to be present with an uneven distribution in the spinal cord.[3] Since the concentration of SP in the dorsal horn and roots was found to be higher than in the ventral horn and roots, the suggestion was made[3] that SP might be the natural neurotransmitter of primary sensory neurons, i.e. the neurons carrying the sensory messages from the peripheral tissues to the central nervous system via synapses in the dorsal horn of the spinal cord. Quite remarkably this hypothesis has so far resisted both the assaults of the years as well as many experimental challenges and SP is today considered as the best candidate for fulfilling the role of neurotransmitter in pathways related to pain.

## II. DISTRIBUTION OF SUBSTANCE P

SP belongs to a family of peptides, the tachykinins, having a widespread distribution in the animal kingdom from vertebrates to invertebrates.[4] All vertebrates so far investigated contain SP. This peptide is present both in the peripheral and central nervous system.[5] The concentration of SP in tissues is determined by radioimmunoassays and its localization by immunofluorescence. Almost all peripheral tissue contain SP-positive nerves in relation to blood vessels or secretory cells. SP is also present in free nerve endings in the skin. Sympathetic ganglia contain SP-positive fibres but no SP-positive cell bodies. SP-fibres are also seen in high numbers along the gastrointestinal tract and particularly in the muscularis mucosae in connnection with the myenteric plexus.

In the central nervous system, the highest concentrations of SP are found in the mesencephalon, brainstem, and spinal cord. Particularly rich areas are the substantia nigra, the substantia gelatinosa, the hypothalamus, amygdala, and globus pallidus.

At an ultrastructural level, SP immunoreactive sites are revealed in dense-core vesicles in nerve endings and in a more diffuse form in the cytoplasm of boutons forming synaptic contacts with dendrites.[6]

## III. PERIPHERAL ACTIONS OF SUBSTANCE P

The peripheral actions of SP consist of a stimulation of smooth muscle contractions as well as vasodilatation.[7] SP contracts the isolated intestine, uterus, ureter, vas deferens, urinary bladder, and trachea, but not the gall bladder. The time-effect relation of contraction and relaxation of these smooth muscles is slower for SP than for acetylcholine, serotonin, or histamine. SP acts directly on the muscle since its contracting activity is not inhibited by cholinergic, serotoninergic, or histaminergic antagonists. Substance P is one of the most potent vasodilatator compounds known. Intravenous administration of a few nanograms of synthetic SP increases the carotid, hepatic, mesenteric, and portal blood flow. Intravenous injections increase the blood flow in skin and muscle.[8]

It is still a controversial point whether SP is responsible for the vasodilatation observed upon antidromic stimulation of sensory nerves. This phenomenon is at the basis of Dale's famous principle,[9] stating that the substance released in the periphery from the sensory nerve endings is identical to the transmitter substance of the central branches of the very same neuron, i.e. the interneuronal transactions of a neuron can only be carried out by a single neurotransmitter. Since SP is present in the nerve terminals along blood vessels in the skin, it is plausible that SP could be released upon antidromic stimulation of sensory nerves and cause vasodilatation. Other properties of SP have also been reported; SP can stimulate salivary secretion,[10] $Na^+$ excretion[11] and increase the concentration of glycine in the brain.[12] *In vitro*, SP stimulates neurite extension in cultures of neuroblastoma cells and chick dorsal ganglia.[13]

## IV. PROPERTIES OF SUBSTANCE P IN THE CENTRAL NERVOUS SYSTEM

If Dale's principle applies to SP, a release of SP in the spinal cord from the central branches of the sensory neurons may be expected. Indeed, SP is released by a calcium-dependent mechanism from the isolated rat spinal cord during electrical stimulation of the dorsal roots.[14]

Moreover, upon application of SP on motorneurons, a depolarization and an increase in membrane conductance takes place.[15] This response is not due to a trans-synaptic effect but to a direct response of the motorneurons since it is also observed when synaptic transmission is blocked by a decrease in extracellular $Ca^{2+}$ or by tetrodotoxine. If one compares the reversal potential of SP-induced depolarization with that of the excitatory postsynaptic potential, one obtains very close values. Such results are consistent with the view that SP is an excitatory transmitter on the motorneurons. The depolarization effect of SP is on a molar basis $10^3$–$10^4$ times stronger than that observed with glutamate and its duration is longer.

When tested on spinal neurons other than motorneurons, SP does not always display the same excitatory properties.[16] In the cuneate nucleus, a region where primary afferent fibres make their synaptic connections, high doses of SP tend to depress their excitability. In small doses, SP improves synaptic transmission. These modulating effects of SP have led to the still-prevailing argumentation that SP might have a regulatory role in neuronal excitability, rather than a role in neurotransmission.

## V. SUBSTANCE P AND THE NEUROTRANSMISSION OF PAIN

SP is highly concentrated in the terminals of small diameter fibres in the substantia gelatinosa of the dorsal horn of the mammalian spinal cord.[17] In this area, SP seems to excite or facilitate the response of cells sensitive to noxious stimuli.[18] In some instances however, systemic administration of SP can also produce analgesia. SP is in fact present in the very same areas of

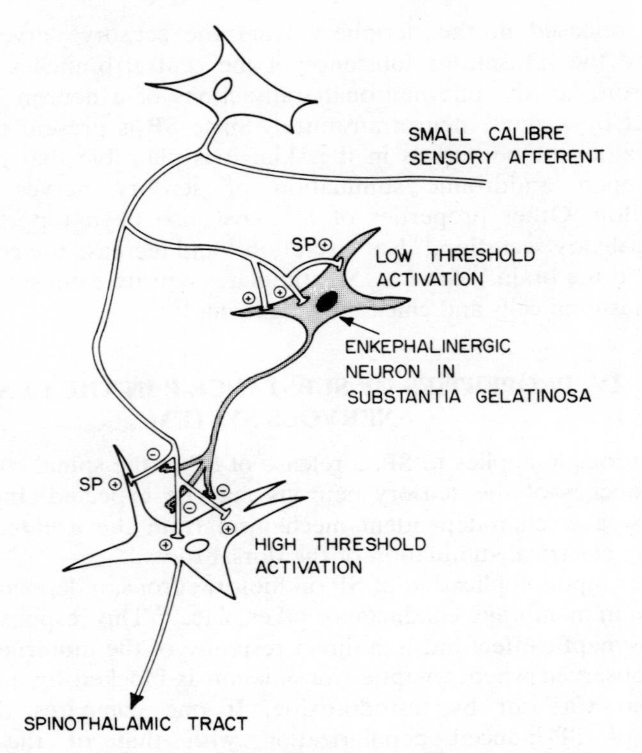


Figure 21.1. Schematic illustration of the interaction between an enkephalinergic neuron in substantia gelatinosa of the spinal cord with a sensory neuron. The transmission of noxious stimuli is regulated by the activation of enkephalinergic neurons by a negative feedback mechanism

the central nervous system where enkephalin containing neurons are found and which are related to pain and analgesia. In the trigeminal nucleus, a mesencephalic structure which can be electrically stimulated to produce strong analgesia, opiate analgesics including a Met-enkephalin analogue were found to suppress the stimulus evoked released of SP.[19] It is suggested (Figure 21.1) that SP in primary afferent terminals would activate spinal neurons sensitive to noxious stimuli as well as enkephalin containing interneurons. These cells would mediate a presynaptic inhibition of SP-containing primary afferent fibres and terminate in that manner the transmission of nociceptive information. Depending on the level of activity in the primary afferent fibres and the respective threshold of activation of spinal neurons, SP could produce either analgesia or pain. Such a model could well respresent the morphological basis of the 'gate theory' of pain.[20]

## VI. THE SUBSTANCE P RECEPTORS

Stimulated by the elucidation of the chemical structure of SP, various studies have been carried out to establish structure–function relationships between synthetic SP analogues.

The peptides are assayed in various systems in which one measures either the contraction of the guinea-pig ileum,[21] the cat blood pressure, or the depolarization of motorneurons in the isolated frog spinal cord. A very good correlation is obtained between these systems. From various studies, it appears that the region of effective interaction of the SP receptor does not accommodate more than the seven amino acid residues corresponding to the C-terminal heptapeptide (Figure 21.2). Indeed, peptides, including SP analogues longer than the C-terminal heptapeptide, display very similar activity.[15] However, modifications or deletions of amino acids in the C-terminal fragment of SP lower or even abolish the activity. This applies in particular to the deletion of the terminal Met·$NH_2$ residue. No substitution of amino acid in the original sequence of SP has been found to substantially improve activity. An interesting compound is obtained upon substitution of the glutamine residue occupying the subsite $S_6$ with a pyroglutamyl moiety. This peptide *p*Glu-Phe-Phe-Gly-Leu-Met·$NH_2$ is among the most active

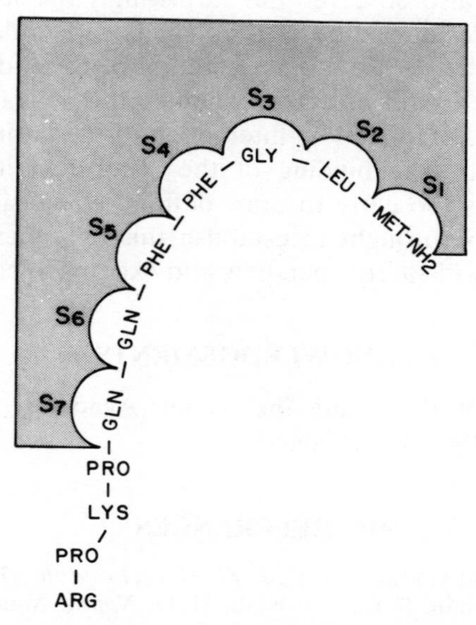


Figure 21.2. A model of the substance P receptor. The active region is composed of seven subsites accommodating the C-terminal heptapeptide

analogues of SP. It displays half-maximal contracting activity on the guinea-pig ileum at a concentration of 2 nM.

So far, the current knowledge about the SP receptor relies exclusively on studies of the biological activity of synthetic SP analogues. Because of the absence of radioactive SP analogues resistant to proteolytic degradation, no binding studies have been carried out successfully. Indeed the half-life of SP and of partial SP sequences in brain homogenates is of the order of 1 min as very effective endopeptidases degrade the peptides.[22] No inhibitor of the SP degrading enzymes has yet been found. The success of binding studies will largely depend on the development of SP analogues which combine the properties of high binding activity together with resistance to proteolytic degradation. Another great challenge with far-reaching pharmacological implications lies in the availability of specific SP antagonists. No such compounds have yet been reported. The only way in which the effects of SP can be transiently blocked is by prior desensitization of the SP receptor. Such a phenomenon is obtained by exposure of the receptor to high concentrations of SP.[7]

It has been reported that SP stimulates adenylate cyclase activity in various areas of the human brain, particularly in the hypothalamus, pineal gland and substantia nigra.[23] SP also increases cyclic AMP level and activates adenylate cyclase in cultured neuroblastoma cells.[13] In the latter case, neurite extension was also observed but surprisingly the stimulatory activity of SP was found to reside in the N-terminal tetrapeptide and not in the C-terminal heptapeptide. In view of the lability of SP and SP analogues in brain homogenates, it is still not clear whether the activation of adenylate cyclase by SP is indeed mediated by interaction of the entire SP molecule to its receptor or only by the binding of the N-terminal tetrapeptide to a separate receptor. It is too early to draw definite conclusions on this issue, but should evidence be brought to establish this hypothesis as a true fact, this other facet of SP will surely open new and exciting avenues of research.

## ACKNOWLEDGEMENTS

Support from the DGRST and the Israel Academy of Sciences and Humanities is gratefully acknowledged.

## VII. REFERENCES

1. von Euler, U. S., and Gaddum, J. H. *J. Physiol. (London),* **72**, 74–87, (1931).
2. Chang, M. M., Leeman, S. E., and Niall, H. D. *Nature New Biol.,* **232**, 86–87, (1971).
3. Lembeck, F. *Naunyn-Schmiedebergs Arch. Exp. Pathol. Pharmakol.,* **219**, 197–213, (1953).
4. Bertaccini, G. *Pharmacol. Rev.,* **28**, 127–177, (1976).

5. Hökfelt, T., Johansson, O., Kellerth, J.-O., Ljungdahl, Å., Nilsson, G., Nygåards, A., and Pernow, B. in *Substance P,* (von Euler, U. S., and Pernow, B., eds.), Raven Press, New York, pp. 117–145, (1977).
6. Pelletier, G., Leclerc, R., and Dupont, A. *J. Histochem. Cytochem.,* **25**, 1373–1380, (1977).
7. Lembeck, F., and Zetler, G. *Int. Rev. of Neurobiol.,* **4**, 159–215, (1962).
8. Burcher, E., Atterhög, J.-M., Pernow, B., and Rosell, S. in *Substance P,* (von Euler, U. S., and Pernow, B., eds.), Raven Press, New York, pp. 261–268, (1977).
9. Dale, H. H. *Proc. Roy. Soc. Lond. (Biol.),* **28**, 319–332, (1935).
10. Lembeck, F., and Starke, K. *Naunyn-Schmiedebergs Arch. Exp. Pathol. Pharmakol.,* **259**, 375–385, (1968).
11. Mills, J., Macfarlane, N., and Ward, P. *Nature (London),* **247**, 108–109, (1974).
12. Stern, P., Catovic, S., and Stern, M. *Naunyn-Schmiedebergs Arch. Exp. Pathol. Pharmakol.,* **281**, 233–239, (1974).
13. Narumi, S., and Maki, Y. *J. Neurochem.,* **30**, 1321–1326, (1978).
14. Otsuka, M., and Konishi, S. *Nature (London),* **264**, 83–84, (1976).
15. Otsuka, M., and Konishi, S. in *Substance P,* (von Euler, U. S., and Pernow, B., eds.), Raven Press, New York, pp. 207–214, (1977).
16. Krnjovic, K. in *Substance P,* (von Euler, U. S., and Pernow, B., eds.), Raven Press, New York, pp. 217–230, (1977).
17. Hökfelt, T., Ljungdahl, Å., Terenius, L., Elde, R., and Nilsson, G. *Proc. Natl. Acad. Sci. U.S.A.,* **74**, 3081–3085, (1977).
18. Henry, J. L. in *Substance P,* (von Euler, U. S., and Pernow, B., eds.), Raven Press, New York, pp. 231–240, (1977).
19. Jessell, T. M., and Iversen, L. L. *Nature (London),* **268**, 549–551, (1977).
20. Melzack, R., and Wall, P. D. *Science,* **150**, 971–979, (1965).
21. Rosell, S., Björkroth, U., Chang, D., Yamaguchi, I., Wan, Y.-P., Rackur, G., Fisher, G., and Folkers, K. in *Substance P* (von Euler, U. S., and Pernow, B., eds.), Raven Press, New York, pp. 83–88, (1977).
22. Blumberg, S., and Teichberg, V. I., *Biochem. Biophys. Res. Comm.,* **90**, 347–354, (1979).
23. Duffy, M. J., Wong, J., and Powell, D. *Neuropharmacology,* **14**, 615–618, (1975).

# Index